W9-CMI-326

Differential Equations

Differential Equations:

An Introduction With Applications

LOTHAR COLLATZ

University of Hamburg
Federal Republic of Germany

Translated by

E. R. DAWSON,

University of Dundee
UK

A Wiley–Interscience Publication

JOHN WILEY & SONS
Chichester · New York · Brisbane · Toronto · Singapore

First published as *Differentialgleichungen*, 6 Aufl., by Lothar Collatz, © 1981. This edition is published by permission of Verlag B. G. Teubner, Stuttgart, and is the sole authorized English translation of the original German edition.

Library of Congress Cataloging-in-Publication Data:

Collatz, L. (Lothar), 1910–
Differential equations.

Translation of: Differentialgleichungen. 6th ed.
'A Wiley–Interscience publication.'
Bibliography: p.
Includes index.
1. Differential equations. I. Title.
QA371.C68213 1986 515.3′5 85–26555

ISBN 0 471 90955 6 (cloth)
ISBN 0 471 91166 6 (paper)

British Library Cataloguing in Publication Data:

Collatz, Lothar
Differential equations: an introduction with particular regard to applications.
1. Differential equations
I. Title
515.3′5 QA371

ISBN 0 471 90955 6 (cloth)
ISBN 0 471 91166 6 (paper)

Printed and Bound in Great Britain

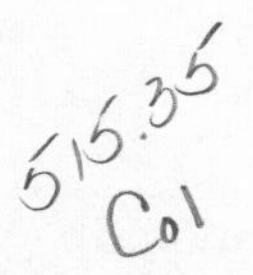

Contents

Preface (to the 5th German edition)

The first edition was an exact reproduction of my lectures during the winter semester of 1945/46 at the Technical University, Hamburg. The second edition was published at the instigation of Professor H. Görder; in it a number of topics were added, some of which are not usually discussed in a normal course of lectures at a Technical University, but which have become of importance for engineers and physicists, so that a short introduction to these fields had to be included in this small volume. To give more than an introduction is, of course, impossible because of the limited space. For further study, reference must be made to larger textbooks.

The new edition differs from the second one, apart from small corrections and additions, mainly by the new, additional sections on systems of linear differential equations with periodic coefficients, which in recent times have acquired ever greater importance, and by the addition of a large number of problems for practice, which have been included in response to often-repeated requests.

The treatment assumes essentially a knowledge of the differential and integral calculus, but otherwise it has been made deliberately full and detailed, especially at the beginning.

Always the descriptive aspect, interpretations from mechanics, and relations with fields of applications have been stressed, in order to induce people, when they encounter differential equations in practice, not to try to steer clear of them but rather to work with them.

A book on differential equations written for a class of readers mainly interested in applications differs from one written 'for mathematicians' mainly in the choice of material, not in the 'rigour' of the presentation. No one would be served by an inexact, schematic presentation. However, it seems justifiable to adopt simple assumptions (for example, about the coefficients appearing in differential equations) which are usually satisfied in practical problems, and thereby to arrive at simpler proofs. By means of matrix theory a transparent theory of linear vibrations can be constructed; so I hope that the book will also be suitable as an introduction for physicists, even though in the space available, eigenvalue problems on quantum physics could not be gone into, and it seemed to be permissible to delete the phrase 'for engineers' which still formed part of the title of the book in its second edition.

Professor Günter Meinardus, Dr. Alfred Meyer and Dr. Rüdiger Nicolovius greatly assisted me with the second edition. They not only participated in the tedious checking of all details of the numerical material, but gave me numerous suggestions for improvements and additions; for example, the combination of clarity and completeness in Chapter 3, section 20, is due to them.

I would like to thank the Teubner Publishers for their good production of the book, and Dr. Erich Behl and Mr. Werner Schütt for their careful proofreading.

Further editions of the book were soon needed after the 1969 publication, and since no errors of a factual kind nor even misprints were known to me, these were unchanged impressions of the third edition.

Hamburg, Summer 1973

Preface to the 6th edition

The purpose of the book is described in the preface to the fifth edition and it has not changed. So only the changes compared with earlier editions are indicated here, without mentioning innumerable smaller additions.

In this sixth edition a number of topics have been included, which, though they could not be presented in great detail, have recently become important; it seems to me useful that the reader should at least be acquainted with their formulation. In particular, the sections on partial differential equations have been reconstructed, because these are becoming increasingly important in applications, although, of course, this book cannot be a substitute for the many good treatises on partial differential equations. In this connection I have chosen problems which have proved useful for their applications and for the numerical methods employed; for numerical methods are constantly gaining significance, because as ever more complicated problems are tackled, the prospect of being able to solve a given problem in 'closed form' is greatly diminished, and in many cases, particularly in non-linear problems, we have to resort to numerical methods. The concept of (pointwise) monotonicity becomes especially important, because, in cases where it is valid, it presents the only method of obtaining, without excessive labour, usable inclusions of the required solution between upper and lower bounds.

Introductions are also given to many problems which have recently become actual, and to methods relating, for example, to free boundary problems, variational equations, incorrectly posed problems, finite-element methods, branching problems, etc.

On the question of modernizing the treatment I deemed it necessary to proceed cautiously. There are quite enough very abstract textbooks, often based on functional analysis, about differential equations, in which, however, applications and concrete aspects feature too little. I have preferred much rather that engineers and scientists should be able to understand the treatment. However, in order that access to modern mathematical literature should not be obstructed, I decided to introduce the fundamental existence and uniqueness theorems twice, once in the classical way, and again in the language of functional analysis; the reader will see that both proofs proceed in the same way.

At this point may I be allowed a word of warning: mathematics—which thrives on its applicability—is moving more and more towards a dangerous isolation in that abstractions are overvalued and actualizations are neglected (often even inexcusably). Abstractions were made in order the better to master the concrete, but now mastery of the concrete is in danger of being lost, as may be susbstantiated in innumerable instances. Often a good engineer is better prepared to deal with a differential equation than a mathematician, and mathematics loses ground. This implies a great and increasingly serious danger for mathematics.

A further modernization concerns the physical units employed. Since the last edition the Government has introduced the international system of units (SI units) into the Federal Republic of Germany, making it compulsory for commercial and official business (by a law passed on 2nd July 1969). This has had to be taken into account in several examples. Instead of the earlier kg-weight or kp the Newton (N) is now used; $1\ \mathrm{N} = 1/9{\cdot}80665$ kp is the new unit of force, and 1 kilonewton $= 1000$ N; the unit of pressure is 1 bar $= 10\ \mathrm{N/cm^2} \simeq 1{\cdot}02\ \mathrm{kp/cm^2}$. I thank my colleague, Mr. K. Wieghardt, for the conversions to the new units.

I also owe particular thanks to Dr. Klaus Meyn, Dr. Klaus Roleff, Mr. Uwe Grothkopf, and Miss Gabriele Theuschl for valuable additions and suggestions for improvement and for their careful proof-reading.

I thank the Teubner Verlag and Dr. Spuhler in particular for their patience over the years and for their excellent (as always) production of the book.

Lothar Collatz
Hamburg
Spring 1980

Preface to the English edition

This English edition is a translation mainly of the sixth German edition, but it also contains additional material, which the publishers invited Professor Collatz to write, on the Laplace and Fourier transforms and their use in solving ordinary and partial differential equations. This new material can be found in Chapter II, §9 and Chapter V, §6. Professor Collatz also took the opportunity to add a short, new section (Chapter V, §1, No.3) on non-linear partial differential equations and the breaking of a wave.

All this new material should make the book even more suitable for use in American and British universities.

I have to thank Mr. R. Sawatzki of the Institut für angewandte Mathematik, University of Hamburg, for his assistance in the proof-reading of the new sections on integral transforms.

E. R. Dawson

Preface to the English edition

Classification of differential equations

1. Definitions and notation

A differential equation is an equation in which the derivatives of one or more functions of one or more independent variables appear. The unknowns to be found are, in ordinary differential equations, functions $y(x)$ of a single independent variable x, and in partial differential equations, functions $u(x, y, \ldots)$ of several independent variables $x, y, \ldots$.

An ordinary differential equation in which only one dependent function $y(x)$ occurs has the general form

$$F[x, y(x), y'(x), y''(x), \ldots, y^{(n)}(x)] = 0.$$

A partial differential equation with, say, two independent variables x, y has, on the other hand, the general form

$$F[x, y, z(x, y), z_x, z_y, z_{xx}, z_{xy}, z_{yy}, \ldots] = 0$$

where the subscripts denote derivatives with respect to x and y, i.e.,

$$z_x = \frac{\partial z}{\partial x}, \quad z_{xy} = \frac{\partial^2 z}{\partial x\, \partial y}, \quad \text{etc.}$$

The order of the highest derivative occurring in a differential equation is called the *order* of the differential equation.

If a differential equation can be written as a polynomial in the required function and its derivatives, then the highest sum of the exponents of the dependent variable and its derivatives which occurs in any one term is called the *degree* of the differential equation. Differential equations of the first degree are also said to be *linear*; in such equations the unknown function and its derivatives occur only to the first power and are not multiplied by one another.

For example, $yy' = x$ is a differential equation of the first order but it is not linear. Further,

$$4x^5y^2 + 2y\left(\frac{dy}{dx}\right)^2 \cos x - 5y + 7 = 0$$

is a first-order differential equation of the third degree, and

$$z_{xx}^2 + z_{yy}^2 = x$$

is a second-order partial differential equation of the second degree.

To *solve* a differential equation of the nth order means *to determine all those n-times continuously differentiable functions which, when substituted with their derivatives into the differential equations, satisfy it identically,* i.e., which make it into an identity.

The *general solution* of an nth-order ordinary differential equation, as we shall see later, contains, as a rule, n free *integration constants* as parameters, which enable us to pick out from the totality of all solutions that solution which satisfies prescribed *initial conditions* or *boundary conditions* of a particular form. Geometrically, the general solution of an nth-order differential equation represents, as does the equation itself, an n-parameter family of curves. The differential equation and its general solution are merely two equivalent, different forms of expression for the same physical or geometrical circumstances; later we shall go into this point further.

2. EXAMPLES OF DIFFERNETIAL EQUATIONS ARISING IN PHYSICS

Example 1. An ordinary differential equation

A mass m hangs on a spring with spring constant c (Fig. I.1), Here c is defined as the constant of proportionality between the extension of the spring and the restoring force K in accordance with 'Hooke's Law' $K = cx$. If the mass m oscillates on the spring, Newton's fundamental law

$$\text{force} = \text{mass} \times \text{acceleration}$$

governs the motion. The restoring force is prefixed by a negative sign since it is opposite to the direction of increasing x, and we obtain the following linear, ordinary differential equation describing the process of oscillation:

$$-cx = m \frac{d^2x}{dt^2},$$

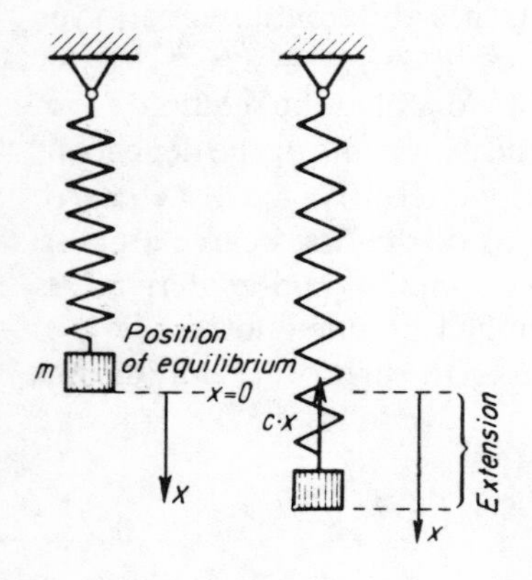

Fig. I.1. Oscillation of a mass on a spring

or, using dots to denote derivatives with respect to time, as is usual in mechanics,

$$\ddot{x} + \frac{c}{m} x = 0.$$

In order the better to characterize the significance of a differential equation, we will anticipate the general solution:

$$x = C_1 \sin \omega (t + C_2);$$

here C_1 and C_2 are the free integration constants, to be determined, say, from prescribed initial conditions. C_1 is the *amplitude*, and C_2 the *phase* of the oscillation, *i.e.*, the time interval from the instant when the mass m last passed through the equilibrium position to the instant from which time is counted.

For conciseness we have written

$$\frac{c}{m} = \omega^2$$

here ω is the *circular frequency,* i.e., the number of oscillations in 2π seconds. If f is the frequency, then we have

$$f = \frac{1}{T} = \frac{\omega}{2\pi}.$$

where T is the *period of an oscillation*. The frequency f increases proportionally with the square root of the spring constant c, and decreases reciprocally with the square root of the mass m. Damping effects have not yet been taken into account here.

Example 2. A system of ordinary differential equations

Let a beam mounted on, say, two supports be loaded by a line load of density $p(x)$, the x-axis lying along the longitudinal axis of the beam (Fig. I.2).

If the function $y(x)$ for the deflection of the beam at the position x is sought, then the following system of two differential equations (not derived here) hold:

$$y'' = -\frac{M(x)}{EJ}, \qquad M'' = -p(x);$$

here $M(x)$ is the 'bending moment', E is the 'modulus of elasticity', and J is the 'second moment of area of the cross-section of the beam about the neutral

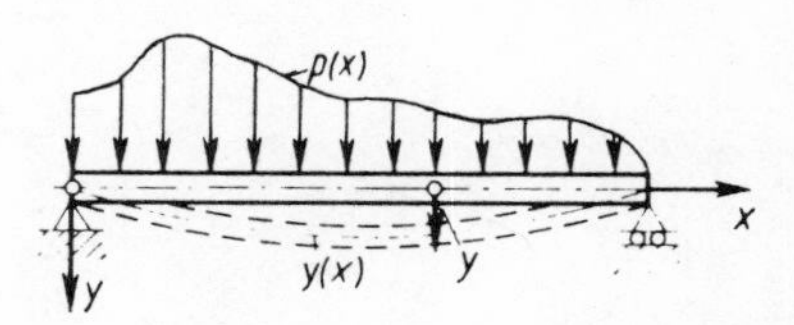

Fig. I.2. Deflection of a loaded beam

axis'. From these two equations for the two unknowns $y(x)$ and $M(x)$ we obtain, by differentiating the first equation twice (assuming EJ to be constant), the 4th-order equation for the unknown function y(x):

$$y^{\text{IV}} = -\frac{M''}{EJ} = \frac{p(x)}{EJ}.$$

Example 3. A 2nd-order partial differential equation

The differential equation for the transverse vibrations of a string is

$$\frac{\partial^2 y}{\partial t^2} = \frac{\mu F}{S}\frac{\partial^2 y}{\partial x^2} \quad \text{or} \quad y_{tt} = a^2 y_{xx},$$

where $y = y(x, t)$ is the deflection of the string at the position x at time t, S is the longitudinal force in the string (assumed to be constant), F is the cross-sectional area, μ the line density, and $a^2 = \mu F/S$. From the solution of this partial differential equation (see Chapter V, §1) it can be seen that a denotes the velocity of transverse waves along the string.

Example 4. A fourth-order partial differential equation

For the deflection of a plate under a 'load-density $p(x, y)$', and with constant 'plate stiffness N', the fourth-order partial differential equation

$$\frac{\partial^4 z}{\partial^4 z} + 2\frac{\partial^4 z}{\partial x^2 \partial y^2} + \frac{\partial^4 z}{\partial y^4} = \frac{p(x, y)}{N}$$

holds, or written symbolically, using the 'Laplacian' operator

$$\Delta \equiv \frac{\partial^2}{\partial x^2} + \frac{\partial^2}{\partial y^2},$$

$$\left(\frac{\partial^2}{\partial x^2} + \frac{\partial^2}{\partial y^2}\right)^2 z(x, y) = \Delta^2 z(x, y) = \Delta\Delta z(x, y) = \frac{p(x, y)}{N}.$$

I

Ordinary differential equations of the first order

§1. THE DIRECTION FIELD AND THE SIMPLEST INTEGRABLE TYPES

3. SOLUTION CURVES IN THE DIRECTION FIELD

As the simplest case it will be assumed that the 'implicit' form of an ordinary differential equation of the first order

$$F[x, y(x), y'(x)] = 0 \tag{I.1}$$

can be solved for y' and be transformed into the 'explicit' form

$$y' = f(x, y), \tag{I.2}$$

in which $f(x, y)$ may be a single-valued function of x and y. That $f(x, y)$ cannot always be expressed in closed form using elementary functions is shown by the counter-example

$$y' = xy - e^{y'}. \tag{I.3}$$

Graphically, $f(x, y)$ signifies a so-called *direction field,* i.e., the differential equation associates a slope with each point x, y. If, therefore, a direction field of small line-elements is plotted (Fig. I.3) the solutions of the differential equation are those curves which 'fit in with' this direction field.

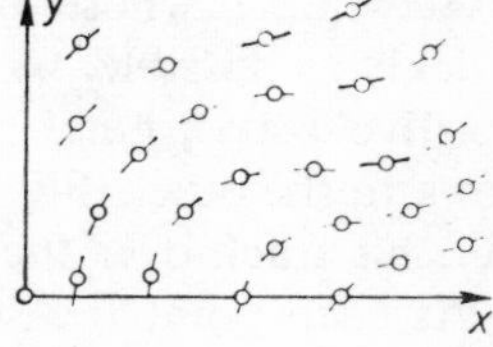

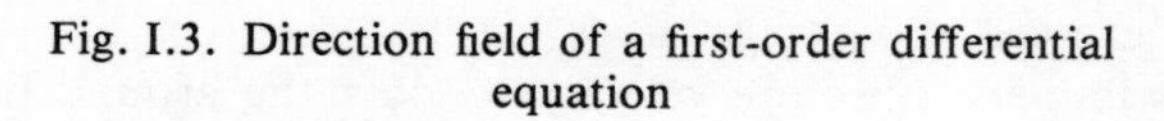
Fig. I.3. Direction field of a first-order differential equation

Example 1.

Let the given differential equation be

$$y' = 1 + x - y. \tag{I.4}$$

To draw the direction field it is more convenient to draw the curves of equal slope, the *isoclines,* rather than the slopes at individual points. The isocline for $y' = 0$ has the equation

$$0 = 1 + x - y \quad \text{or} \quad y = x + 1.$$

This is the 'curve' with horizontal line elements in Fig. I.4. Similarly, for $y' = 1$ the straight line $y = x$ is obtained. As the figure shows, this is, in fact, one solution of the differential equation, a fact which can be verified by putting $y = x$ in the differential equation, since the equation then becomes the identity

$$1 \equiv 1 + x - x.$$

With the further isoclines for $y' = 2$ and $y' = -1$ (see Fig. I.4), the course of the solution curves can already be approximately made out (cf. (I.9)).

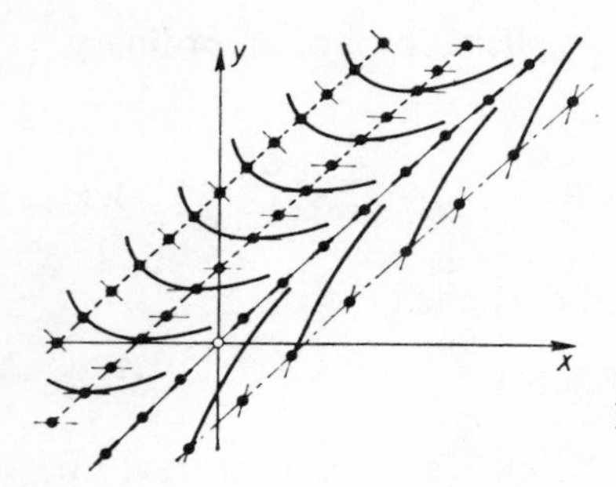

Fig. I.4. Direction field of the differential equation $y' = 1 + x - y$

Example 2. A velocity field

The velocity distribution in a stream of width $2a$ flowing with velocity v_0 in the middle can be assumed to be a parabolic function of the distance x from the middle.

$$v_F = v_0\left(1 - \frac{x^2}{a^2}\right).$$

At the 'banks', i.e., at the boundary points $x = \pm a$, the rate of flow is $v_F = 0$.

A swimmer whose own speed $v_E = \text{const.}$ swims so as to cross the stream as quickly as possible, by heading relatively to the stream in the direction of the positive x-axis, i.e., always perpendicular to the direction of flow, from one bank to the other (Fig. I.5). The question asked is, what is his absolute track, i.e., his track over the ground.

As each point the swimmer's absolute velcoity v_A over the ground is the

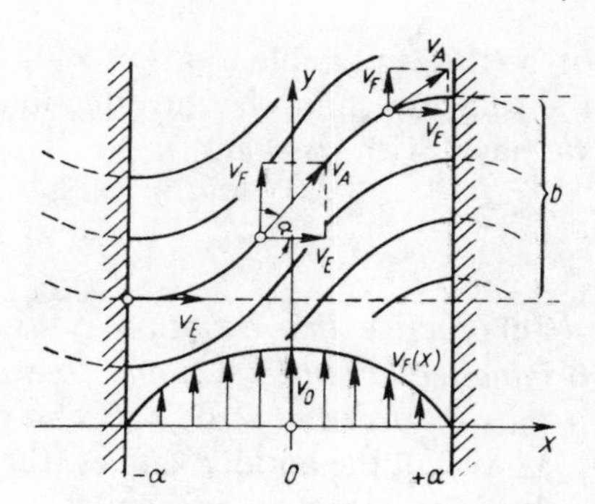

Fig. I.5. A direction field: swimming in a flowing stream

vector sum of v_E and v_F; let α be the angle made by v_A with the positive x-axis (Fig. I.5). The differential equation for the curve of the swimmer's track can now be set up:

$$\tan\alpha = \frac{\mathrm{d}y}{\mathrm{d}x} = y' = \frac{v_F}{v_E} = \frac{v_0}{v_E}\left(1 - \frac{x^2}{a^2}\right).$$

Since this differential equation is particularly simple (y does not occur explicitly), it can be integrated immediately:

$$y = \frac{v_0}{v_E}\left(x - \frac{x^3}{3a^2}\right) + C.$$

The free parameter C, i.e., the integration constant, determines the swimmer's starting-point. The family of curves consists of cubical parabolas. The swimmer's total drift is obtained as the difference

$$b = y(a) - y(-a) = \frac{v_0}{v_\mathrm{E}}\left(2\mathrm{a} - \frac{2a^3}{3a^2}\right) = \frac{4}{3}\cdot\frac{v_0}{v_E}\cdot a.$$

4. Separation of the variables

Following these simple integrations we come to the first real method of solving differential equations, viz., *separation of the variables.* This method can be applied to equations of the form

$$y'(x) = \frac{\mathrm{d}y}{\mathrm{d}x} = \frac{f(x)}{g(y)} = f(x)\varphi(y). \tag{I.5}$$

Suppose $f(x)$ is continuous in an interval $a \le x \le b$ (denoted more concisely by $[a, b]$), and that $g(y)$ is continuous in an interval $[c, d]$. We integrate both sides of the equation

$$g(y)y' = f(x)$$

with respect to x from a chosen point x_0 up to a variable point x (both points are to lie in $[a, b]$), and on the left we introduce y as a new variable instead of x (this is possible if $y'(x) \neq 0$, since then $y(x)$ is montone in $[x_0, x]$). By the rule governing the introduction of a new variable (the substitution rule), we have, with $y_0 = y(x_0)$,

$$\int_{y_0}^{y} g(y)\, dy = \int_{x_0}^{x} f(x)\, \mathrm{d}x. \tag{I.6}$$

In practice this equation is usually set up by multiplying formally the original equation by dx, *and formal separation of the variables x, y to bring it into the form* $g(y)\,\mathrm{d}y = f(x)\,\mathrm{d}x$, *and then integrating both sides.*

If we call the antiderivatives (the indefinite integrals, which are determined only up to arbitrary constants)

$$\int g(y)\,\mathrm{d}y = G(y)$$

and

$$\int f(x)\,\mathrm{d}x = F(x),$$

then, under the above assumptions, the general solution of the differential equation is

$$G(y) = F(x) + C.$$

By differentiating, it can be verified immediately that these functions really are solutions of (I.5).

Example 1. An electrical circuit

An electromotive force U is applied by means of a switch S to an inductance L in series with an ohmic resistance R (Fig. I.6) The growth of current $J(t)$ after switching on is required.

From Faraday's law of induction and Kirchhoff's theorem we have

$$U = RJ + L\,\frac{\mathrm{d}J}{\mathrm{d}t} \tag{I.7}$$

(the individual resistances of the elements are either neglected or are included in R). Equation (I.7) asserts that U is equal to the sum of the voltage drops. Separation of the variables $J(t)$ and t gives

$$\int L\,\frac{\mathrm{d}J}{U - RJ} = \int \mathrm{d}t$$

$$-\frac{L}{R}\ln|U - RJ| = t + C$$

$$U - RJ = \pm\,\mathrm{e}^{-R(t+C)/L}.$$

So the general solution is

$$J(t) = \frac{U}{R} \mp \frac{1}{R}\,\mathrm{e}^{-R(t+C)/L}.$$

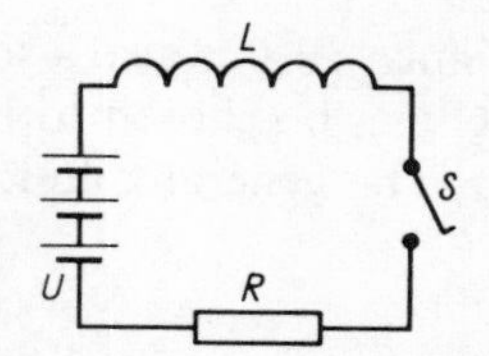

Fig. I.6. An electrical circuit

From this set of solutions we select a particular solution by using the initial condition that at the moment of switching on, $t = 0$, the current strength $J(t) = 0$. The condition $J(t) = 0$ can be satisfied only if the negative sign is chosen, and then C is determined by

$$e^{-(RC/L)} = U.$$

Hence (see Fig. I.7)

$$J(t) = \frac{U}{R}(1 - e^{-Rt/L}) \tag{I.8}$$

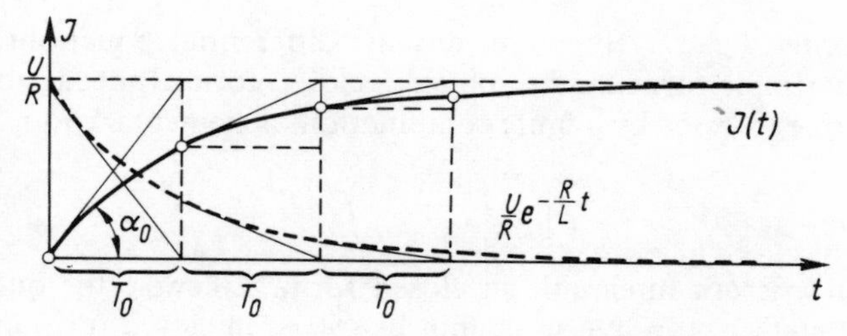

Fig. I.7. Growth of current after switching on

Example 2.

In (I.4)

$$y' = 1 + x - y$$

the variables cannot be separated immediately, but can be if a substitution is made for the quantity which is preventing separation, namely, with

$$x - y(x) = z(x), \qquad \text{i.e., } 1 - y' = z',$$

$$-z' = -\frac{dz}{dx} = z.$$

We can now separate and integrate:

$$-\int \frac{dz}{z} = \int dx$$

$$-\ln|z| = x + C_1$$

$$z = \pm e^{-x-C_1} = \underbrace{\pm e^{-C_1}}_{C} e^{-x} = Ce^{-x}.$$

Since $z = 0$ is also a solution of the differential equation $z' = -z$, the constant C can be chosen to be any real number.

The general solution for y is therefore

$$y = x - C\,e^{-x}. \tag{I.9}$$

We test this solution by differentiating:

$$y' = 1 + Ce^{-x} \equiv 1 + x - x + C\,e^{-x}.$$

The differential equation does become an identity.

This simple test should always be carried out as a matter of course after solving any not entirely simple differential equation, for only the test can ensure absence of error. We omit the test here quite often, but only for lack of space; the reader should always carry it out, particularly of course, when he himself has worked out an example. Even an experienced manipulator may make errors.

A note on other types. By no means all differential equations given in comparatively simple form can be solved in closed form. Indeed, not even all integrals can be expressed by a finite combination of elementary functions e.g.,

$$\left(\int_0^x \frac{\sin x}{x}\,dx\right),$$

i.e., $xy' = \sin x$ is not integrable in closed form. Likewise the quite simply constructed equation $y' = x + y^2$ cannot be solved in closed form in terms of elementary functions, but leads to 'cylinder functions', i.e., Bessel functions.

A pure integration (a *quadrature*) is regarded as a simpler problem than solving a differential equation. So a differential equation is regarded as having been solved if it has been reduced to a quadrature problem, i.e., graphically considered, to finding an area.

5. The similarity differential equation

Equations of the form

$$y' = f\left(\frac{y}{x}\right) \tag{I.10}$$

will here be called similarity differential equations. In the literature they are often called homogeneous equations, but we want to keep the term 'homogeneous' to convey a different meaning.

To determine the isoclines, let y' be put equal to a constant a. This yields straight lines $y = Ax$ through the origin; on each of these lines the prescribed slope $y' = f(A)$ is constant.

If a solution $y(x)$ fits in with the direction field (Fig. I.8), then every curve obtained from it by a similarity transformation with respect to the origin also

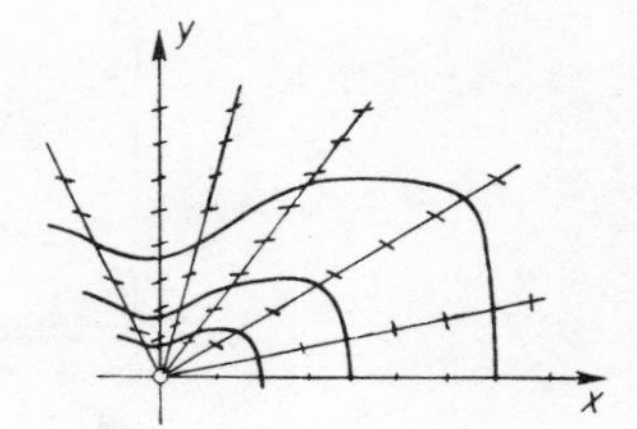

Fig. I.8. Direction field of a similarity differential equation

fits in with the direction field. (But it must not be concluded that all the solution curves of a similarity differential equation are similar to one another. A simple counter-example is provided by the differential equation

$$y' = \frac{x}{2y},$$

which has the solutions $x^2 - 2y^2 = c$. No two of the three solutions for $c = -1, 0, +1$ are similar to one another.)

Method of solution. To solve (I.10) the substitution

$$\frac{y(x)}{x} = z(x), \quad \text{i.e.,}\ y = xz,\ y' = z + xz'$$

suggests itself. In the new differential equation

$$z + xz' = f(z)$$

separation of the variables leads to our objective:

$$\int \frac{\mathrm{d}x}{x} = \int \frac{\mathrm{d}z}{f(z) - z} + C_1$$

or with $C_1 = -\ln|C|$

$$\ln|Cx| = \int \frac{\mathrm{d}z}{f(z) - z}. \tag{I.11}$$

After the integration just the reverse transformation has to be carried out. (See (I.64) for a worked-out example.) If the denominator in (I.11) has a zero ξ, *i.e.*, if $f(\xi) - \xi = 0$, then in addition to the above solution the solution $y = \xi x$ also arises.

6. Simple Cases Reducible to the Similarity Differential Equation

A particular case of a similarity differential equation is

$$\frac{\mathrm{d}y}{\mathrm{d}x} = y' = \frac{ax + by}{ax + \mathrm{d}y} = \frac{a + b(y/x)}{c + d(y/x)}.$$

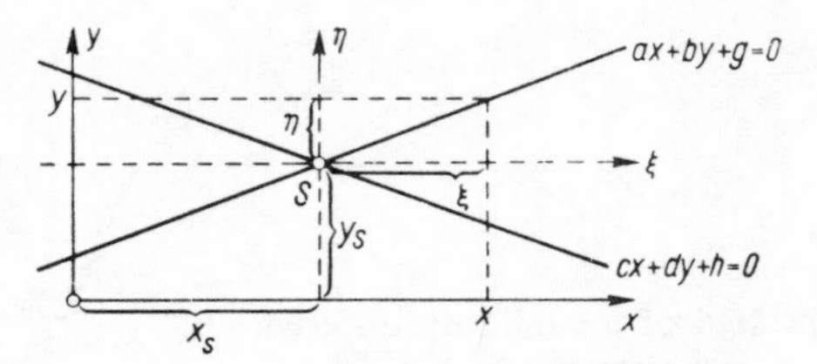

Fig. I.9. Diagram for the differential equation

$$y' = (ax + by + g)/(cx + dy + h)$$

The more general differential equation

$$\frac{\mathrm{d}y}{\mathrm{d}x} = y' = \frac{ax + by + g}{cx + dy + h} \tag{I.13}$$

can also be reduced to this form by a suitable transformation of co-ordinates. The numerator and denominator when equated to zero each represent a straight line (Fig. I.9)

$$ax + by + g = 0$$

and

$$cx + dy + h = 0.$$

Suppose these lines intersect in a point S, *i.e.*, suppose that the determinant

$$ad - bc \neq 0.$$

If their equations are transformed into a new co-ordinate system ξ, η with its origin at the point of intersection $S x = x_s, y = y_s$ of the straight lines and the transformation relations

$$x = x_s + \xi, \qquad y = y_s + \eta$$

hold, then the constants g and h drop out, and under the transformation the differential equation becomes

$$\frac{\mathrm{d}\eta}{\mathrm{d}\xi} = \frac{a + b(\eta/\xi)}{c + d(\eta/\xi)}.$$

We thus arrive at the form of equation (I.12) already discussed.

But if $ad - bc = 0$ and if b and d are not simultaneously both zero (say $b \neq 0$), then the differential equation can be brought into an integrable form by putting

$$ax + b \cdot y(x) + g = z(x)$$

Introduction of polar co-ordinates. Similarity differential equations can often be greatly simplified by transforming them into polar co-ordinates. (Fig. I.10). The transformation equations for the co-ordinates and the differentials

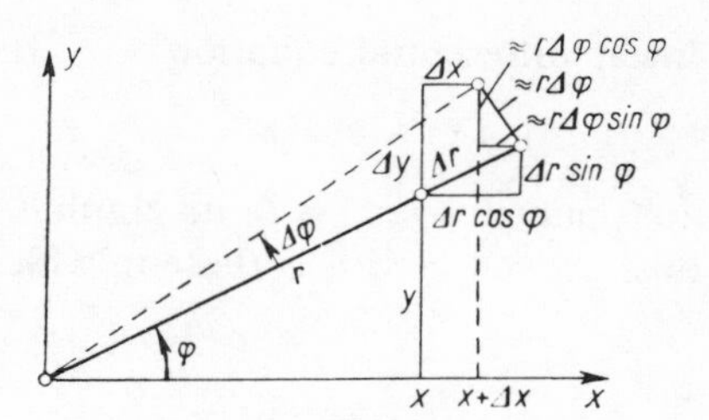

Fig. I.10. Illustrating the introduction of polar co-ordinates

are

$$\left.\begin{aligned} x &= r\cos\varphi \\ y &= r\sin\varphi \end{aligned}\right\} \qquad \left.\begin{aligned} \mathrm{d}x &= \mathrm{d}r\cos\varphi - r\sin\varphi\,\mathrm{d}\varphi \\ \mathrm{d}y &= \mathrm{d}r\sin\varphi + r\cos\varphi\,\mathrm{d}\varphi \end{aligned}\right\} \tag{I.14}$$

$$r = \sqrt{x^2 + y^2}$$

$$\varphi = \text{arc tan}\,\frac{y}{x}$$

(Cf. the worked example of equation (I.64) in §7.)

§2. THE LINEAR DIFFERENTIAL EQUATION OF FIRST ORDER

7. Homogeneous and Inhomogeneous Equations; the Trivial Solution

Another integrable type of differential equation is the so-called *linear differential equation* (cf. No. 1)

$$y'(x) + p(x)y = r(x). \tag{I.15}$$

$p(x)$ and $r(x)$ are given functions, perhaps continuous; for example, $y' + x^2y = \mathrm{e}^x$ is such a linear equation.

If y' still has a function $q(x)$ as a factor, the whole equation is to be divided by this function, provided that $q(x) \neq 0$.

If the right-hand side $r(x)$ vanishes identically, the equation is said to be *homogeneous.* Thus $y' + x^2y = 0$ is the homogeneous equation corresponding to the above example.

But if the right-hand side does not vanish identically, i.e., if $r(x)^2 \not\equiv 0$, then the differential equation is said to be *inhomogeneous.* In such an equation $r(x)$ is called the 'perturbing function' or the 'perturbing term'.

The nomenclature *homogeneous and inhomogeneous* is analogous to that used for linear systems of algebraic equations. Homogeneous linear systems of equations in algebra always have the so-called 'trivial' solution (all the unknowns are zero), but the 'non-trival solutions' are of special interest.

The homogeneous linear differential equation

$$y'(x) + p(x)y = 0 \tag{I.16}$$

also has a trivial solution, namely, $y(x) \equiv 0$; its graph is the x-axis. But there are 'non-trivial solutions' as well, and it is these in which we are much more interested.

8. Solution of a homogeneous equation

Separation of the variables in (I.16) gives

$$\int \frac{dy}{y} = -\int p(x)\,dx + C_1. \tag{I.17}$$

Let $P(x)$ denote an antiderivative of the integral on the right; then

$$\ln|y| = -P(x) + C_1.$$

Hence

$$y(x) = C_2 \exp\left(-\int^x p(\xi)\,d(\xi)\right), \qquad y = C_2 e^{-P(x)}. \tag{I.18}$$

For the sake of clarity the letter ξ has been used as the variable of integration to distinguish it from the upper limit x of the integral. It does not matter what lower limit x_0 is chosen; often 0 is convenient (of course, x_0 must lie in the domain of the function $p(x)$).

Example

$$y' + x^2 y = 0.$$

The solution is

$$y = C \exp\left(-\int_0^x \xi^2\,d\xi\right) = C \exp\left(-\frac{1}{3}x^3\right)$$

or, taking x_0 as the lower limit and writing $y_0 = y(x_0)$,

$$y(x) = y_0 \exp\left(-\int_{x_0}^x \xi^2\,d\xi\right) = y_0 \exp\left(\frac{1}{3}(x_0^3 - x^3)\right).$$

For $C = 0$ or $y_0 = 0$ we obtain the trivial solution $y(x) \equiv 0$.

9. Solution of the inhomogeneous equation

The homogeneous equation has the general solution $y = C\,e^{-P(x)}$ with C as an arbitrary constant. We wish to try to account for the 'distortion' of the solution by the 'perturbing term' by putting $y = C(x)\,e^{-P(x)}$, i.e., we transform the differential equation (I.15) into a new one for the function $C(x)$, hoping that the new equation will be simpler. Thus in place of the constant C a function $C(x)$, which is to be determined, appears; the method is therefore known as the method of *variation of the constants* (*Lagrange*). From

$$y(x) = C(x) \exp\left(-\int^x p(\xi)\,\mathrm{d}\xi\right) \tag{I.19}$$

it follows that

$$y'(x) = C'(x) \exp\left(-\int^x p(\xi)\,\mathrm{d}\xi\right) + C(x) \exp\left(-\int^z p(\xi)\,\mathrm{d}\xi\right)(-p(x)).$$

Substitution into the differential equation (I.15) gives

$$C'(x) = r(x) \exp\left[\int^z p(\xi)\,\mathrm{d}\xi\right].$$

By integrating we obtain immediately

$$C(x) = C_2 + \int^x r(\eta) \exp\left[\int^n p(\xi)\,\mathrm{d}\xi\right] \mathrm{d}\eta.$$

So the general solution of the inhomogeneous linear differential equation (I.15) is

$$y(x) = \exp\left[-\int^x p(\xi)\,\mathrm{d}\xi\right]\left\{C + \int^x r(\eta) \exp\left[\int^\eta p(\xi)\,\mathrm{d}\xi\right] \mathrm{d}\eta\right\}. \tag{I.20}$$

The method of variation of the constants can also be applied to linear differential equations of higher order (No. II.16).

Example 1.

To illustrate the effect of the 'perturbing function' we compare the direction fields of the differential equations

$$\begin{aligned} y' - y &= 0 \qquad &&\text{(homogeneous linear)} \\ y' - y &= x \\ y' - y &= \sin x \end{aligned} \quad \Bigg\} \ \text{(inhomogeneous linear)}$$

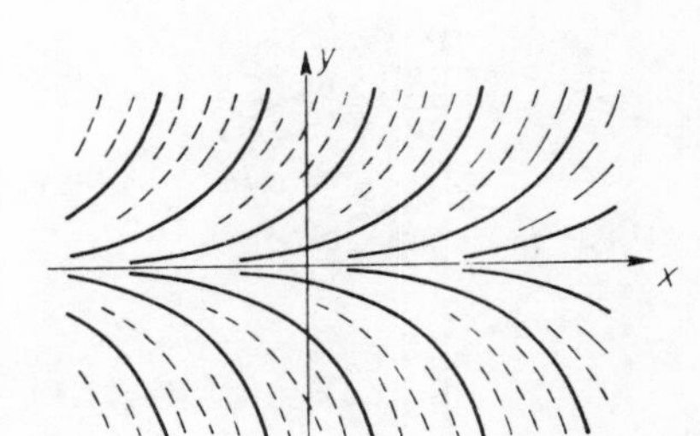

Fig. I.11. Direction field of the differential equation $y' = y$

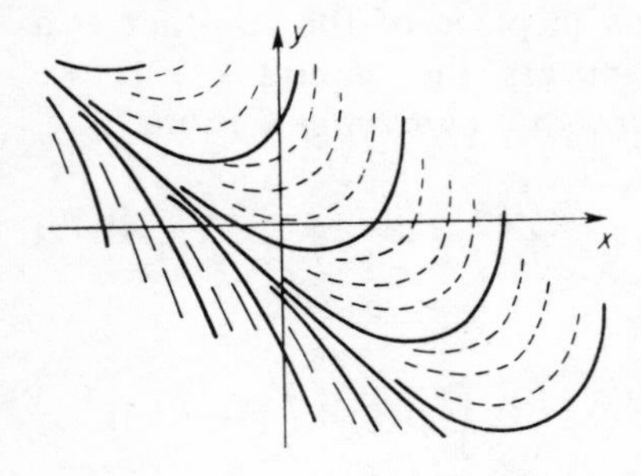

Fig. I.12. Direction field of the differential equation $y' = y + x$

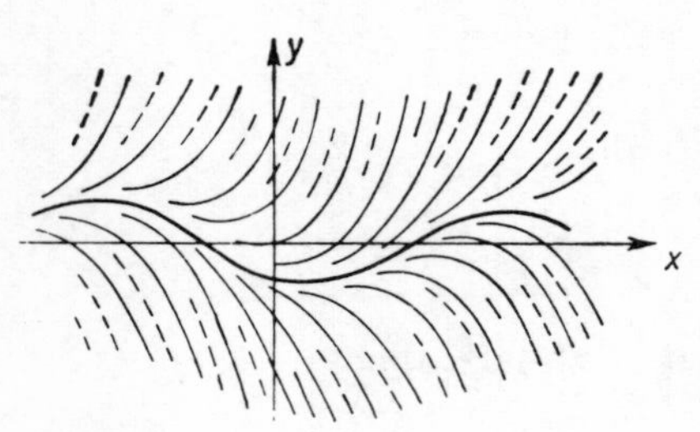

Fig. I.13. Direction field of the differential equation $y' = y + \sin x$

Their general solutions are:

$$y = C\,e^x, \qquad \text{Fig. I.11}$$

$$y = C\,e^x - 1 - x, \qquad \text{Fig. I.12}$$

$$y = C\,e^x - \tfrac{1}{2}\sin x - \tfrac{1}{2}\cos x, \qquad \text{Fig. I.13}$$

Example 2. Bending stress in a bent rod

In a more exact calculation of the bending stresses in a simple rod under flexure, namely, by considering the normal stress $\sigma = \sigma(v)$, which depends on the distance v from the neutral axis passing through the centroid of the rod's cross-sectional area (Fig. I.14), we obtain the differential equation

$$\frac{d\sigma}{dv} - \frac{\mu}{r - v}\sigma + \frac{\mu\alpha}{r - v} = 0.$$

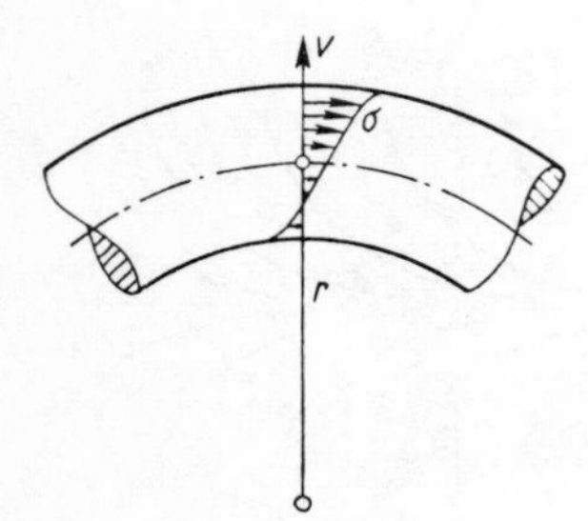

Fig. I.14. Distribution of bending stress in a rod under flexure

Here, r is the radius of curvature of the rod's axis, μ is a constant of elasticity, and α is a constant depending on the particular normal force, the bending moment, and the geometrical dimensions.

From the general solution (I.20) of a linear differential equation we can immediately find the solution for this case too. Making the appropriate changes in notation

$$x \sim v, \qquad y \sim \sigma, \qquad p(x) \sim -\frac{\mu}{r-v}, \qquad r(x) \sim \frac{-\mu\alpha}{r-v}, \qquad e^{P(x)} \sim (r-v)^{\mu},$$

we have

$$\sigma(v) = (r-v)^{-\mu}\left[C - \mu\alpha \int^{v} (r-\eta)^{\mu-1}\, d\eta\right] = \alpha + C(r-v)^{-\mu}.$$

Here we know that in the fibre $v = 0$ only the normal stress σ_0 acts, and so, for $v = 0$ we have $\sigma = \sigma_0$. Hence we can determine the constant C so as to pick out from the general solution the special case under consideration:

$$C = (\sigma_0 - \alpha)r^{\mu}.$$

Hence finally

$$\sigma(v) = \alpha + (\sigma_0 - \alpha)\left(1 - \frac{v}{r}\right)^{-\mu}.$$

§3. THE BERNOULLI DIFFERENTIAL EQUATION

10. Reduction to a linear differential equation

An equation similar to (I.15) but with $r(x)$ multiplied by y^n is known as a 'Bernoulli differential equation':

$$y'(x) + p(x)y = r(x)y^n; \qquad (I.21)$$

(here we assume that $n \neq 1$ and $n \neq 0$; for in either of these cases, (I.21) is merely a linear differential equation).

We make a general substitution

$$y^k(x) = z(x), \qquad y = z^{1/k}, \qquad y' = \frac{z'}{k} z^{(1-k)/k}$$

with $k \neq 0$, and try to determine the most suitable k. Multiplying (I.21) by $z^{(k-1)/k}$ we obtain

$$\frac{z'}{k} + pz = rz^{(k-1+n)/k};$$

the exponent of z on the right-hand side of the new equation can be reduced to 0 by choosing $k = 1 - n$, and then the differential equation becomes a linear inhomogeneous equation.

Thus

$$y^{1-n}=z,\ \frac{z'}{1-n}+pz=r.$$

is the appropriate substitution for solving a Bernoulli differential equation by first reducing it to a linear differential equation.

Example

$$y'+\frac{y}{x}=x^2y^2.$$

With

$$y^{1-n}=\frac{1}{y}=z,\qquad y'=-z'z^{-2},$$

it follows that

$$-z'z^{-2}+\frac{z^{-1}}{x}=x^2z^{-2},\quad \text{or}\quad z'-\frac{z}{x}=-x^2.$$

As in §2 we first integrate the corresponding homogeneous differential equation

$$z'-\frac{z}{x}=0,\qquad \int\frac{\mathrm{d}z}{z}=\int\frac{\mathrm{d}x}{x}+C_1,$$

$$\ln|z|=\ln|x|+\ln|C_2|,\qquad (C_1=\ln|C_2|),\qquad z=C_2x.$$

'Variation of the constants'
$z=C_2(x)x,\qquad z'=C_2'(x)xw+C_2(x)$
gives for the inhomogeneous equation

$$C_2'(x)=-x,\qquad C_2(x)=-\frac{x^3}{2}+C,$$

$$z=\left(-\frac{x^2}{2}+C\right)x=-\frac{x^3}{2}+Cx,$$

and the required general solution for y is

$$y=\frac{1}{Cx-x^3/2}.$$

11. The Riccati Differential Equation

As a further type, we mention the so-called 'Riccati differential equation'

$$y'(x)+p(x)y+r(x)y^2=q(x). \tag{I.22}$$

If one particular solution of this is known, say

$$y=u(x)$$

then (I.22) can be reduced to a Bernoulli differential equation by putting

$$y = u(x) + v(x).$$

This gives

$$\underbrace{u' + pu + ru^2}_{=q} + v' + (p + 2ur)v + rv^2 = q.$$

Thus v is indeed given by the Bernoulli differential equation

$$v' + \{p(x) + 2u(x)r(x)\}v = -r(x)v^2.$$

Example

$$y' - (1 - 2x)y + y^2 = 2x.$$

By trial we find the solution $y \equiv 1$ and so we put $y = 1 + v(x)$ to obtain the Bernoulli equation

$$v' + (1 + 2x)v = -v^2$$

and then put $v = 1/z$. The equation for z is

$$z' - (1 + 2x)z = 1$$

and by (I.20) it has the general solution

$$z = e^{x(1+x)}\left(C + \int^x e^{-\xi(1+\xi)}\, d\xi\right).$$

The integral appearing here cannot be evaluated in closed form (cf. *Gauss's error integral* $\int e^{-x^3}\, dx$).

With $y = 1 + 1/z$, the general solution of the original equation reads

$$y = 1 + \frac{e^{-x(1+x)}}{C + \int^x e^{-\xi(1+\xi)}\, d\xi}.$$

Tables or numerical quadrature formulae may be used to evaluate the integral.

§4. INTEGRATING FACTORS

12. Exact differential equations

Let a differential equation $y' = \phi(x, y)$ be written in the form

$$f(x, y)\, dx + g(x, y)\, dy = 0.$$

This is called an *exact* differential equation if

$$f(x, y) = \frac{\partial z}{\partial x}, g(x, y) = \frac{\partial z}{\partial y},$$

and then $z(x, y) = \text{const} = c$ is the general solution in implicit form. Now it is well known that the mixed second derivatives of a function $z(x, y)$ are identical with one another provided they are continuous:

$$\frac{\partial^2 z}{\partial x \partial y} = \frac{\partial^2 z}{\partial y \partial x},$$

hence we obtain a simple necessary condition for a differential equation to be exact, viz.,

$$\frac{\partial f}{\partial y} = \frac{\partial g}{\partial x}.$$

However, cases where a differential equation satisfies this condition immediately rarely occur.

13. Integrating Factors

Sometimes, however, a differential equation can easily be changed into an exact differential equation by multiplying it throughout by a suitable, so-called *integrating factor* $\mu(x, y)$. We therefore require of $\mu(x, y)$ that for

$$\mu(x, y) f(x, y)\, \mathrm{d}x + \mu(x, y) g(x, y)\, \mathrm{d}y = 0$$

the condition

$$\frac{\partial}{\partial y}(\mu f) = \frac{\partial}{\partial x}(\mu g) \tag{I.23}$$

shall be satisfied, or, differentiating out,

$$\frac{\partial \mu}{\partial x} g - \frac{\partial \mu}{\partial y} f + \mu\left(\frac{\partial g}{\partial x} - \frac{\partial f}{\partial y}\right) = 0.$$

At first sight it seems as if the difficulties have been increased, because now, instead of an ordinary differential equation, we have to solve a partial differential equation for the function $\mu(x, y)$. But sometimes a particular solution of this partial differential equation can be guessed, and any one solution (which must, of course, not vanish identically) is all that is needed.

Example

$$y' = \frac{\mathrm{d}y}{\mathrm{d}x} = -\frac{xy^3}{1 + 2x^2y^2}$$

or

$$xy^3\, \mathrm{d}x + (1 + 2x^2y^2)\, \mathrm{d}y = 0.$$

Here

$$f(x, y) = xy^3 \quad \text{and} \quad g(x, y) = 1 + 2x^2y^2.$$

The equation (I.23) for μ here is

$$\frac{\partial \mu}{\partial y} xy^3 + \mu 3xy^2 = \frac{\partial \mu}{\partial x}(1 + 2x^2y^2) + \mu 4xy^2.$$

It is the expression in the bracket which is particularly awkward here. To make it vanish, we try to get by with

$$\frac{\partial \mu}{\partial x} = 0$$

i.e., to make μ depend only on y. The equation for μ then reduces to the simple form

$$y \frac{\mathrm{d}\mu}{\mathrm{d}y} = \mu \quad \text{with the solution } \mu = Cy.$$

Since only one solution is needed, we take $C = 1$. If we now multiply the differential equation by the integrating factor $\mu = y$ to get

$$xy^4 \,\mathrm{d}x + (y + 2x^2y^3)\,\mathrm{d}y = 0,$$

then it follows from $\frac{\partial z}{\partial x} = xy^4$ by integration that

$$y = y^4 \int x \,\mathrm{d}x + w_1(y) = \tfrac{1}{2} x^2y^4 + w_1(y).$$

The function $w_1(y)$ appears instead of an integration constant because we have regarded y as constant during the integration. Similarly

$$\frac{\partial z}{\partial y} = y + 2x^2y^3$$

integrates to give

$$z = \tfrac{1}{2} y^2 + \tfrac{1}{2} x^2y^4 + w_2(x).$$

Comparing the two solutions for z shows that we must have

$$w_1(y) = \tfrac{1}{2} y^2 + \text{const} \quad \text{and} \quad w_2(x) = \text{const} = C_1.$$

Hence the general solution of our original equation reads

$$y^2 + x^2y^4 = C.$$

In general the determination of a particular solution of the partial differential equation for μ may be very difficult. Equations which are solvable by means of an integrating factor can often be solved more easily in a different way. Our example above could be regarded as a Bernoulli differential equation for the function $x(y)$ inverse to the required function $y(x)$ (this artifice is often worthwhile):

$$\frac{\mathrm{d}x}{\mathrm{d}y} + \frac{2}{y} x = -\frac{1}{y^3} x^{-1},$$

Here the rôles of x and y have been interchanged. With

$$x^2 = z; \qquad \frac{dx}{dy} = \frac{1}{2z^{1/2}} \frac{dz}{dy}$$

we obtain the differential equation

$$\frac{dz}{dy} + \frac{4}{y} z = -\frac{2}{y^3},$$

the solution of which is

$$z = x^3 = Cy^{-4} - y^{-2}$$

and hence

$$y^2 + x^2 y^4 = C,$$

as above.

§5. PRELIMINARIES TO THE QUESTIONS OF EXISTENCE AND UNIQUENESS

14. SINGLE-VALUED AND MULTI-VALUED DIRECTION FIELDS

In differential equations with multi-valued functions $y' = f(x, y)$, i.e., with a multi-valued slope, *e.g.*,

$$y' = x \pm \sqrt{2 + x^2 + y^2},$$

the general solution consists of several families of solutions; in the above example, it consists of two one-parameter families of solutions, since with each point there are associated two directions, because of the alternative signs before the square root (Fig. I.15).

An example of infinitely-many-valued fields is the one-parameter family of tangents to a sine curve (Fig. I.16); at each point there are indeed infinitely many, but still only definite discrete slopes which satisfy the differential equation.

The equation of the tangent at a point (ξ, sin ξ) is, as shown in Fig. I.16,

$$y - \sin \xi = (x - \xi)\cos \xi.$$

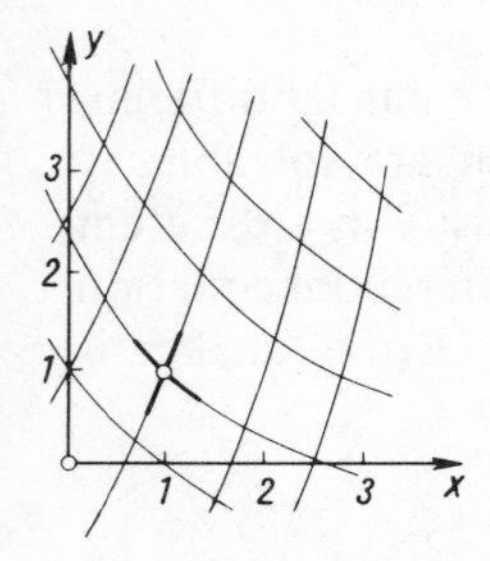

Fig. I.15. The two-valued direction field for the differential equation

$$y' = x \pm \sqrt{2 + x^2 + y^2}$$

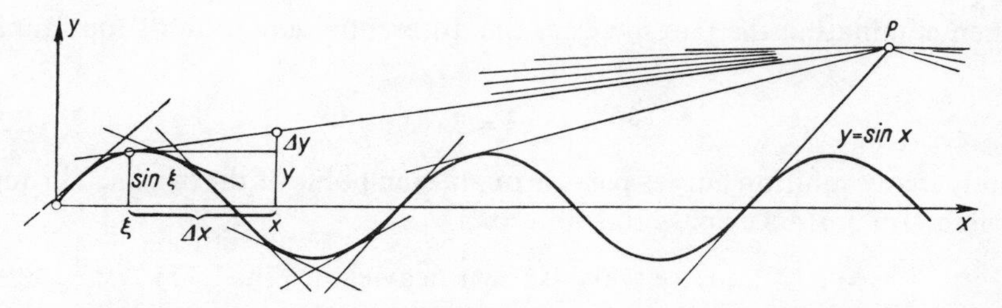

Fig. I.16. The infinitely-many-valued direction field of the tangents to a sine curve

Eliminating the parameter ξ by using

$$\cos \xi = y', \qquad \sin \xi = \sqrt{1 - y'^2}$$

we obtain the differential equation

$$y - \sqrt{1 - y'^2} = xy' - y' \operatorname{arc\,cos} y'.$$

15. NON-UNIQUENESS OF THE SOLUTION

For physical applications it is important to know sufficient conditions for uniqueness of the solution of a differential equation.

That it is not sufficient for the slope function $y' = f(x, y)$ to be single-valued and continuous to ensure uniqueness of the solution is shown by the following example. For the 1-parameter family of cubical parabolas, Fig. I.17,

$$y = (x - c)^3$$

we obtain, by differentiating

$$y' = 3(x - c)^2$$

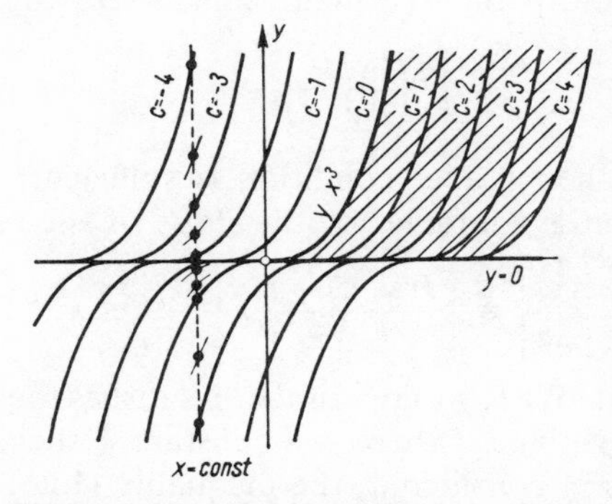

Fig. I.17. Continuity of the slope function $f(x, y)$ is not sufficient to ensure uniqueness of the solution

and then eliminating the parameter c, the differential equation of the family, viz.,

$$y' = 3\sqrt[3]{y^2} = 3y^{2/3}. \tag{I.24}$$

Infinitely many solution curves pass through each point of the x-axis. Through the origin, for instance, pass the solutions

$$y = x^3 \quad \text{and} \quad y \equiv 0 \quad \text{(shown heavier in Fig. I.17)}$$

and also the solutions which consist of a bit of the x-axis and half of a cubical parabola joined to it:

$$y = \begin{cases} 0 & \text{for } 0 \leqslant x \leqslant a \\ (x-a)^3 & \text{for } \qquad x \geqslant a, \end{cases}$$

where $a > 0$ can be chosen arbitrarily.

All these solutions satisfy the differential equation and cover the 'funnel' of solutions associated with the origin $x = y = 0$, which is shaded in Fig. I.17. (Cf. L. Bieberbach, *Einführung in die Theorie der Differentialgleichungen im reellen Gebiet*, Berlin, 1965, p. 44.) Thus even with a single-valued continuous function $f(x, y)$, branchings are possible.

16. The Lipschitz condition; its stronger and weaker forms

The non-uniqueness in the last example is due to the fact that in the differential equation (I.24) the function $y' = f(x, y)$ increases 'too rapidly' with y. It will turn out that, in single-valued direction fields, uniqueness of solution is ensured if the absolute value of

$$\frac{\partial f(x, y)}{\partial y}$$

remains bounded, i.e., if there is a constant k, a so-called 'Lipschitz constant' such that, in the region of the x, y-plane considered, the condition

$$\left|\frac{\partial f(x, y)}{\partial y}\right| \leq k. \tag{I.25}$$

holds. Actually, a rather weaker condition is sufficient; we can replace the slope of the tangent in a y-interval by the slope of the secant (Fig. I.18):

$$\left|\frac{\Delta f}{\Delta y}\right| \equiv \left|\frac{f(x, y) - f(x, y^*)}{y - y^*}\right| \leq k. \tag{I.26}$$

Here x is held constant. We say that a function satisfies the *Lipschitz condition* in a region of the x, y-plane if there is a constant k such that, for all values of x, y, y^* in the region considered, the inequality (I.26) is satisfied.

For cubical parabolas

$$y = (x-c)^3 \quad \text{with} \quad y' = f(x, y) = 3y^{2/3}$$

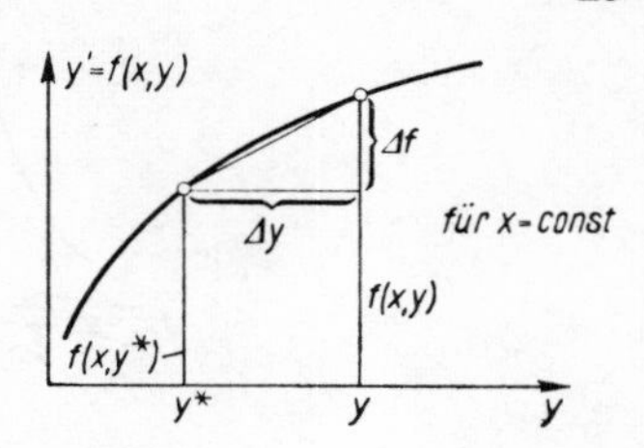

Fig. I.18. Illustrating the Lipschitz condition

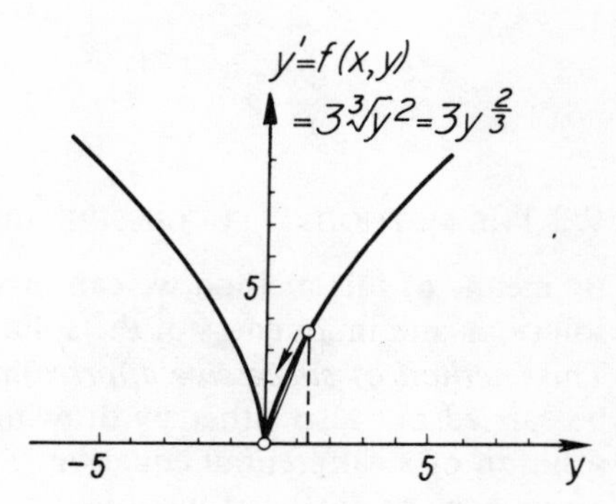

Fig. I.19. The Lipschitz condition is infringed

the Lipschitz condition in both its forms is infringed at the point $y = 0$: the derivative is not bounded (Fig. I.19):

$$\left|\frac{\partial(3y^{2/3})}{\partial y}\right| = \left|\frac{2}{\sqrt[3]{y}}\right| \to \infty \quad \text{as} \quad y \to 0,$$

and the slopes of the secants remain unbounded:

$$\left|\frac{f(x,0) - f(x,y)}{0 - y}\right| = \left|\frac{3}{\sqrt[3]{y}}\right|.$$

That the first form of the Lipschitz condition demands more than the second is shown, for instance, by the example in Fig. I.20 for the differential equation $y' = f(x, y) = |y|$ with the solutions $y = 0$ and $|C|\,e^{x}, -|C|\,e^{-x}$ (Fig. I.21). For $y' = f(x, y) = |y|$ (see Fig. I.20) a slope $\partial f/\partial y$ is not defined at the point $y = 0$, and so the Lipschitz condition in its first form (I.25) is not satisfied; but it does hold in its second form (I.26), since the slope of the secant is always defined and is never more than 1 in absolute value, and hence it is bounded. The differential form of the condition demands more, but usually it can be more easily checked.

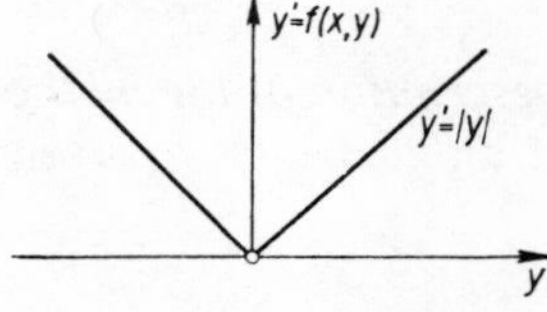

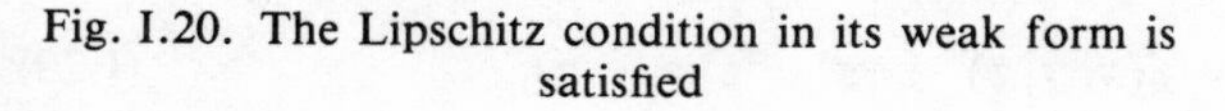
Fig. I.20. The Lipschitz condition in its weak form is satisfied

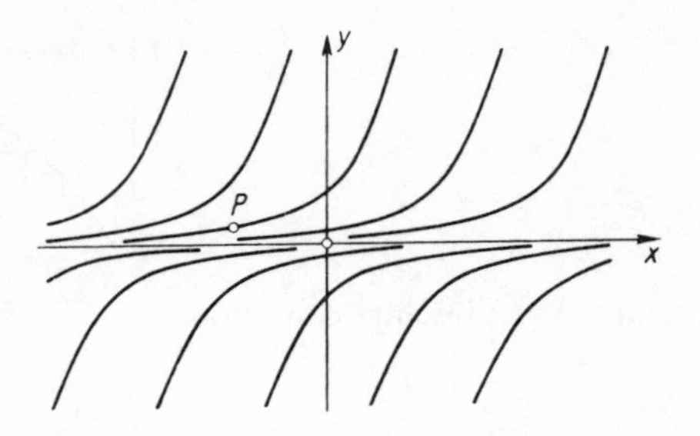

Fig. I.21. For the differential equation
$y' = |y|$

17. THE METHOD OF SUCCESSIVE APPROXIMATIONS

By means of this method we can prove a general theorem on the existence of solutions and uniqueness of the solution if the Lipschitz condition is satisfied. This *method of successive approximations* is an iterative method, which can be carried out also either by drawing or by calculation, for constructing the solution of a differential equation. Starting from an arbitrary function $y_0(x)$ a sequence of further functions

$$y_1(x), y_2(x), \ldots, y_n(x)$$

is formed, which converges towards the required solution function.

The argument is of particularly wide applicability because we can always reduce any differential equation of higher order to a system of 1st-order equations, and so we can obtain propositions about differential equations of any order.

Differential equations of higher order and systems of differential equations. If we have, for example, the differential equation

$$y^{(n)} = f[x, y, y', y'', \ldots, y^{(n-1)}] \tag{I.27}$$

we can regard the individual derivatives as functions of x:

$$y = y_{(1)}(x) \qquad y' = y_{(2)}(x), \qquad y'' = y_{(3)}(x), \ldots, \qquad y^{(n-1)} = y_{(n)}(x).$$

Thus instead of the nth-order differential equation, we can consider the system of n first-order differential equations

$$\left.\begin{aligned} y'_{(1)} = y_{(2)}, \qquad y'_{(2)} = y_{(3)}, \qquad y'_{(3)} = y_{(4)}, \ldots, \qquad y'_{(n-1)} = y_{(n)} \\ y'_{(n)} = y^{(n)} = f(x, y_{(1)}, y_{(2)}, y_{(3)}, \ldots, y_{(n)}) \end{aligned}\right\}. \tag{I.28}$$

Description of the method. In this section, we restrict ourselves, for the sake of an elementary presentation, to a system of $n = 2$ differential equations

$$y' = f(x, y, z) \qquad z' = g(x, y, z), \tag{I.29}$$

but we emphasize that the arguments can be carried through for any n (cf. No. 20 on this point). We choose two arbitrary functions $y_0(x)$ and $z_0(x)$ as the initial functions or the 'zeroth approximations', for example, by using the initial conditions

$$y = y_0, \qquad z = z_0 \quad \text{for } x = x_0,$$

we could take the straight lines

$$y_0(x) = \text{const} = y_0, \qquad z_0(x) = \text{const} = z_0.$$

If the rough shape of the solution curves can already be seen from the sort of problem under consideration, we can start off from better approximation functions, but their curves must always pass through the starting point $x = x_0$, $y = y_0$, $z = z_0$.

We now put $y_0(x)$ and $z_0(x)$ into the right-hand sides of the differential equations (I.29) and determine the next approximations $y_1(x)$ and $z_1(x)$ from

$$y'_1(x) = f[x, y_0(x), z_0(x)]. \qquad z'_1(x) = g[x, y_0(x), z_0(x)].$$

By integrating and using the initial conditions we obtain the first approximations $y_1(x)$ and $z_1(x)$:

$$\left.\begin{aligned} y_1(x) &= y_0 + \int_{x_0}^{x} f[\xi, y_0(\xi), z_0(\xi)]\,\mathrm{d}\xi, \\ z_1(x) &= z_0 + \int_{x_0}^{x} g[\xi, y_0(\xi), z_0(\xi)]\,\mathrm{d}\xi, \end{aligned}\right\}$$

Using these we repeat the process to obtain the second approximations $y_2(x)$ and $z_2(x)$, and so on; see Fig. I.22.

We can express this in another way. We introduce an 'operator T' which transforms every pair of continuous functions $u_1(x)$, $u_2(x)$ into a new pair of continuous functions $v_1(x)$, $v_2(x)$ according to the prescription

$$\begin{Bmatrix} v_1(x) \\ v_2(x) \end{Bmatrix} = T\begin{Bmatrix} u_1(x) \\ u_2(x) \end{Bmatrix} = \begin{Bmatrix} y_0 \\ z_0 \end{Bmatrix} + \int_{x_0}^{z} \begin{Bmatrix} f[\xi, u_1(\xi), u_2(\xi)] \\ g[\xi, u_1(\xi), u_2(\xi)] \end{Bmatrix} \mathrm{d}\xi.$$

The reader will, no doubt, immediately understand this way of writing the rule: in each curved bracket one is to take either always the top element or always the bottom element. But perhaps it becomes clearer if we use vector notation:

$$U = \begin{pmatrix} u_1(x) \\ u_2(x) \end{pmatrix}; \qquad V = \begin{pmatrix} v_1(x) \\ v_2(x) \end{pmatrix}; \qquad F = \begin{pmatrix} f \\ g \end{pmatrix};$$

$$U_n = \begin{pmatrix} y_n(x) \\ z_n(x) \end{pmatrix}; \qquad Y_0 = \begin{pmatrix} y_0 \\ z_0 \end{pmatrix}.$$

This will also prepare us for the case discussed in No. 20, where there are s functions, which we denote by $y_{(1)}, y_{(2)}, \ldots, y_{(s)}$.

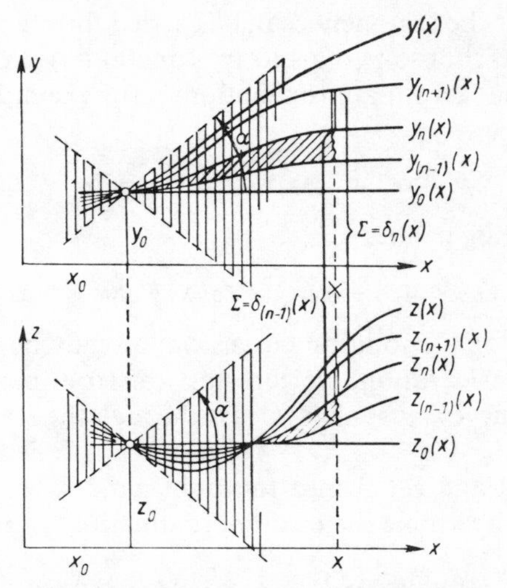

Fig. I.22. The method of successive approximations

In this vector notation the prescription defining operator T is

$$V = TU = Y_0 + \int_{x_0}^{x} F(\xi, U(\xi))\, d\xi \tag{I.30}$$

The equations determing the $(n+1)$th approximation are

$$\left.\begin{aligned} y_{n+1}(x) &= y_0 + \int_{x_0}^{x} f[\xi, y_n(\xi), z_n(\xi)]\, d\xi, \\ z_{n+1}(x) &= z_0 + \int_{x_0}^{x} g[\xi, y_n(\xi), z_n(\xi)]\, d\xi. \end{aligned}\right\} \quad (n = 0, 1, 2, \ldots)$$

or

$$\begin{Bmatrix} y_{n+1}(x) \\ z_{n+1}(x) \end{Bmatrix} = T \begin{Bmatrix} y_n(x) \\ z_n(x) \end{Bmatrix}, \qquad (n = 0, 1, 2, \ldots)$$

or

$$U_{n+1} = TU_n.$$

If by chance we had put in the actual solution functions as our first approximations, then the process would simply reproduce these; i.e., we would find

$$y_1(x) = y_2(x) = \ldots = y(x) \quad \text{and} \quad z_1(x) = z_2(x) = \ldots = z(x).$$

Thus

$$\begin{Bmatrix} y(x) \\ z(x) \end{Bmatrix} = T \begin{Bmatrix} y(x) \\ z(x) \end{Bmatrix}, \tag{I.32}$$

i.e., the pair $y(x), z(x)$ is a 'fixed point' of the operator T.

§6. THE GENERAL EXISTENCE AND UNIQUENESS THEOREM

18. The existence theorem

For the system of differential equations

$$y' = f(x, y, z); \qquad z' = g(x, y, z)$$

'Peano's theorem'—which we quote here without proof—states that the existence of a solutions system $y(x)$, $z(x)$ is assured provided that the functions $f(x, y, z)$ and $g(x, y, z)$ are continuous and bounded.

But to prove the *uniqueness* we require rather more, for example, the 'Lipschitz conditions', which in our case can be written

$$\left.\begin{aligned} |f(x,y,z)-f(x,y^*,z^*)| &\le k\{|y-y^*|+|z-z^*|\} \\ |g(x,y,z)-g(x,y^*,z^*)| &\le k\{|y-y^*|+|z-z^*|\}. \end{aligned}\right\} \qquad \text{(I.33)}$$

In this form, they resemble the first terms of the Taylor expansion for functions of two independent variables and they correspond exactly to the condition in the simpler form used earlier in (I.26)

$$|f(x,y)-f(x,y^*)| \le k\,|y-y^*|.$$

Theorem

In the system of differential equations

$$y' = f(x, y, z), \qquad z' = g(x, y, z)$$

let the functions f and g be continuous, be bounded by

$$|f(x, y, z)| \le M, \qquad |g(x, y, z)| \le M \qquad \text{(I.34)}$$

and let the Lipschitz conditions (I.33) *be satisfied. Let these conditions hold for all x, y, y^*, z, z^* which lie in the domain*

$$|x-x_0| < h, \qquad |y-y_0| < hM, \qquad |z-z_0| < hM,$$
$$|y^*-y_0| < hM, \qquad |z^*-z_0| < hM.$$

Then there is in the interval $|x-x_0| < h$ one and only one system of functions $y(x)$, $z(x)$ which satisfies the given system of differential equations and the initial conditions $y(x_0) = y_0$, $z(x_0) = z_0$. The method of successive approximations (from §5, equation (I.31) *provides sequences of functions $y_n(x)$, $z_n(x)$ which converge uniformly to this system of functions.*

[Note. If (I.33) and (I.34) hold in a box $|x-x_0| \le a, |y-y_0| \le b, |z-z_0| \le c$, then we can take as h the smallest of the 3 numbers $a, b/M, c/M$.]

For the solution curves the conditions for boundedness of the functions

$$|y'| = |f(x, y, z)| \le M,$$
$$|z'| = |g(x, y, z)| \le M,$$

mean that the slopes are bounded, that they do not exceed a limiting angle $\delta = \tan^{-1} M$, see Fig. I.22; all the approximations and the solutions lie within the shaded sectors if $y_0(x)$ and $z_0(x)$ belong to these regions.

We investigate first only the existence of the solution, and its uniqueness later. If we write down (I.31) for the subscripts $n-1$ and n, then by subtraction we obtain for the differences

$$y_{n+1}(x) - y_n(x) = \int_{x_0}^{x} \{f[\xi, y_n(\xi), z_n(\xi)] - f[\xi, y_{n-1}(\xi), z_{n-1}(\xi)]\}\, d\xi,$$

$$z_{n+1}(x) - z_n(x) = \int_{x_0}^{x} \{g[\xi, y_n(\xi), z_n(\xi)] - g[\xi, y_{n-1}(\xi), z_{n-1}(\xi)]\}\, d\xi.$$

We can estimate the integrals, because the absolute value of a definite integral of a function is at most equal to the definite integral over the same interval of the absolute value of the function. Hence (for $x \geq x_0$)

$$\begin{aligned} |y_{n+1}(x) - y_n(x)| &\leq \int_{x_0}^{x} |f(\xi, y_n, z_n) - f(\xi, y_{n-1}, z_{n-1})|\, d\xi, \\ |z_{n+1}(x) - z_n(x)| &\quad \text{similarly,} \end{aligned} \tag{I.35}$$

or, using the Lipschitz condition (I.33),

$$\begin{aligned} |y_{n+1}(x) - y_n(x)| &\leq k \int_{x_0}^{x} \{|y_n - y_{n-1}| + |z_n - z_{n-1}|\}\, d\xi, \\ |z_{n+1}(x) - z_n(x)| &\leq \quad \text{the same bound.} \end{aligned}$$

Instead of proving the convergence of the two integrals separately we can prove, more simply, the convergence of their sum, which we denote by $\delta_n(x)$.

We therefore add the two inequalities

$$\underbrace{|y_{n+1}(x) - y_n(x)| + |z_{n+1}(x) - z_n(x)|}_{\delta_n(x)} \tag{I.36}$$

$$\leq 2k \int_{x_0}^{x} \{|y_n - y_{n-1}| + |z_n - z_{n-1}|\}\, d\xi \leq 2k \int_{x_0}^{x} \delta_{n-1}(\xi)\, d\xi$$

and show that the sum of all the $\delta_n(a)$ converges as $n \to \infty$.

For $y_1(x)$ (and similarly for $z_1(x)$) it follows from (I.31) that

$$|y_1(x) - y_0(x)| = \left| \int_{x_0}^{x} f(\xi, y_0, z_0)\, d\xi \right| \leq M|x - x_0|,$$

(we write the absolute value $|x - x_0|$, because the argument can be carried out for $x < x_0$ in exactly the same way; the solution can also be continued 'towards the left'); therefore

$$\delta_0(x) = |y_1(x) - y_0(x)| + |z_1(x) - z_0(x)| \leq 2M|x - x_0|.$$

Further, we have that

$$\delta_1(x) \leq 2k \int_{x_0}^{x} \delta_0(\xi)\,\mathrm{d}\xi \leq 2M\,2k \int_{x_0}^{x} (\xi - x_0)\,\mathrm{d}\xi = 2M\,2k\,\frac{|x - x_0|^2}{2!}$$

at first for $x \geq x_0$; the same bound holds for $x < x_0$. From this we can already see the rule for forming the general sum $\delta_n(x)$ of the differences, viz.,

$$\delta_n(x) \leq 2M(2k)^n \frac{|x - x_0|^{n+1}}{(n+1)!} = \frac{M}{k}\,\frac{(2k\,|x - x_0|)^{n+1}}{(n+1)!}, \tag{I.37}$$

which can easily be proved by induction, as follows.

If we assume this bound to be correct for the subscript n, then it follows from (I.36) that

$$\delta_{n+1} \leq 2k \int_{x_0}^{x} \delta_n(\xi)\,\mathrm{d}\xi \leq 2M(2k)^{n+1} \int_{x_0}^{x} \frac{|\xi - x_0|^{n+1}}{(n+1)!}\,\mathrm{d}\xi$$

$$= 2M(2k)^{n+1} \frac{|x - x_0|^{n+2}}{(n+2)!}$$

again first for $x \geq x_0$; but the same bound is also obtained for $x < x_0$. Thus the bound asserted by (I.37) for $n+1$ has been obtained; and since (I.37) holds for $n = 1$, it must therefore hold for all n.

We now consider a series $R(x)$, the partial sums of which are precisely the $y_n(x)$, and which will turn out to be the required solution function $y(x)$:

$$R(x) = y_0(x) + (y_1 - y_0) + (y_2 - y_1) + \cdots.$$

We shall prove that this series is *absolutely convergent,* i.e., that the series of its absolute values converges

$$|R(x)| \leq |y_0(x)| + |y_1 - y_0| + |y_2 - y_1| + \cdots.$$

This again can be estimated by the series

$$|R(x)| \leq |y_0(x)| + \delta_0(x) + \delta_1(x) + \delta_2(x) + \cdots.$$

If $2kh = \zeta$, then also

$$2k\,|x - x_0| \leq \zeta,$$

and we obtain as the 'majorant' the series, convergent for all ζ, for the function $\mathrm{e}^{\zeta} - 1$, namely,

$$|R(x)| \leq |y_0(x)| + \frac{M}{k}\underbrace{\left\{\frac{\zeta}{1!} + \frac{\zeta^2}{2!} + \frac{\zeta^3}{3!} + \cdots\right\}}_{\mathrm{e}^{\zeta} - 1}.$$

The series under investigation therefore converges absolutely for all x such that $|x - x_0| < h$. Hence the $y_n(x)$ tend to a limit function $R(x)$, and likewise $z_n(x) \to S(x)$. Moreover, the *convergence is uniform,* because the terms of the series are bounded by numbers, *independent of x,* which were terms of a convergent series. The uniformity of the convergence ensures, according to simple theorems in the theory of convergence, that the limit functions $R(x)$, $S(x)$ are continuous, and we may interchange the passage to the limit $n \to \infty$ and the integration:

$$\int_{x_0}^{x} f[\xi, y_n(\xi), z_n(\xi)]\, d\xi \to \int_{x_0}^{x} f[\xi, R(\xi), S(\xi)]\, d\xi$$

and similarly for the integral with $g(x, y, z)$; and so the limit functions $R(x), S(x)$ satisfy the equations

$$R(x) = y_0 + \int_{x_0}^{x} f[\xi, R(\xi), S(\xi)]\, d\xi, \qquad S(x) = z_0 + \int_{x_0}^{x} g[\xi, R(\xi), S(\xi)]\, d\xi$$

and therefore they are differentiable (because the integrands are continuous), and they satisfy the differential equations $y' = f(x, y, z)$, $z' = g(x, y, z)$. Thus $R(x)$ and $S(x)$ are solutions of the initial-value problem, and the existence of solutions of the initial-value problem has been proved.

19. Proof of uniqueness

The uniqueness of the solution system $y(x)$, $z(x)$ is proved by showing that the assumption that there is a second solution system $Y(x)$ and $Z(x)$ satisfying the differential equations leads to a contradiction. (In the proof we shall, for the sake of simplicity, consider only the 'right-hand' side $x \geq x_0$).

Suppose, then, that the differential equations

$$\begin{aligned} y' &= f(x, y, z) \\ z' &= g(x, y, z) \end{aligned} \quad \text{and} \quad \begin{aligned} Y' &= f(x, Y, Z) \\ Z' &= g(x, Y, Z) \end{aligned}$$

hold. If we write the differences of the two solutions as

$$y(x) - Y(x) = \eta(x), \qquad z(x) - Z(x) = \zeta(x).$$

then from

$$\begin{aligned} \eta' &= y' - Y' = f(x, y, z) - f(x, Y, Z), \\ \zeta' &= z' - Z' = g(x, y, z) - g(x, Y, Z) \end{aligned}$$

we obtain by integration

$$\eta(x) = \int_{x_0}^{x} \{f(\xi, y, z) - f(\xi, Y, Z)\}\, d\xi, \qquad \zeta(x) \text{ similarly.}$$

As above, using the Lipschitz conditions (I.33) and writing

$$|\eta(x)| + |\zeta(x)| = \delta(x) \quad \text{for } x \geq x_0$$

we obtain for the absolute values the inequalities

$$|\eta(x)| \leq \int_{x_0}^{x} |f(\xi, y, z) - f(\xi, Y, Z)|\, d\xi$$

$$\leq k \int_{x_0}^{x} \{|y - Y| + |z - Z|\}\, d\xi = k \int_{x_0}^{x} \{|\eta(\xi)| + |\zeta(\xi)|\}\, d\xi$$

$$= k \int_{x_0}^{x} \delta(\xi)\, d\xi.$$

Exactly the same bound holds for $|\zeta(x)|$, and hence we obtain for $\delta(x)$

$$\delta(x) \leq 2k \int_{x_0}^{x} \delta(\xi)\, d\xi.$$

If D is the maximum absolute values of δ in the interval $x_0 \leq x \leq x_0 + h^*$ and is attained at the point x_m say, see Fig. I.23, then

$$\delta(x) \leqq 2k \int_{x_0}^{x} \delta(\xi)\, d\xi \leq 2k \int_{x_0}^{x_0+h^*} \delta(\xi)\, d\xi \leq 2kh^* D.$$

Since this holds for every x and, in particular, for x_m, we also have

$$D \leq 2kh^* D.$$

Nothing has so far been assumed about h^*. If we now choose $h^* < 1/(2k)$, then, provided that $D \neq 0$, we have $D < D$. Thus there is a contradiction unless $\delta(x) \equiv 0$.

The interval $x_0 - h \leq x \leq x_0 + h$ can now be covered by a finite number of intervals of length h^*, and so uniqueness holds throughout the entire interval of length $2h$.

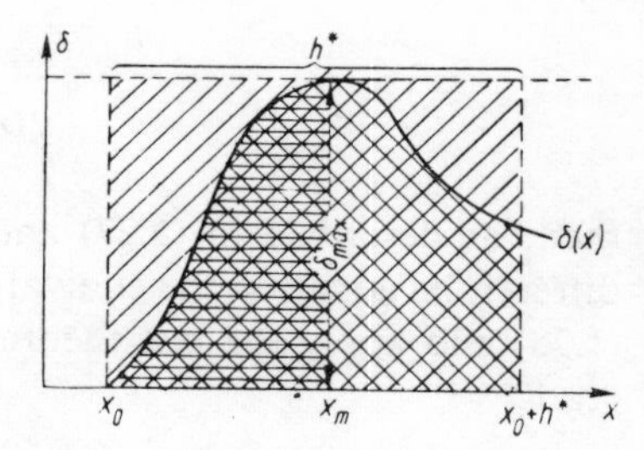

Fig. I.23. For the proof of uniqueness

20. SYSTEMS OF DIFFERENTIAL EQUATIONS AND A DIFFERENTIAL EQUATION OF *n*th ORDER

The proofs in Nos. 18 and 19 can be extended without any difficulty to the case of a system of s first-order differential equations for s functions ($s \geq 2$) to give the result:

For a given system of s first-order differential equations

$$\frac{dy_{(j)}(x)}{dx} = f_j[x, y_{(1)}(x), y_{(2)}(x), \ldots, y_{(s)}(x)] \quad (\textit{for } j = 1, 2, 3, \ldots, s) \quad (I.38)$$

for s unknown functions $y_{(1)}(x), y_{(2)}(x), y_{(3)}(x), \ldots, y_{(s)}(x)$ *with the initial conditions*

$$y_{(1)}(x) = y_{(1)0}, \qquad y_{(2)}(x) = y_{(2)0}, \ldots, \qquad y_{(s)}(x) = y_{(s)0} \quad \text{for } x = x_0 \tag{I.39}$$

the existence and uniqueness of the solution system is assured, provided that the functions f_j *appearing on the right-hand sides of the differential equations are continuous in the domain*

$$|x - x_0| < h, |y_{(j)} - y_{(j)0}| < hM_{(j)}$$

and satisfy the conditions

$$|f_j(x, y_{(1)}, y_{(2)}, \ldots, y_{(s)}| \leq M_{(j)} \quad \text{(the boundedness condition)} \tag{I.40}$$

and

$$\begin{aligned} &|f_j[x, y_{(1)}, y_{(2)}, \ldots, y_{(s)}] - f_j[x, \overset{*}{y}_{(1)}, \overset{*}{y}_{(2)}, \ldots, \overset{*}{y}_{(s)}]| \\ &\quad \leq \mathrm{k}\{|y_{(1)} - \overset{*}{y}_{(1)}| + \cdots + |y_{(s)} - \overset{*}{y}_{(s)}|\}. \end{aligned} \quad \text{(Lipschitz condition).} \tag{I.41}$$

An *n*th-order differential equation

Given an *n*th-order differential equation

$$y^{(n)} = f[x, y(x), y'(x), y''(x), \ldots, y^{(n-1)}(x)] \tag{I.42}$$

and the initial conditions at $x = x_0$

$$\begin{aligned} y(x_0) &= y_0 \\ y'(x_0) &= y_0' \\ \ldots \quad &\ldots \\ y^{(n-1)}(x_0) &= y_0^{(n-1)}, \end{aligned}$$

then, as described in (I.27) and (I.28), we can associate with the *n*th-order differential equation the equivalent system of *n* 1st-order differential equations (I.28) and apply the last theorem to the latter. Here the Lipschitz condition reads

$$\begin{aligned} &f[x, y, y', y'', \ldots, y^{(n-1)}] - f[x, y^*, y'^*, y''^*, \ldots, y^{(n-1)*}]| \\ &\quad \leq k\{|y - y^*| + |y' - y^*| + \ldots + |y^{(n-1)} - y^{(n-1)^*}|\}. \end{aligned} \tag{I.43}$$

If in a neighbourhood, say $|x - x_0| \leq a$, $|y^{(q)} - y_0^{(q)}| \leq b$ *(for* $q = 0, 1, \ldots, n-1$*), of the point* $x_0, y_0, y_0', \ldots, y_0^{(n-1)}$ *the function* f *satisfies the conditions of continuity, boundedness* $|f| \leq M, |y^{(q)}| \leq M$ *(for* $q = 0, 1, \ldots, n-1$*), and the Lipschitz condition, then the existence and uniqueness of the solution of the nth-order differential equation is assured in the interval* $|x - x_0| \leq \min(a, b/M)$.

Example of non-uniqueness

For the differential equation of second order and second degree

$$y''(1+y')=1 \quad \text{or} \quad y''=f(x,y,y')=\frac{1}{1+y'}$$

the conditions of boundedness and continuity are already infringed for variable y'.

When $y'=-1$, uniqueness is infringed; for example, both the curves given by the equations

$$y=1-x\pm\tfrac{1}{3}\sqrt{(2x)^3}$$

go through the point $x=0$, $y=1$ with the slope $y'=-1$.

As a particularly important case we now discuss the

Linear *n*th-order differential equation

If the right-hand side of the differential equation (I.42) is linear in y and its derivatives, i.e., if the differential equation is of the form

$$\sum_{v=0}^{n} p_v(x)y^{(v)}(x)=r(x)$$

with $p_n(x)=1$, and if we rewrite the differential equation as a system

$$y'_{(q)}=y_{(q+1)} \quad (q=1,\ldots,n-1), \qquad y'_{(n)}=f=r(x)-\sum_{v=0}^{n-1} p_v(x)y_{(v+1)},$$

then the partial derivatives of f with respect to the $y_{(q)}$ are precisely the coefficient functions$-p_{q-1}(x)$ for $q=1,2,\ldots,n$. If we assume these to be continuous and bounded in an interval $[A,B]$, which might extend to infinity in one or both directions,

$$|p_q(x)|\leq P \text{ for } q=0,1,\ldots,n-1 \quad \text{and} \quad |r(x)|\leq P \quad \text{with } P\geq 1,$$

then P can be used as the Lipschitz constant k. Bounds M for $|y_{(q+1)}|$ and $|f|$ are also required. Let us consider, say, the box

$$|x-x_0|\leq a, \qquad |y^{(q)}-y_0^{(q)}|\leq b \quad \text{for } q=0,1,\ldots,n-1.$$

where, of course, $|x-x_0|\leq a$ must be contained in $[A,B]$ and b must be ≥ 1. If β is the largest of the absolute values $|y_0^{(q)}|$ of the given initial values, then $|y^{(q)}|\leq\beta+b$ for $q=0,1,\ldots,n-1$, and we can choose

$$M_{(j)}=M=(n+1)P(\beta+b).$$

The existence and uniqueness of the solution in an interval

$$|x-x_0|\leq h, \quad \text{where } h=\min(a,b/M),$$

is then assured. We now show that the solution exists and is unique in the whole interval $[A, B]$. The number b can be chosen arbitrarily large, and in particular to be $> \beta$, so that $b/M \geq 1/(2(n+1)) = \alpha$; since a is subjected only to the restriction that $|x - x_0| \leq a$ is contained in $[A, B]$, we can continue the solution uniquely from x_0 over an interval of finite, fixed length α (which must not, of course, extend beyond $[A, B]$. We can repeat this, so that, for a finite interval $[A, B]$, we reach the end of the interval after a finite number of steps (in this process, the number b may have to be chosen afresh from step to step, but this does not affect α).

21. Some fundamental concepts in functional analysis

The considerations in Nos. 18 to 20 can be made clearer and simpler by using some of the abstract ideas of functional analysis. The most important concepts, which have proved of value in many complicated studies in differential equations, are those of a Banach space, operators and their fixed points; we shall explain these ideas, adapting each to the case of differential equations.

Linear space

A set R of elements is called a 'linear space' if in R the operations of addition of elements and multiplication of elements by numbers from a field K are defined, these operations being governed by the rules of vector algebra. Illustration: in this section K can be taken to be the field of real numbers, i.e., if an element f belongs to R, then so does $c \cdot f$, where c is any real number; further, if f, g are any 2 elements of R, then the sum $f + g$ also belongs to R; we can operate with the elements of R as we do with vectors v, which we can also add and multiply by real numbers.

Thus we already have simple examples of a linear space:

(I) Vectors $v = (v_1, \ldots, v_n)$ with components v_θ in the n-dimensional space of real points.

(II) The most important example, however, is the set of functions $f(x)$ which are defined on an interval $J = [a, b]$ of the real x-axis, and for which $f + g$ and $c \cdot f$ have their usual meanings in elementary analysis.

(III) We can also (for a fixed k) regard k functions $f_1(x), \ldots, f_k(x)$ defined on J as a single element of a linear space.

Every linear space has a 'null element' θ with the properties

$$f - f = \theta, \; c \cdot \theta = \theta, \; 0 \cdot f = \theta$$

for all $f \in R$ and for all c in K.

Normed space

A linear space R is said to be 'normed' if to every element f in R there corresponds a non-negative number $\| f \|$ called its 'norm', such that, for any

f, g in R and any number c of the field K (here, in particular, for any real number c), the following conditions are satisfied:

$$\left.\begin{array}{lll} \text{1. Definiteness:} & \|f\| = 0 \quad \text{if and only if } f = \theta \\ \text{2. Homogeneity:} & \|cf\| = |c| \cdot \|f\| \\ \text{3. Triangle inequality:} & \|f+g\| \leq \|f\| + \|g\|. \end{array}\right\} \tag{I.44}$$

The homogeneity implies that $\|-f\| = \|f\|$ (for $c = -1$).

The norm is, in a sense, a measure of the 'size' of the element concerned.

With every pair of elements f, g can be associated a 'distance'

$$\rho(f, g) = \|f - g\|. \tag{I.45}$$

Examples of a normed space

1. The space R^n already mentioned of n-dimensional real vectors v with the components $v_1, \ldots, v_n$ is a normed space, in which we can introduce various norms:

(A) the 'Euclidean' norm

$$\|v\| = \left[\sum_{\nu=1}^{n} v_\nu^2\right]^{1/2}$$

is the Euclidean length of the vector v. The triangle inequality asserts that in every triangle with sides of length $\|f\|$, $\|g\|$, $\|f+g\|$ no side can be longer than the sum of the lengths of the other two sides (Fig. I.24).

(B)
$$\|v\| = \sum_{\nu=1}^{n} |v_\nu| \quad \text{and} \quad \|v\| = \max_\theta |v_\theta|$$

also satisfy the conditions for a norm.

2. From the space previously mentioned of functions defined on a real interval $J = [a, b]$, we can pick out the sub-space $C[J]$ of continuous functions $f(x)$ and define on it the norm

$$\|f\| = \max_J |f(x)| \tag{I.46}$$

as the greatest distance from the null function, considering the graph of f (Fig. I.25). Here too the triangle inequality holds (see Fig. I.26). In this inequality,

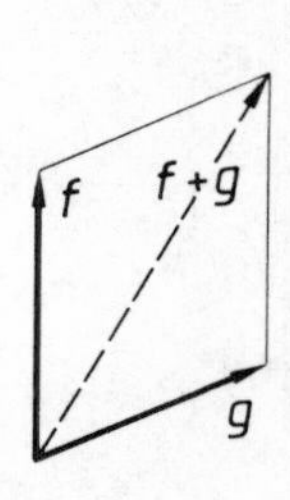

Fig. I.24. For the triangle inequality

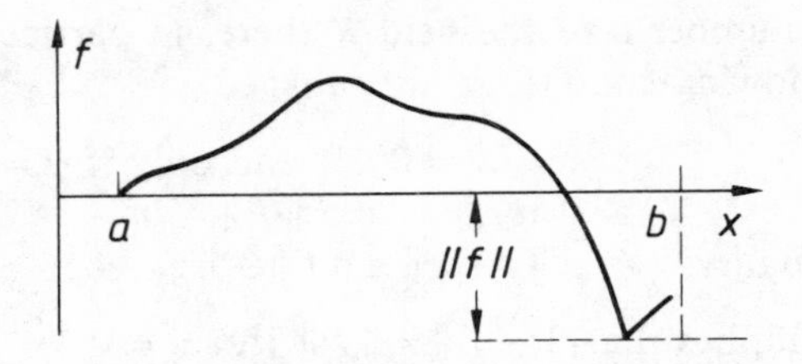

Fig. I.25. The maximum norm

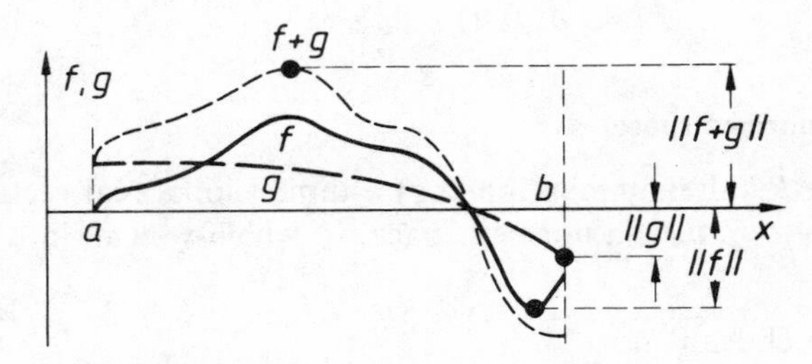

Fig. I.26. The triangle inequality with the maximum norm

the equality sign holds if there is a point $x = \xi$ at which $|f|$ and $|g|$ both assume their maximum values values and f and g have the same sign.

Another possible norm, rather more general, is

$$\|f\| = \max_J p(x)\,|f(x)| \tag{I.47}$$

where $p(x)$ is a positive continuous function in J, chosen once and for all. If $p(x) \equiv 1$, this is the same as the norm (I.46).

The distance introduced in (I.45) between the functions f and g also has an obvious interpretation as the distance between the graphs of f and g measured in the direction of ordinates, Fig. I.27 (in the so-called 'Hausdorff distance', distances are measured in the usual geometrical way).

Similary, for a multi-dimensional, closed, bounded domain B in the space

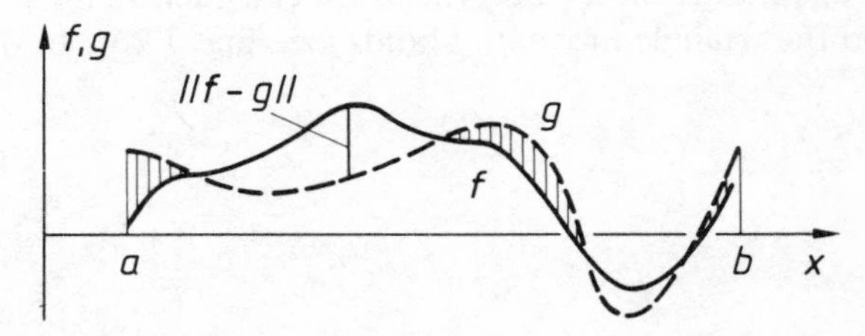

Fig. I.27. The distance between two functions f and g

$(x_1, \ldots, x_n)$ we can consider the space $C[B]$ of continuous functions $f(x) = f(x_1, \ldots, x_n)$ with the following norm

$$\|f\| = \max_{x \in B} |f(x)|.$$

3. Let $k > 1$ be a fixed natural number and consider a system $\Phi(x)$ of k functions $f_{(1)}(x), \ldots, f_{(k)}(x)$ which all belong to C[J] (cf. the previous example). Then possible norms are

$$\|\Phi\| = \max_{\varkappa} \max_{x \in J} |f_{(\varkappa)}(x)| \quad \text{and} \quad \|\Phi\| = \sum_{\varkappa=1}^{k} \max_{x \in J} |f_{(\varkappa)}(x)|. \tag{I.48}$$

Completeness

With the distance (I.45) we have a 'topology', and we can introduce the well-known topological concepts from point-set theory such as neighbourhood, limit-point, convergence, etc. A sequence $f_1, f_2, \ldots$ of elements is said to be 'convergent' if there is an element f such that

$$\lim_{n \to \infty} \|f - f_n\| = 0. \tag{I.49}$$

The sequence is said to be 'Cauchy convergent' if the distance between f_m and f_n always converges to zero when m and n increase to infinity independently of one another:

$$\lim_{m,n \to \infty} \|f_m - f_n\| = 0. \tag{I.50}$$

In the elementary theory of convergence of sequences of real numbers, these two types of convergence are equivalent to one another; but they are revealed to be different if we take the space R to be the set of rational numbers, addition and multiplication by numbers c to be in the classical sense (but with the factors c restricted to be rational numbers), and we take the norm to be the absolute value, $\|f\| = |f|$. For example, let f be the sequence of rational numbers obtained by terminating at successive places the decimal expansion of π:

$$f_1 = 3, \qquad f_2 = 3.1, \qquad f_3 = 3.14 \qquad f_4 = 3.141, \text{ etc.}$$

This sequence of numbers is Cauchy convergent in the sense of (I.50); but it is not convergent in this space R in the sense of (I.49), because there is no element in R which satisfies (I.49), since π, which is an irrational number, does not belong to R. We are therefore justified in making the following definition (which is given here in a very specialized form).

Definition of completeness

A (proper or improper) subset D of a normed space R is said to be complete *if for every Cauchy convergent sequence f_n in D there is an element $f \in D$ such that* (I.49) *holds.*

(Expressed briefly, the limit elements of convergent sequences must themselves belong to the subset D.)

Example 1

The space R^n already mentioned several times of n-dimensional vectors v with real (even irrational) components $v_1, \ldots, v_n$ is complete.

2. For the space $C[J]$ of continuous functions $f(x)$ on J with the norm (I.46), Cauchy convergence of a sequence of functions $f_1(x), f_2(x), \ldots$ means that, for every $\varepsilon > 0$, there is a number N such that

$$|f_m(x) - f_n(x)| < \varepsilon \quad \text{for } m > N \text{ and for all } x \in J, \tag{I.51}$$

i.e., there is uniform convergence. Now a classical theorem in real analysis asserts that every sequence $f_\nu(x)$ of continuous functions which is uniformly convergent in a closed interval J converges to a limit-function which is continuous in J; so the limit-element also belongs to R. Thus the space $C[J]$ with the norm (I.46) (and also with the norm (I.47)) is complete.

Every complete normal space is called a 'Banach space'.

Definition of an operator

Let R_1, R_2 be normed spaces and let D be a (proper or improper) subset of R_1. If to every element $u \in D$ there corresponds a unique element $v \in R_2$, we can write

$$v = Tu \tag{I.52}$$

and call T a transformation or mapping with the domain D (and a range W in R_2).

Definition

The operator T is said to be 'Lipschitz-bounded' if there is a constant K such that

$$\|Tu_1 - Tu_2\| \leq K\|u_1 - u_2\| \quad \textit{for all } u_1, u_2 \textit{ in } D. \tag{I.53}$$

A Lipschitz-bounded operator T is said to be 'contractive' or a 'contraction' if there is a constant $K < 1$ such that (I.53) *holds.*

Definition

Let $R_1 = R_2$, i.e., the operator T maps elements in $D \subseteq R_1$ into R_1 again. Every element in D which is mapped into itself by T, i.e., such that

$$u = Tu \tag{I.54}$$

holds, is called a 'fixed point' of T.

A fundamental example

Just as we went from the initial-value problem (I.29) with two differential equations to a fixed-point equation $U = TU$ with the operator T defined by formulae (I.30), so we can also proceed with s unknown functions $y_1(x), \ldots, y_s(x)$ in an interval $J = [a, b]$, and rewrite the initial-value problem with the differential equations (I.38) and the given initial values $y_{(\sigma)}(x_0)$ in vector form (for brevity we write y_σ instead of $y_{(\sigma)}$

$$U = \begin{pmatrix} u_1(x) \\ \vdots \\ u_s(x) \end{pmatrix}, \qquad V = \begin{pmatrix} v_1(x) \\ \vdots \\ v_s(x) \end{pmatrix}, \qquad Y_0 = \begin{pmatrix} y_1(x_0) \\ \vdots \\ y_s(x_0) \end{pmatrix},$$

$$F = F(x, Y) = \begin{pmatrix} f_1 \\ \vdots \\ f_s \end{pmatrix} = \begin{pmatrix} f_1(x, y_1, \ldots, y_s) \\ \vdots \\ f_s(x, y_1, \ldots, y_s) \end{pmatrix} \tag{I.5}$$

as the operator equation (I.30) and ask whether there is a fixed point $V = TU = U$.

If T has a fixed point $V = U$, then differentiation of (I.30) yields

$$U' = F(x, U(x)), \qquad U(x_0) = Y_0,$$

i.e., U solves the initial-value problem specified and, conversely, every solution of the initial-value problem is a fixed point of the operator T.

Let $[a, b]$ be an interval of the real x-axis which contains x_0 in its interior, and let $Z(x)$ be an arbitrary continuous function with components $z_1(x), \ldots, z_s(x)$ which is defined on $[a, b]$ and has values lying in the domain of definition of $F(x, Y)$. Let W denote a cube in the s-dimensional $z_i, \ldots,$ z_s-space with its centre at Y_0 and its edges parallel to the co-ordinate axes, the length of the edges being $2h$:

$$W = \{ (z_\sigma; |z_\sigma - y_{\sigma 0}| \leq h \text{ for } \sigma = 1, \ldots, s\} \quad (\text{with } y_{\sigma 0} = y_\sigma(x_0))$$

Let the norm corresponding to (I.47) be used

$$\| V \| = \max_{1 \leq \sigma \leq s} \max_{x \in [a, b]} (p(x) \, | v_\sigma(x) |). \tag{I.56}$$

Let Q be the $(s + 1)$-dimensional box $a \leq x \leq b, z_\sigma \in W$. Since Q is closed, the continuous functions f_τ are bounded in it, by as constant M, say:

$$|f_\sigma(x, z_v)| \leq M \quad \text{for } \sigma = 1, \ldots, s, \quad \text{and} \quad (x, z_v) \in Q.$$

There is now a Lipschitz constant L such that

$$| f_\sigma(x, Z) - f_\sigma(x, \hat{Z})| \leq L \sum_{v=1}^{8} | z_v - \hat{z}_v | \quad \text{for } \sigma = 1, \ldots, s, (x, z_v), (x, \hat{z}_v) \in Q, \tag{I.57}$$

where the $\hat{z}_\nu$ are the respective components of $\hat{Z}$. A Lipschitz condition like (I.57) is satisfied, for example, if the functions f_σ have continuous partial derivatives with respect to all the y_ν. Now let the interval J and the corresponding Q be restricted to the (proper or improper) sub-intervals $\tilde{J}$ and $\tilde{Q}$:

$$\tilde{J} = \{x, \mid x - x_0 \mid \leq d\}; \qquad d = \min\left\{\mid x_0 - a \mid, \ \mid y_0 - b \mid, \frac{h}{M}\right\}$$

$$\tilde{Q} = \{(x, z_\sigma), x \in \tilde{J}, \mid z_\sigma - y_{\sigma 0} \mid \leq h\}.$$

We have now to show that the operator T maps the box $\tilde{Q}$ into itself. Let Z be defined in $\tilde{Q}$; then for the σth component of TZ we have

$$|(TZ)_\sigma - y_{\sigma 0}| \leq \left| \int_{x_0}^{x} f_\sigma(\xi, Z(\xi))\, \mathrm{d}\xi \right| \leq M \left| \int_{x_0}^{x} \mathrm{d}\xi \right| \leq Md \leq h,$$

and so the values of TZ again lie in $\tilde{Q}$. We next check whether T is a contraction operator. For 2 vectors Z, $\hat{Z}$ we have, by (I.57),

$$\| TZ - T\hat{Z} \| = \max_{1 \leq \sigma \leq s} \left[\max_{x \in J} p(x) \left| \int_{x_0}^{x} f_\sigma(\xi, Z(\xi))\, \mathrm{d}\xi - \int_{x_0}^{x} f_\sigma(\xi, \hat{Z}(\xi))\, \mathrm{d}\xi \right| \right]$$

$$\leq \max_{x \in J} Lp(x) \left| \int_{x_0}^{x} \sum_{\upsilon=1}^{s} \mid z_\upsilon - \hat{z}_\upsilon \mid \mathrm{d}\xi \right| \leq Ls \left(\max_{x \in J} p\,(x) \cdot \left| \int_{x_0}^{x} \frac{\mathrm{d}\xi}{p(\xi)} \right| \right) \cdot \| Z - \hat{Z} \|,$$

since by (I.56)

$$| z_\upsilon(\xi) - \hat{z}_\upsilon(\xi) | \leq \frac{1}{p(\xi)} \| Z - \hat{Z} \|.$$

Further we can now choose $p(x)$ to be a positive continuous function on J. For

$$p(x) = e^{-\alpha} \mid x - x_0 \mid \quad \text{with } \alpha > Ls$$

we have

$$p(x) \left| \int_{x_0}^{x} \frac{\mathrm{d}\xi}{p(\xi)} \right| = \frac{1}{\alpha} (1 - e^{-\alpha | x - x_0 |}) \leq \frac{1}{\alpha}.$$

Hence it follows altogether that

$$\| TZ - T\hat{Z} \| \leq K \| Z - \hat{Z} \| \quad \text{with } K = \frac{Ls}{\alpha} < 1.$$

Therefore the operator T is indeed a contraction in the domain considered.

22. BANACH'S FIXED-POINT THEOREM AND THE EXISTENCE THEOREM FOR ORDINARY DIFFERENTIAL EQUATIONS

We have now introduced the concepts needed for formulating a general fixed-point theorem, from which the existence theorem for ordinary differential equations can easily be deduced.

Fixed-point theorem (First formulation)

Let D be a complete (proper or improper) sub-domain of a linear normed space R, and let D be mapped into itself by a contraction operator T. Then there is in D precisely one element u which satisfies the equation (I.54) $u = Tu$.

For numerical computation a second formulation of the theorem is more convenient; in this we do not need to know in advance whether the given subset D is complete, and we obtain in addition an explicit bound for the error.

Fixed-point theorem (Second formulation)

Let D be a sub-domain of a complete normed space, on which a contraction operator T with a Lipschitz constant $K < 1$ is defined. From a given element u_0 of D we can calculate $u_1 = Tu_0$. Let the 'ball' S of elements q such that

$$\| q - u_1 \| \leq \frac{K}{1-K} \| u_1 - u_0 \| \tag{I.58}$$

be contained in D. Then there is in D precisely one element $\hat{u}$ which satisfies the equation (I.54) $u = Tu$. *The iterative process*

$$u_{n+1} = Tu_n \qquad (n = 0, 1, 2, \ldots) \tag{I.59}$$

can be carried out repeatedly without limit, and the error bound

$$\| u_m - \hat{u} \| \leq \frac{K^m}{1-K} \| u_1 - u_0 \| \qquad (m = 1, 2, 3, \ldots). \tag{I.60}$$

holds. Thus $\{u_n\}$ converges to $\hat{u}$ as $n \to \infty$.

Proof. Let u_0 be in D. With the first formulation, all the elements computed according to (I.59) again lie in D. With the second formulation, we prove by induction over n (for $n \geq m$) by using the triangle inequality that the following inequalities hold:

$$\left.\begin{aligned} &\| u_{n+1} - u_n \| \leq K^n \| u_1 - u_0 \| \\ &\| u_{n+1} - u_m \| \leq \| u_{n+1} - u_n \| + \| u_n - u_{n-1} \| + \cdots + \| u_{m+1} - u_m \| \\ &\qquad \leq \| u_1 - u_0 \| \sum_{v=m}^{n} K^v \leq \frac{K^m}{1-K} \| u_1 - u_0 \|. \end{aligned}\right\} \tag{I.61}$$

From the last inequality it follows for $m = 1$ that u_{n+1} lies in the ball S, i.e., in the domain of definition of T, and therefore that the iteration can be carried out repeatedly without limit. The inequalities (I.61) therefore hold for both formulations. Further, it follows from the last inequality in (I.61) that

$$\lim_{n,\, m \to \infty} \| u_n - u_m \| = 0.$$

So the sequence $\{u_n\}$ is Cauchy convergent and since, by hypothesis, the space

R is complete, it follows that there is a limit element $\hat{u}$:

$$\lim_{n\to\infty} \| u_n - \hat{u} \| = 0. \tag{I.61*}$$

With the first formulation, $\hat{u}$ lies in D because D is complete. With the second formulation, $\hat{u}$ lies in R because R is complete. But in fact $\hat{u}$ lies in the ball S, because it follows from the last inequality in (I.61) that

$$\| \hat{u} - u_m \| \le \| \hat{u} - u_{n+1} \| + \| u_{n+1} - u_m \| \le \|\hat{u} - u_{n+1} \| + \frac{K^m}{1-K} \| u_1 - u_0 \| .$$

If we let $n \to \infty$ here, then, because of (I.61*), we obtain the error bound (I.60). Putting $m = 1$ in this, we see that $\hat{u}$ also belongs to the ball S defined by (I.58).

Finally the triangle inequality gives, with (I.59),

$$\begin{aligned} 0 \le \| \hat{u} - T\hat{u} \| &\le \|\hat{u} - u_{n+1} \| + \| u_{n+1} - T\hat{u} \| \\ &= \| \hat{u} - u_{n+1} \| + \| Tu_n - T\hat{u} \| \le \| \hat{u} - u_{n+1} \| + K \| u_n - \hat{u} \|. \end{aligned}$$

The right-hand side tends to 0 as $n \to \infty$, and therefore

$$\| \hat{u} - T\hat{u} \| = 0 \quad \text{and} \quad \hat{u} = T\hat{u}.$$

hence $\hat{u}$ is a fixed-point of T.

To prove the uniqueness, let us assume that the fixed-point equation (I.54) has two solutions in D with $u = Tu$, $v = Tv$; then

$$\| u - v \| = \| Tu - Tv \| \le K \| u - v \|$$

or

$$\| u - v \| (1 - K) \le 0, \quad \text{i.e., } \| u - v \| = 0, u = v.$$

Hence there can be only one solution of (I.54) in D.

We now easily obtain the general

Existence and uniqueness theorem for systems of differential equations

In the initial-value problem (I.38), (I.39) *let the given functions* $f_j(x, y_{(1)}, \ldots, y_{(s)})$ *satisfy the boundedness condition* $|f_j| \le M$ *and the Lipschitz condition* (I.41) *in a box*

$$Q: \qquad a \le x \le b, \qquad | y_{(j)} - y_{(j)0} | \le h.$$

Now let $\tilde{Q}$ *be a box, which may be smaller than* Q:

$$\tilde{Q}: | x - x_0 | \le d, \qquad | y_{(j)} - y_{(j)0} | \le h, \qquad d = \min \left(| x_0 - a |, |x_0 - b |, \frac{h}{M} \right).$$

Then the initial-value problem has precisely one solution in the interval $J: | x - x_0 | \le d$ *and this solution lies entirely in* $\tilde{Q}$ *for* x *in* J.

To prove this we have only to adduce the results of the last two sections. Using the vector notation (I.55) we have that the initial-value problem is equivalent to determining the fixed point $U = TU$ of the operator T of (I.30). As was shown in No. 21, the operator T is a contraction in the box $\tilde{Q}$. The fixed-point theorem in its first formulation then guarantees the existence and uniqueness of a solution U.

§7. SINGULAR LINE-ELEMENTS

Without going more deeply into the difficult general theory of singular line-elements we want to acquaint the reader with the typical phenomena by means of some simple examples. Since in applications singular points and other exceptional phenomena appear more often than one might at first suppose, it is important to obtain at least a rough idea of the general picture. A more detailed theory cannot be given here.

23. Regular and Singular Line-Elements. Definitions and Examples

Suppose the differential equation

$$F(x, y, y') = 0.$$

is given. We ask when its expression in the explicit form

$$y' = G(x, y)$$

is possible and when this form is uniquely determined.

By well-known theorems on implicit functions, solubility for y' at a point in question is ensured if $\partial F(x, y, y')/\partial y' \neq 0$. Points where $\partial F/\partial y' = 0$ have therefore to be considered separately.

The three numbers x, y, y' determine a *line-element;* (we may visualize this as a small part of a straight line going through the point x, y with slope y'). We used such line-elements earlier in §1 to make clear the idea of the direction-field of a differential equation. A line-element of the given differential equation $F(x, y, y') = 0$ is said to be *regular* if a neighbourhood about the point x, y can be specified such that through every point of this neighbourhood there passes just *one* solution curve of the differential equation; otherwise the line-element is said to be *singular*.

Without the assumption of continuity of $F(x, y, y')$ the condition

$$\frac{\partial F \cdot (x, y, y')}{\partial y'} = 0 \tag{I.62}$$

is neither necessary nor sufficient as a criterion for the existence of singular line-elements, i.e., this condition in certain circumstances may not include all the singular points, and it can also give too many. We shall give examples of this. Thus (I.62) can be regarded only as a rough and ready rule.

If, however, continuity of $F(x, y, y')$ is assumed, then the following is true; only at a point x, y, y' where (I.62) holds can the corresponding line-element be singular, and if (I.62) does hold, the line-element usually is singular.†

A singular integral or a singular solution is to be understood to mean an integral curve which contains *only* singular line-elements.

Examples for singular solutions. Example 1. Cubical parabolas

With the family of curves (No. 15, Fig. I.17)

$$y = (x - c)^3$$

with the differential equation

$$y' = 3\sqrt[3]{y^2}$$

or, implicitly,

$$F(x, y, y') \equiv y'^3 - 27y^2 = 0,$$

the condition (I.62) yields $y' = 0$, or, substituted in

$$F(x, y, y'), \qquad -27\, y^2 = 0.$$

Hence, it follows that $y = 0$ is a singular solution.

Example 2. Tangents to a parabola and their envelope

The tangents to the parabola

$$\eta = \xi^2$$

(ξ and η are the co-ordinates of a point on the parabola, x and y are those of a point on a tangent) have the equation

$$\tan \alpha = 2\xi = \frac{y - \xi^2}{x - \xi}.$$

From the family of tangents to the parabola (with ξ as a parameter)

$$y = 2\xi x - \xi^2.$$

with

$$y' = 2\xi$$

we obtain by eliminating ξ the differential equation

$$F(x, y, y') \equiv \tfrac{1}{4} y'^2 - xy' + y = 0.$$

†A precise presentation of the facts can be found. e.g., in H. V. Mangold–K. Knopp: *Einführung in die höheren Mathematik,* vol. 3, 10th ed.; Stuttgart 1958, p. 584ff.; and E. Kamke, *Differentialgleichungen reeller Funktionen,* 3rd ed., Leipzig 1956, p. 115f.

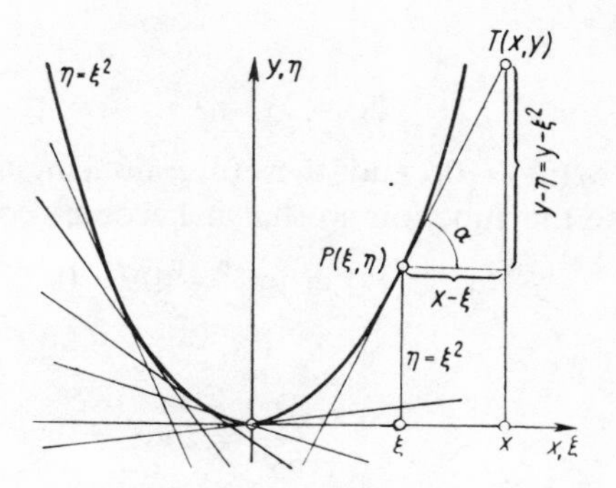

Fig. I.28. The parabola as the envelope of its tangents

The rule of thumb (I.62) yields

$$\frac{\partial F}{\partial y'} = \tfrac{1}{2}\, y' - x = 0.$$

Elimination of y' from $F = 0$ and $F_{y'} = 0$ gives, in fact, $y = x^2$, that is, the parabola is the singular solution, and here it is the envelope of the tangents, see Fig. I.28.

Example 3. The semi-cubical parabola (Neil's parabola) and cusps

However, the rule of thumb (I.62) does not always give the envelope. Let us examine the family of semi-cubical parabolas, Fig. I.29,

$$y = (x - c)^{2/3}.$$

We have

$$y' = \tfrac{2}{3}(x - c)^{-1/3} = \tfrac{2}{3}\, y^{-1/2}$$

or

$$F(x, y, y') \equiv y'^{\,2} - \frac{4}{9y} = 0.$$

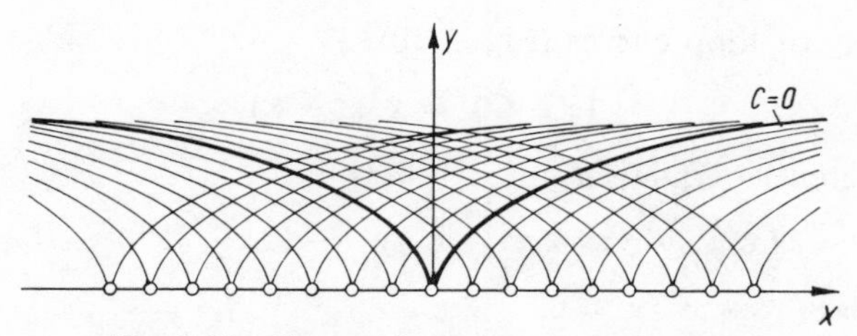

Fig. I.29. Cusps of semi-cubical parabolas

Using

$$F_{y'} = 2y' = 0$$

we would obtain $4/(9y) = 0$, and therefore no singular line-elements. If, however, we rewrite the function so that it becomes *continuous*

$$F^*(x, y, y') \equiv yy'^2 - 4/9 = 0,$$

then

$$\frac{\partial F^*}{\partial y'} = 2yy' = 0$$

is satisfied for $y = 0$ as well as for $y' = 0$, and it yields, in fact, the locus of all the singular points, i.e., the *discriminant locus.* However, here $y = 0$ is not a solution of the differential equation. One has always to investigate further whether the discriminant loci are at the same time solutions of the differential equation, and whether they are contained in the general solution or not.

Example 4. Only regular line-elements and yet $\partial F/\partial y' = 0$ is satisfied

For the set of straight lines inclined at 45° to the x-axis, $y = x + C$, the differential equation $y' = 1$ holds. But if we write it differently as

$$F(x, y, y' \equiv (y' - 1)^2 = 0,$$

then

$$Fy' = 2(y' - 1) = 0$$

yields as the discriminant loci the straight lines

$$y' = 1, \ y = x + C.$$

i.e., the same straight lines as we started from.

According to this, all line-elements would appear to be exceptional elements; they are, however, regular. So caution is indeed needed in using the rule of thumb (I.62).

Example 5. Curves of the third degree with double points

For the family of loop curves (Fig. I.30)

$$(y - C)^2 = x^2(1 - x)$$

with the differential equation

$$F(x, y, y') \equiv 4(x^2 - x^3)y'^2 - (2x - 3x^2)^2 = 0$$

we have $F_{y'} = 8(x^2 - x^3)y' = 0$,
from which we get either $y' = 0$
(i.e., the maxima and minima),

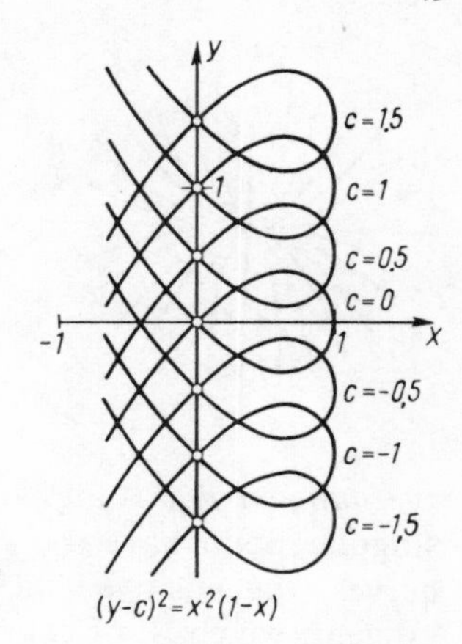

Fig. I.30. Curves of the third degree with double points

or $x^2(1-x)=0$, i.e., the crossing points and the points $x=1$ where the tangents are vertical.
(In examples 2, 3 and 5 the direction-fields are many-valued, and so we must, according to the definitions initially given, say that all the solutions are singular.)

24. Isolated singular points

Typical cases are provided by the differential equation

$$F(x, y, y') \equiv xy' - ay = 0. \tag{I.63}$$

From $F_{y'} = x = 0$ we obtain $y = 0$, i.e., the origin. To obtain the geometrical interpretation we solve the differential equation by separation of the variables:

$$\int \frac{dy}{y} = \int \frac{a}{x}\,\mathrm{d}x + C_1; \qquad \ln|y| = a \ln|x| + C_1; \qquad y = Cx^a.$$

We obtain different families of curves according to the value of a. $a=1$ gives straight lines $y = Cx$. We obtain a *node of the first kind,* also called a *star,* Fig. I.31. Every half-line through the node is a tangent to a solution curve.

$a=2$ yields parabolas $y = Cx^2$ and a *node of the second kind*, also called a

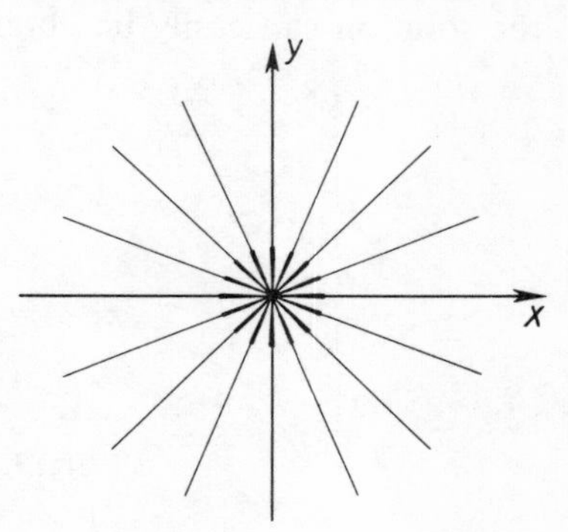

Fig. I.31. Node of the first kind, or star

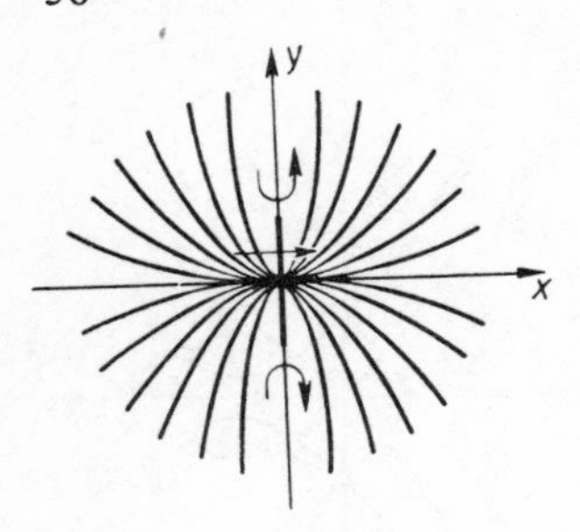

Fig. I.32. Node of the second kind; two-tangent node

two-tangent node, Fig. I.32. The parabolic solution curves leading into the singular point have the same tangent (the x-axis). The two special solution curves (the positive and negative parts of the y-axis) also have the same tangent, which is, of course, different from the common tangent of the other solution curves.

$a = -1$ yields the hyperbolas $y = C/x$ and a saddle point (Fig. I.33). There are exactly four solution curves which lead into the saddle point; two of them have a common tangent, and the other two likewise have a common tangent, which is different from the first one.

For the differential equation

$$F(x, y, y') \equiv yy' + x = 0,$$

(I.62) yields the origin, $x = 0$, $y = 0$, The solutions

$$x^2 + y^2 = C$$

form a family of circles. The singular point $(0, 0)$ is called a *centre or vortex point*. Through every point of a certain neighbourhood of the centre passes a closed *Jordan curve* as the solution curve (here a circle) which contains the centre in its interior (Fig. I.34).

For the similarity differential equation

$$y' = \frac{x + y}{x - y} = \frac{1 + y/x}{1 - y/x} \tag{I.64}$$

the solution can easily be obtained, as shown in §1.

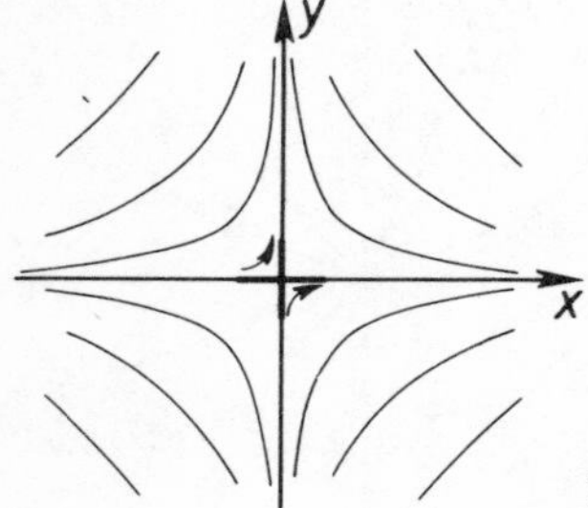

Fig. I.33. A saddle point

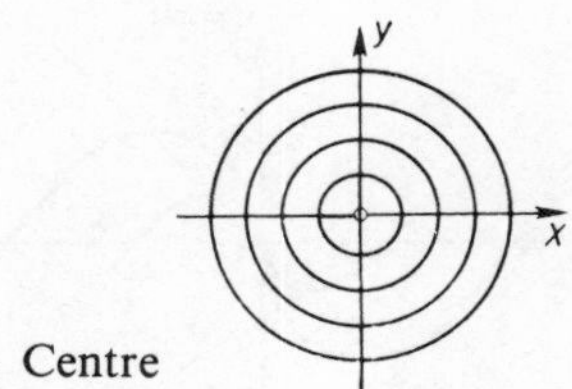

Fig. I.34. Centre

Putting

$$y(x) = xz(x), \; y' = z + xz'$$

we find the differential equation for z

$$z + xz' = \frac{1+z}{1-z},$$

$$\frac{dx}{x} = \frac{1-z}{1+z^2}\, dz,$$

and its solution

$$\ln|x| = \arc\tan z - \tfrac{1}{2}\ln(1+z^2) + \ln C_1,$$

$$\ln C_2 x \sqrt{1+\frac{y^2}{x^2}} = \ln C_2 \sqrt{x^2+y^2} = \arc\tan\frac{y}{x}.$$

This equation suggests introducing polar co-ordinates (I.14):

$$\ln C_2 r = \varphi \quad \text{or} \quad r = C\,e^{\varphi}.$$

These are logarithmic spirals (Fig. I.35). We could also have transformed the original differential equation using (I.14) to obtain

$$\frac{dr}{d\varphi} = r' = r,$$

again giving the solution $r = C e^{\varphi}$.

The point $x = 0$, $y = 0$ is a *focus*, which all the solution curves approach asymptotically. No solution curve leads into the focus in a definite direction, but the solution curves cut any straight line through the focus infinitely often; Fig. I.35.

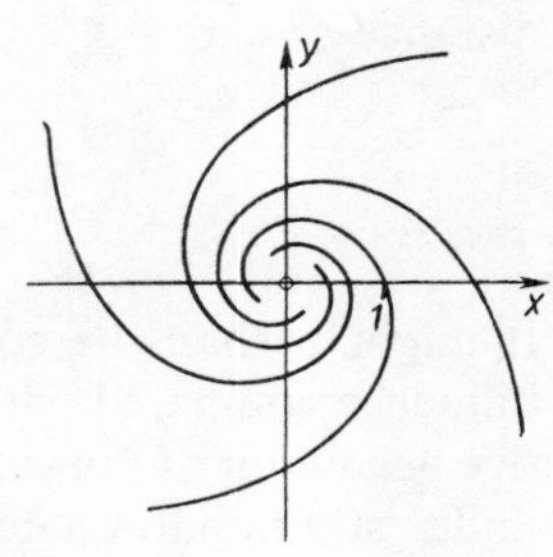

Fig. I.35. A focus

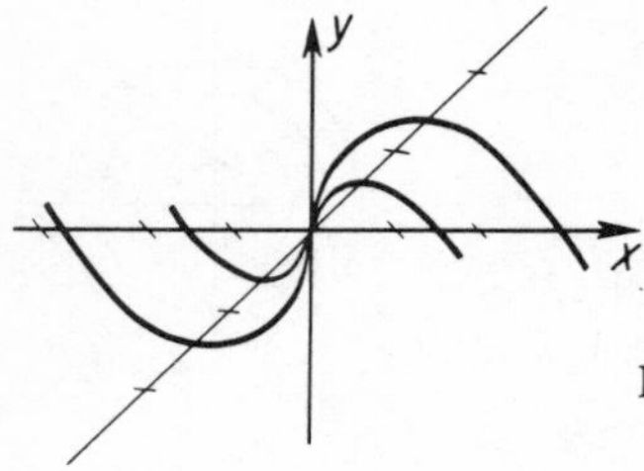

Fig. I.36. Node of the third kind or one-tangent node

Finally, there is still a *node of the third kind,* or a *one-tangent node.* In this case all the solution curves leading into the node have the same tangent. A simple example of this case is provided by the similarity differential equation

$$y' = \frac{y}{x} - 1$$

with the solutions (Fig. I.36)

$$y = C\,|\,x\,| - x \ln |\,x\,|\,.$$

25. On the theory of isolated singular points

Here we shall only give some results from this theory without proof. In applications we often encounter differential equations of the type

$$y' = \frac{G(x, y)}{F(x, y)}, \tag{I.65}$$

where F and G have a common zero at the point $x = x_0$, $y = y_0$ but behave regularly in a neighbourhood of this zero. Let us assume that $x_0 = y_0 = 0$ (this can always be made so by shifting the co-ordinate system), so that we can bring the differential equation into the form

$$y'(x) = \frac{cx + dy + g(x, y)}{ax + by + f(x, y)} \tag{I.66}$$

with $f(0, 0) = g(0, 0) = 0$. It is further assumed that in the neighbourhood of the zero the functions f and g are 'of a smaller order of magnitude' than the distance $r = \sqrt{x^2 + y^2}$, i.e., that in an arbitrary approximation to the zero we have

$$\lim_{r\to 0} \frac{f}{r} = \lim_{r\to 0} \frac{g}{r} = 0. \tag{I.67}$$

It might perhaps be conjectured that the functions f and g would have no influence compared with the linear terms $ax + by$ and $cx + dy$ on the qualitative behaviour of the solution curves near the null point, but this is correct only under additional assumptions.

First of all, (I.66) without the f and g, that is,

$$y' = \frac{cx + dy}{ax + by} \tag{I.68}$$

being a similarity differential equations, can be discussed fully without difficulty.† Here we merely sketch how the general picture can also be obtained in the following way (described in more detail, e.g., in W. W. Stepanow: *Lehrbuch der Differentialgleichungen,* Berlin 1956, p. 71ff.

We seek a transformation $\xi = \alpha x + \beta y$, $\eta = \gamma x + \delta y$, which will bring the differential equation into the form

$$\frac{d\eta}{d\xi} = \frac{\gamma(ax + by) + \delta(cx + dy)}{\alpha(a\chi + by) + \beta(cx + dy)} = \frac{\lambda\eta}{\mu\xi} = \frac{\lambda(\gamma x + \delta y)}{\mu(\alpha x + \beta y)};$$

this equation is of the form (I.63), and from it we can gain a clear picture of the nature of the solutions in the vicinity of the zero. First, if the numerator assumes the desired form $\lambda\eta$, we obtain a linear homogeneous system of equations for γ and δ:

$$\gamma(a - \lambda) + \delta c = 0, \ \gamma b + \delta(d - \lambda) = 0, \tag{I.69}$$

which has a non-trivial solution if and only if the determinant vanishes:

$$\Phi(\lambda) = \begin{vmatrix} a - \lambda & b \\ c & d - \lambda \end{vmatrix} = 0. \tag{I.70}$$

Exactly the same determinant arises if we determine α and β so that the denominator takes the desired form $\mu\xi$.

Let $\Phi(\lambda) = 0$ have the roots λ_1, λ_2. It will be assumed that the determinant $D = ad - bc \neq 0$, for if $D = 0$ the denominator in (I.68) vanishes or we get $y' = \text{const}$. $D \neq 0$ implies that $\lambda_{1,2} \neq 0$. The general solution of the transformed equation then reads

$$\eta = c \cdot |\xi|^{\lambda_2/\lambda_1}, \tag{I.71}$$

with c as the integration constant. We now have to distinguish various cases, depending on whether the roots λ_1, λ_2 are real or complex, simple or coincident, and in the real case whether they have the same or different signs. The roots are real if and only if the discriminant

$$\Delta = 4bc + (a - d)^2 \quad \text{is} \geq 0; \tag{I.72}$$

if Δ is < 0, we can choose γ and δ as the complex conjugates of α and β, $\gamma = \bar{\alpha}, \delta = \bar{\beta}$, and then remain among the reals by making the further transformation $\xi^* = (\xi + \eta)/2$, $\eta^* = (\xi - \eta)/(2i)$. The results for the several

†A detailed discussion can be found, e.g., in K. H. Weise, *Gewöhnliche Differentialgleichungen,* Göttingen 1966; a more precise discussion in L. Bieberbach, *Einführung in die Theorie der Differentialgleichungen im reellen Gebiet,* Berlin 1965. The classification goes back essentially to H. Poincaré: Mémoire sur les courbes définies par une équation differentielle, *Journ. Math. pur. et appl. (3)* 7 (1881), 375–422.

cases are given in the following table (after Bieberbach. loc.cit.), where, under 'normal form', y and x are again used instead of η and ξ, or η^* and ξ^*. One must further distinguish between the cases where in the system of equations for γ, δ and α, β all or not all the coefficients vanish.

N:	D	Δ	Roots	Standard form	Type
1		$\Delta > 0$	λ_1, λ_2 real, $\lambda_1 \neq \lambda_2, \lambda_1\lambda_2 > 0$	$y' = \dfrac{\lambda_2 y}{\lambda_1 x}$	Two-tangent node
2		$\Delta = 0$	$\lambda_1 = \lambda_2 = \lambda$ real	$y' = \dfrac{y}{x}$	Star
3	$D > 0$			$y' = \dfrac{y}{x + y}$	One-tangent node
4		$\Delta < 0$	$\lambda = \mu \pm i\upsilon$, $\mu \neq 0$, $\upsilon \neq 0$, μ, θ real	$y' = \dfrac{-\upsilon x + \mu y}{\mu x + \upsilon \xi}$	Focus
5			$\lambda = \pm i\upsilon$, $\lambda \neq 0$, υ real	$y' = -\dfrac{x}{y}$	Centre
6	$D < 0$	$\Delta > 0$	λ_1, λ_2 real, $\lambda_1\lambda_2 < 0$	$y' = \dfrac{\lambda_2 y}{\lambda_1 x}$	Saddle point

The theory for the more general differential equation (I.66) is more difficult. Under the assumption of smallness of f and g (I.67) it can be shown (see Bieberbach, loc. cit. 'dominance of the linear termsd') that the solution curves of (I.66) in the cases 1 to 4 have the same qualitative behaviour near the null point as those for the corresponding abbreviated equation (I.68). But an example can be given to show that the type of the singular point does not always stay the same. For the differential equation $y' = -x/y$ there is a centre at the origin (type No. 5); the changed equation $y' = (-x + yr)/(y + xr)$ can be put into polar co-ordinates immediately by using (I.14); we obtain the simple equation

$$\frac{dr}{d\varphi} + r^2 = 0$$

with the solutions

$$r = \frac{1}{\varphi + c},$$

where c is an arbitrary constant; the solutions are spiral shaped, and the origin is now a focus.

26. CLAIRAUT'S AND d'ALEMBERT'S DIFFERENTIAL EQUATIONS

The family of solution curves of *Clairaut's equation* consists of a curve and the totality of its tangents (cf. the earlier examples of tangents to a sine curve in §5, Fig. I.16, and tangents to a parabola, Fig. I.28).

Let a curve be given by the equation $y = f(x)$, where f is a continuously differentiable function for x in the interval $a \leq x \leq b$. The tangent at the point ξ then has the equation (Fig. I.37)

$$\tan \alpha = f'(\xi) = \frac{y - f(\xi)}{x - \xi}. \tag{I.73}$$

From this and $y' = f'(\xi)$ we can eliminate ξ. Let $\xi = g(y')$ be the function inverse to $y' = f'(\xi)$. Then (I.73) can be rewritten as

$$y = xy' + \{f[g(y')] - g(y')y'\}.$$

The expression in braces represents a new function $\varphi(y')$. Thus the standard form for the differential equation for the tangent of a curve reads

$$y = xy' + \varphi(y'). \tag{I.74}$$

Conversely, if a *Clairaut* differential equation like this is given, with a continuously differentiable function $\varphi(y')$, we can obtain its solutions by the dodge of first differentiating the equation. The y' then drops out:

$$y' = y' + xy'' + \frac{\mathrm{d}\varphi}{\mathrm{d}y'}\, y''$$

or

$$y''\left(x + \frac{\mathrm{d}\varphi}{\mathrm{d}y'}\right) = 0.$$

This equation is satisfied either for $y'' = 0$ or for

$$x + \frac{\mathrm{d}\varphi}{\mathrm{d}y'} = 0.$$

The first alternative gives after two integration the straight lines $y = C_1 x + C_2$ as the general solution. The second parameter C_2 is attributable to the increase in the order of the differential equation because of the differentiation. C_1 and

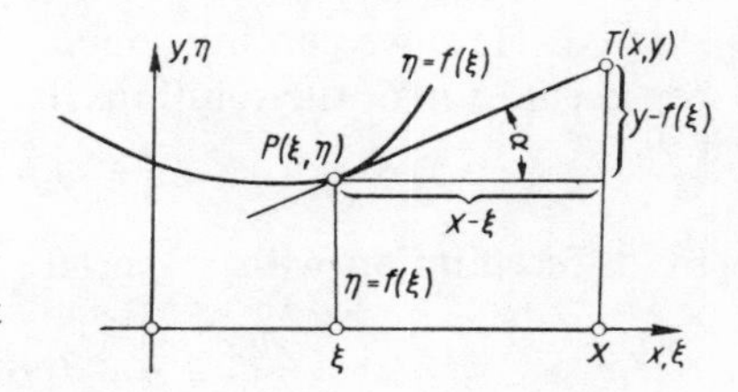

Fig. I.37. For the derivation of the Clairaut differential equation.

C_2 cannot be chosen independently of one another, but only so that the differential equation (I.74) is satisfied, i.e., since $y' = C_1$, so that

$$C_2 = \varphi(C_1).$$

Hence, $y' = 0$ leads to the one-parameter family of solutions

$$y = C_1 x + \varphi(C_1).$$

In other words, in Clairaut's equation we have only to replace y' by C_1.

The second factor in the differential equation is precisely the partial derivative with respect to y' of the left-hand side of the differential equation

$$F(x, y, y') \equiv xy' + \varphi(y') - y = 0$$

$$\frac{\partial F}{\partial y'} = x + \frac{\mathrm{d}\varphi}{\mathrm{d}y'} = 0.$$

It therefore yields the *envelope* of the family of straight lines, provided this family has an envelope. If, in fact, φ is twice continuously differentiable and if $\varphi''(y) \neq 0$, then the envelope does exist (cf. Mangold–Knopp: *Einführung in die höhere Mathematik,* vol. II, 11th ed., Stuttgart 1958, p. 521. Actually, there is a further condition necessary to the existence of the envelope, viz., that

$$\begin{vmatrix} F_x & F_y \\ F_{xy'} & F_{yy'} \end{vmatrix} \neq 0;$$

but here this condition is fulfilled automatically, because

$$\begin{vmatrix} y' & -1 \\ 1 & 0 \end{vmatrix} = 1.$$

The Clairaut differential equation is a particular case of d'Alembert's differential equation

$$y = xf(y') + \varphi(y') \tag{I.75}$$

with $f(y') \equiv y'$. Let f and g be be continuous in a certain interval of y'. For some value $y' = c$ the equation $f(c) = c$ is satisfied, and so the straight line $y = xc + \varphi(c)$ is a solution of the differential equation. If we disregard the straight lines $x = \text{const.}$, the equation (I.75) is the most general differential equation having rectilinear isoclines. By drawing the isocline field a general picture of the shape of the solution curves can thus be easily obtained. However, the case where $f(y') \neq y'$ in a certain y'-interval still has to be considered. Here we can introduce $y' = p$ as a parameter and seek a parametric representation of the solutions in the form $x = x(p)$, $y = y(p)$. It follows from

$$y = x/(p) + \varphi(p)$$

by differentiation with respect to x that

$$\frac{\mathrm{d}y}{\mathrm{d}x} = p = f(p) + x\frac{\mathrm{d}f}{\mathrm{d}p}\frac{\mathrm{d}p}{\mathrm{d}x} + \frac{\mathrm{d}\varphi}{\mathrm{d}p}\frac{\mathrm{d}p}{\mathrm{d}x}.$$

Hence a linear differential equation for $x(p)$ is obtained:

$$\frac{dx}{dp}(p - f(p)) = x\frac{df}{dp} + \frac{d\varphi}{dp}. \tag{I.76}$$

The factor $p - f(p)$ is assumed to be non-zero. After solving this linear differential equation we have the required parametric representation of the solution:

$$x = x(p), \qquad y = x(p)f(p) + \varphi(p). \tag{I.77}$$

Example

$$y = -xy' + y'^2.$$

Here

$$f(p) = -p, \varphi(p) = p^2,$$

and (I.76) reads

$$2p\frac{dx}{dp} = -x + 2p$$

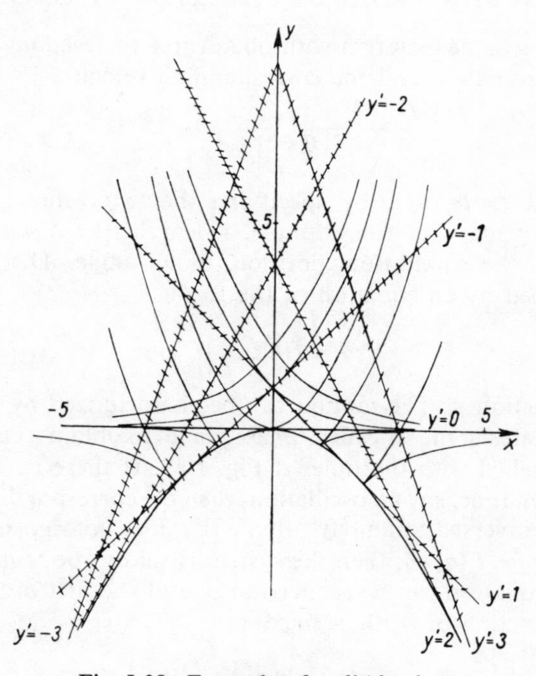

Fig. I.38. Example of a d'Alembert differential equation

with the solution

$$x(p) = \frac{c}{\sqrt{|p|}} + 2p/3.$$

Then, in accordance with (I.77)

$$y(p) = -c\sqrt{|p|} + p^2/3.$$

Figure I.38 shows some of the isoclines and solution curves. Equation (I.62) leads here to

$$-x + 2y' = 0$$

and elimination of y' from this equation and the differential equation yields in this case the equation $y = -x^2/4$ for the cusps of the solutions curves. For a given c the cusp is at the point

$$x = 3 \cdot \sqrt[3]{\frac{c^2}{6}}, \qquad y = -\frac{9}{4}\sqrt[3]{\frac{c^4}{36}}.$$

In addition to the given solutions there is also the solution $y \equiv 0$, which is obtained from the zero $p = 0$ of the equation $f(p) = -p = p$.

27. Oscillations with one degree of freedom. Phase curves

The state of a mechanical system with one degree of freedom is fixed by the value of its co-ordinate q and the corresponding velocity

$$\dot{q} = \frac{dq}{dt}$$

at a time-instant $t = t_0$, i.e., by specifying the two values $q_0 = q(t_0)$ and $\dot{q}_0 = \dot{q}(t_0)$; here q may be, for example, a length (the distance during the oscillations from the equilibrium position) or an angle. The motion of the system is described by an equation of the form

$$\ddot{q} = f(t, q, \dot{q}) \tag{I.78}$$

with a given function f. The motion can be characterized by plotting q and $v = \dot{q}$ as co-ordinates in the so-called 'phase plane' to obtain a curve, the points of which are labelled with the times t; Fig. I.39. If there is, for example, a periodic process in time, say an oscillation, then the corresponding phase curve is closed and is traversed 'infinitely often'. If f does not depend explicitly on the time, i.e., if $f = F(q, \dot{q})$, then the system is said to be 'autonomous' and the differential equation can be reduced to one of the first order. For, it v is regarded as a function of q, then since

$$\ddot{q} = \frac{dv}{dt} = \frac{dv}{dq}\frac{dq}{dt} = v\frac{dv}{dq}$$

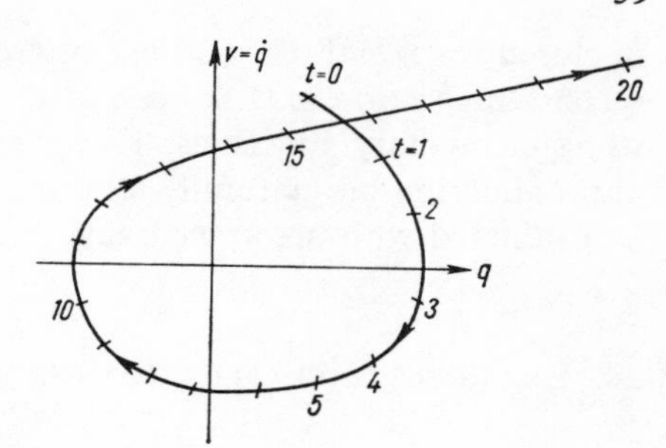

Fig. I.39. A phase curve

the equation (I.78) becomes

$$\frac{\mathrm{d}v}{\mathrm{d}q} = \frac{F(q, v)}{v}. \tag{I.80}$$

By this first-order differential equation a family of curves (the 'phase curves') is determined in the phase plane, and for the discussion of these curves the theory of singular points is often of great use, since (I.80) has the same form as (I.65). If, in particular, F depends on q alone or on v alone, then the differential equation (I.80) can be integrated immediately by separation of the variables; for example, if

$$\ddot{q} = F(q, v) = g(q), \tag{I.81}$$

then letting

$$G(q) = \int_0^q g(s)\,\mathrm{d}s \tag{I.82}$$

be an antiderivative of $g(q)$, equation (I.80) on integration gives

$$\frac{v^2}{2} = G(q) + c \tag{I.83}$$

as the equation of the family of phase curves. If q_0, v_0 are the prescribed initial values, then

$$\tfrac{1}{2}(v^2 - v_0^2) = G(q) - G(q_0) \quad \text{or} \quad c = \tfrac{1}{2} v_0^2 - G(q_0). \tag{I.84}$$

If desired, the time lapse can be determined by a further integration:

$$\frac{\mathrm{d}q}{\mathrm{d}t} = v = \pm\sqrt{2(G(q) + c)} \tag{I.85}$$

$$t - t_0 = \pm \int_{q_0}^{q} \frac{\mathrm{d}q}{\sqrt{2(G(q) + c)}}. \tag{I.86}$$

This formula can be used, for example, to determine the period of oscillation as a function of the deflection.

In non-linear oscillations we can often carry out the first integration (I.83) in closed form and discuss the circumstances in the phase plane, although the second integration in (I.86) can often no longer be carried out in closed form using elementary functions. However, numerical and graphical methods of approximation are naturally still available; in particular, integrals can easily be evaluated with great accuracy.

28. EXAMPLES OF OSCILLATIONS AND PHASE CURVES

Example 1. A pendulum with large deflections

The position of a 'simple pendulum' of length l and mass m oscillating under gravity is defined by φ, the angle of inclination to the equilibrium position; φ corresponds to the preceding q. In this ideal pendulum let the 'string' be rigid and of negligible mass. If g is the acceleration of gravity, the equation of motion (moment of external forces = moment of inertia times the angular acceleration) reads (Fig. I.40)

$$-lmg\sin\varphi = ml^2\ddot{\varphi}$$

or

$$\ddot{\varphi} = -\frac{g}{l}\sin\varphi = g(\varphi). \tag{I.87}$$

From (I.83) we obtain immediately for the phase curves the equation

$$v^2 = \frac{2g}{l}\cos\varphi + C. \tag{I.88}$$

First we draw v^2 as a function of φ, Fig. I.41, and thence obtain a general view of the phase curves, Fig. I.42.

We obtain 3 different types of curves:

1. the curves of type A correspond to the proper oscillations of the simple pendulum (swinging between $-\varphi_0$ and $+\varphi_0$);
2. a limit curve of type B. To the point P_0 corresponds $\varphi_0 = 0$, $v_0 = \sqrt{2g/l}$. If the pendulum in its equilibrium position $\varphi_0 = 0$ receives an impulse mv_0 with this value of v_0, then it swings up to the position of unstable

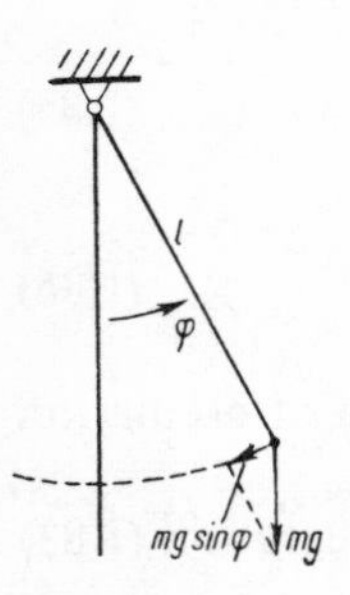

Fig. I.40. Pendulum with large deflections

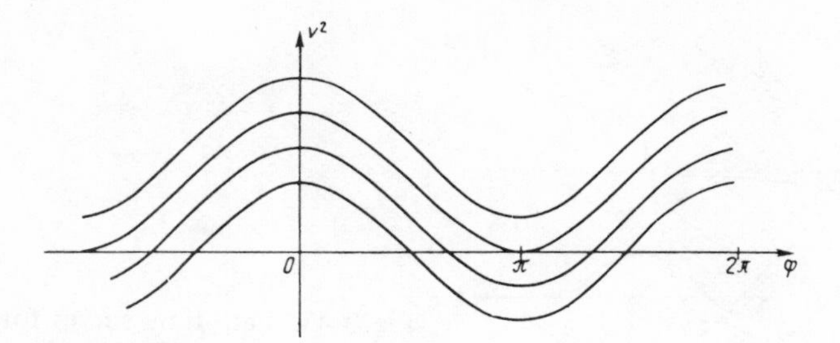

Fig. I.41. Auxiliary curves for the phase curves

equilibrium $\varphi = \pi$, but it reaches it only after an infinite time, as can be seen from the integral (I.86);

3. the curves of type C correspond to complete revolutions.

Singular points occur for the differential equation

$$\frac{dv}{d\varphi} = -\frac{g}{l}\frac{\sin\varphi}{v} \tag{I.89}$$

if $v = 0$ and $\sin\varphi = 0$, i.e., for point of type P_1, Fig. I.42,

$$\varphi = 2m\pi, \qquad v = 0 \quad (m = 0, \pm 1, \ldots)$$

and for points of type P_2

$$\varphi = (2m+1)\pi, \qquad v = 0 \quad (m = 0, \pm 1, \ldots);$$

P_1 is a centre, and P_2 a saddle point.

The determination of the time elapsed in the motion is here no longer possible in closed form with elementary functions, but only with elliptic functions; we can, however, obtain the shape of the q, t-curves, Fig. I.43, from a discussion of (I.86). The calculation, which leads to the elliptic integrals, may be set out briefly for oscillations of type A as follows.

If we choose as the initial conditions, say, $\varphi(0) = \varphi_0$, $v(0) = 0$, and write

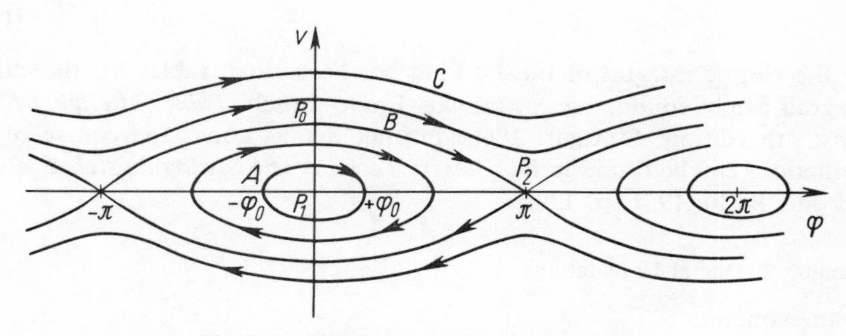

Fig. I.42. Phase curves for the pendulum

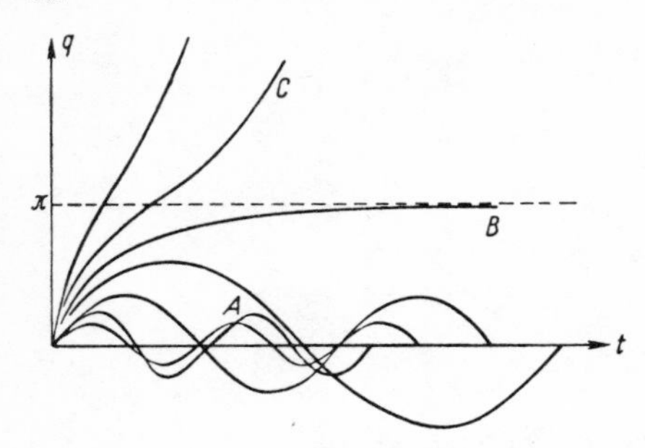

Fig. I.43. Path-time curves for the pendulum

$s^2 = g/l$, then (I.88) gives

$$v^2 = 2s^2(\cos\varphi - \cos\varphi_0) = \left(\frac{\mathrm{d}\varphi}{\mathrm{d}t}\right)^2;$$

since $\dot{\varphi} < 0$ for small t the following equation corresponds to formula (I.86) with the minus sign

$$t = \frac{1}{s\sqrt{2}}\int_{\varphi}^{\varphi_0}\frac{\mathrm{d}u}{\sqrt{\cos u - \cos\varphi_0}} = \frac{1}{2s}\int_{\varphi}^{\varphi_0}\frac{\mathrm{d}u}{\sqrt{\sin^2(\varphi_0/2) - \sin^2(u/2)}},$$

because $\frac{1}{2}\cos u = \frac{1}{2} - \sin^2 u/2$ We now substitute into the integral

$$\sin(u/2) = r, \qquad \sin(\varphi_0/2) = k, \qquad \sin(\varphi/2) = \rho,$$

and thus obtain

$$st = \int_{\rho}^{k}\frac{\mathrm{d}r}{\sqrt{(1-r^2)(k^2-r^2)}}.$$

If further we now put $r = kz$, then

$$st = \int_{\rho/k}^{1}\frac{\mathrm{d}z}{\sqrt{(1-k^2z^2)(1-z^2)}} = -F(k, \arcsin\rho/k) + K(k). \tag{I.90}$$

Here the usual notation

$$F(k,\varphi) \equiv \int_0^{\sin\varphi}\frac{\mathrm{d}z}{\sqrt{(1-z^2)(1-k^2z^2)}} = \int_0^{\varphi}\frac{\mathrm{d}\psi}{\sqrt{1-k^2\sin^2\psi}}, \qquad K(k) = F\left(k, \frac{\pi}{2}\right) \tag{I.91}$$

for the elliptic integral of the 1st kind has been used; tables for the elliptic integrals can be found, e.g., in Jahnke–Emde–Lösch, *Tables of Higher Functions,* 7th edition, Stuttgart 1966. Further details about the course of the oscillations can be found in K. Klotter: *Technische Schwingungstehre,* vol. 1, 2nd ed., Berlin 1951, p. 137.

Example 2. Damped oscillations

To the equation

$$\ddot{q} + k\dot{q} + cq = 0 \tag{I.92}$$

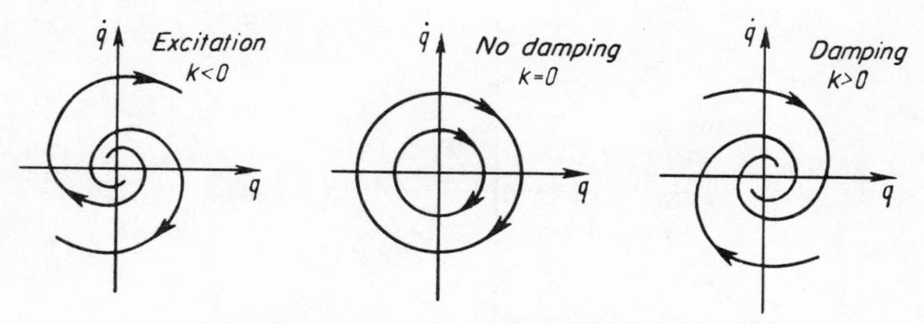

Fig. I.44. Phase curves with excitation and with damping

for the oscillations studied in more detail in Chapter II, 15 there corresponds, according to (I.80), the first-order differential equation

$$\frac{\mathrm{d}v}{\mathrm{d}q} = -\frac{kv + cq}{v}.$$

Depending on the sign of k we obtain in the phase plane for small $|k|$ the curves shown in Fig. I.44: we see that for $k > 0$ there is stability; after a small displacement from the equilibrium position, a spiral is described in the phase plane on which the null-point (the equilibrium position) is again approached. In contrast, with $k < 0$, (excitation instead of damping), the equilibrium position is unstable; after a small perturbation, an outward running spiral is described.

Example 3. Coulomb (dry) friction

When there is dry friction we have the differential equation

$$\ddot{q} + cq + k\,\mathrm{sgn}\,\dot{q} = 0;$$

where

$$\mathrm{sgn}\,\dot{q} = \begin{cases} 1 \text{ for } \dot{q} > 0 \\ -1 \text{ for } \dot{q} < 0 \\ 0 \text{ for } \dot{q} = 0. \end{cases} \tag{I.93}$$

Here, the corresponding phase equation reads

$$\frac{\mathrm{d}v}{\mathrm{d}q} = \frac{-cq \mp k}{v}, \tag{I.94}$$

the upper sign ($-k$) holding in the upper half-plane ($v > 0$), and the lower sign ($+k$) holding in the lower half-plane ($v < 0$), Fig. I.45. If the scales for the q-axis and the q-axis are suitably chosen, semi-circular arcs are obtained (with other scales, semi-ellipses), these arcs having centre A in the upper half-plane, and centre B in the lower half-plane. In a complete oscillation from a maximum q_0 to the minimum q_1 and to the next maximum q_2 the amplitude has

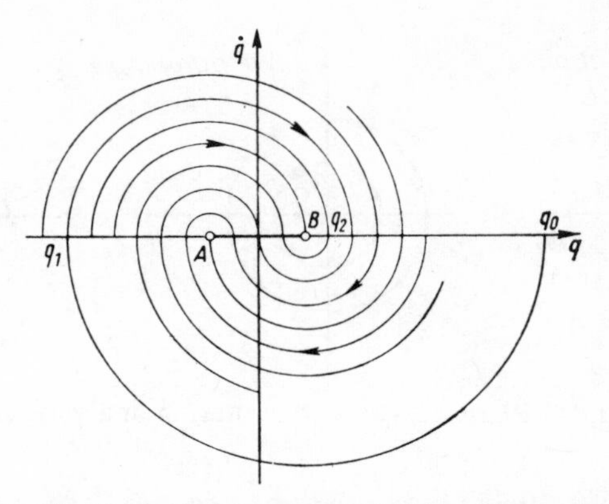

Fig. I.45. Phase curves with dry friction

decreased by the fixed amount $2Q$, where Q corresponds to the interval AB. Thus the amplitudes decrease each time by Q until, after a finite number of oscillations, a phase point on the interval AB is reached. The system then remains at rest, even though q does not have the value 0.

Example 4. A non-linear oscillator

A mass m is arranged between two springs, of spring constant c, as shown in Fig. I.46. In the equilibrium position the mass and the two springs lie in a straight line. We consider oscillations of the mass perpendicular to this straight line. Let x_0 be the initial deflection at time $t = 0$ and let $v_0 = \dot{x}_0 = 0$ be the initial velocity. The restoring force here is not proportional to the deflection; in fact, the restoring force when the spring extension is

$$\lambda = \sqrt{l^2 + x^3} - l$$

is

$$K = 2c(\sqrt{l^2 + x^2} - l)\cos\alpha$$

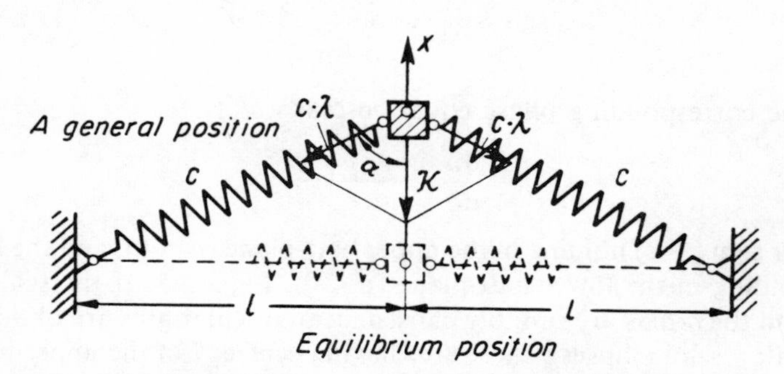

Fig. I.46. Example of a non-linear oscillation system; it is also non-linearizable

where

$$\cos\alpha = \frac{x}{\sqrt{l^2 + x^2}}$$

and the equation of motion is

$$m\ddot{x} = 2c\left(-x + \frac{lx}{\sqrt{l^2 + x^2}}\right). \tag{I.95}$$

The spring characteristic has the form shown in Fig. I.47. Equation (I.95) has the same form as (I.81) and can therefore be integrated once, as in (I.83):

$$\tfrac{1}{2}m\dot{x}^2 = 2c\left(-\frac{x^2}{2} + l\sqrt{l^2 + x^2}\right) + C_1.$$

The second integration, which led to equation (I.86) here gives

$$t - t_0 = \pm\int \frac{\mathrm{d}x}{\sqrt{\frac{4c}{m}\left(-\frac{x^2}{2} + l\sqrt{l^2 + x^2} + C_2\right)}}.$$

This in an elliptic integral. It can be evaluated graphically, for example. If, however, the oscillations are small, i.e., if $x \ll l$, we can get further using elementary means, albeit with certain approximations. We use the binomial theorem to expand the root appearing in (I.95) as the series

$$\frac{1}{\sqrt{l^2 + x^2}} = \frac{1}{l}\left(1 + \frac{x^2}{l^2}\right)^{-1/2} = \frac{1}{l}\left(1 - \frac{1}{2}\frac{x^2}{l^2} + \frac{1\cdot 3}{2\cdot 4}\frac{x^4}{l^4} - \dots\right).$$

The series converges if $x^2/l^2 < 1$, and by hypothesis this condition is satisfied. Substitution in the equation of motion yields the new differential equation

$$m\ddot{x} = 2c\left(-x + \frac{lx}{l}\left[1 - \frac{1}{2}\frac{x^2}{l^2} + \frac{1\cdot 3}{2\cdot 4}\frac{x^4}{l^4} - \dots\right]\right).$$

If we terminate the series after the second term we have as an approximation

$$\ddot{x} = -kx^3; \tag{I.96}$$

where $c/ml^2 = k$; from (I.83) again, and bringing in the initial conditions $t = 0$: $x = x_0$, $\dot{x} = 0$, we have

$$\frac{1}{2}\dot{x}^2 = \frac{k}{4}(x_0^4 - x^4).$$

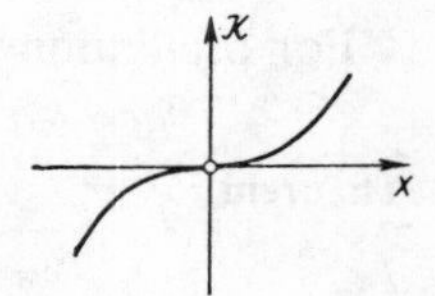

Fig. I.47. Non-linear spring characteristic

Separation of the variables yields

$$\pm \int_{x_0}^{x} \frac{\mathrm{d}x}{\sqrt{\frac{k}{2}(x_0^4 - x^4)}} = \int_0^t = t.$$

(for $x_0 > 0$ the lower sign, minus, is to be taken.)

On the left there is again an elliptic integral, which cannot be solved in closed form elementarily. However, important facts, such as the dependence of the period of oscillation on the deflection, can be deduced from the equation without integration. For the period T of oscillation (because of the symmetry, we can consider just $T/4$) we thus obtain

$$\frac{T}{4} = \int_0^{x_0} \frac{\mathrm{d}x}{\sqrt{\frac{k}{2}(x_0^4 - x^4)}} \quad \text{(for } t = T/4 \text{ we have } x = 0\text{)}$$

or putting $x = x_0 u$; $\mathrm{d}x = x_0\, \mathrm{d}u$

$$T = 4\sqrt{\frac{2}{k}}\,\frac{1}{x_0}\int_0^1 \frac{\mathrm{d}u}{\sqrt{1-u^4}}.$$

The definite (improper but convergent) integral is a fixed constant. If we include all the constants of the system together in k^*, then

$$T = \frac{1}{f} = \frac{k^*}{x_0}.$$

Thus here the period of oscillation is inversely proportional to the initial deflection x_0, i.e., the frequency f is directly proportional to the initial deflection x_0.

29. Periodic oscillations of an autonomous undamped system with one degree of freedom

A real-valued function $x(t)$ defined for all real values of t is said to be 'oscillatory' if it has infinitely many zeros T_ν ($\nu = 1, 2, \ldots$); it is 'non-oscillatory' if it has at most a finite number of zeros. Functions $x(t)$ which do not vanish identically but which have an uncountable infinity of zeros will not be considered here.

A function $x(t)$ is said to be 'periodic' if there is a positive number p, the 'period', such that

$$x(t+p) = x(t) \quad \text{for all real } t. \tag{I.97}$$

For oscillation equations of the form (I.81) the following theorem holds.

Theorem

Let

$$\ddot{x} + g(x) = 0 \tag{I.98}$$

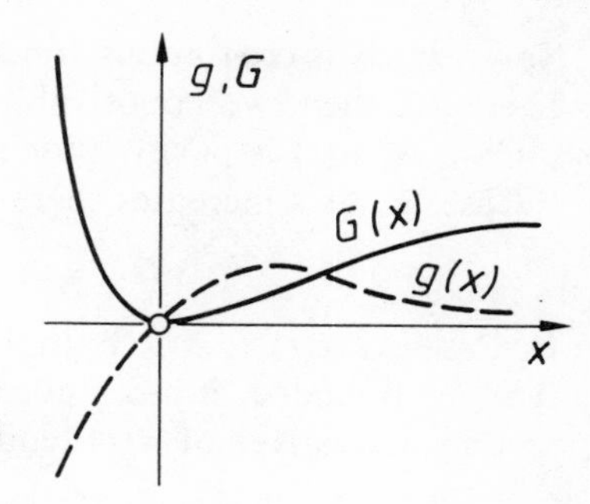

Fig. I.48. The restoring-force term and its antiderivative

be the differential equation for oscillations of an autonomous system with one degree of freedom and with a given (non-linear) restoring-force term $g(x)$. Let $g(x)$ satisfy the conditions (see Fig. I.48):

(1) *$g(x)$ is continuous for all real x and $xg(x)$ is 'positive definite'* (i.e., *$xg(x) > 0$ for all $x \neq 0$).*

(2) *$G(u) = \int_0^u g(s)\,\mathrm{d}s$ increases beyond all bounds as $u \to \pm\infty$.* (I.99)
Then every non-identically-zero solution $x(t)$ of the differential equation is periodic in t.

Proof. I. First consider the solution $x(t)$ of the differential equation (I.98) with the initial values

$$x(0) = 0, \qquad \dot{x}(0) = v_0 > 0$$

where v_0 is a given positive initial velocity. With $v = v(t) = \dot{x}(t)$ the equation becomes

$$v\frac{\mathrm{d}v}{\mathrm{d}x} = -g(x)$$

or

$$v\,\mathrm{d}v = -g(x)\,\mathrm{d}x$$

which we integrate over the interval $(0, t)$. As in (I.83) (but there with a different sign before the g) we obtain (apart from a mass factor) the energy integral

$$\tfrac{1}{2}v^2 - \tfrac{1}{2}v_0^2 = -G(x)$$

or

$$G(x) \leq \tfrac{1}{2}v_0^2.$$

$G(x) = G(x(t))$ is therefore bounded above and, by hypothesis (2), $x(t)$ also remains bounded above, i.e., there is a constant K, which can depend on v_0, such that

$$x(t) \leq K \quad \text{for all } t.$$

Now $v(t)$ is a continuous function which has the positive value v_0 at $t = 0$. There are then two conceivable cases:

Case A. $v(t)$ is positive for all values of $t \geq 0$;

Case B. As t increases there is a first zero of $v(t)$, say at t_1, and

$$v(t_1) = 0,\ v(t) > 0 \quad \text{for } 0 \leq t < t_1. \tag{I.100}$$

In Case A, $x(t)$ is strictly monotone increasing for all $t > 0$ and, since $x(t)$ remains bounded, it has a positive limit value x as $t \to +\infty$ and the first and second derivatives of $x(t)$ tend to 0 as $t \to +\infty$;

$$\lim_{t\to\infty} x(t) = \hat{x}, \qquad \lim_{t\to\infty} \dot{x}(t) = 0, \qquad \lim_{t\to\infty} \ddot{x}(t) = 0.$$

But if we let $t \to +\infty$ in the differential equation (I.98), we obtain

$$0 = \lim_{t\to+\infty} (\ddot{x} + g(x)) = \lim_{t\to+\infty} \ddot{x}(t) + \lim_{t\to+\infty} g(x) = 0 + g(\hat{x}),$$

and therefore
$$g(\hat{x}) = 0$$

contradicting hypothesis (1) that $xg(x) > 0$. Hence, Case A cannot occur, and there is a $t_1 > 0$ satisfying (I.100).

II. The solution $x(t)$ with the initial conditions $x(0) = 0$, $\dot{x}(0) = v_0$ is now continued by 'reflection' in the abscissa $t = t_1$ (Fig. 1.49) into a function $x^*(t)$

$$x^*(t) = x(2\,t_1 - t) \quad \text{for } t_1 \leq t \leq 2t_1 = t_2.$$

$x^*(t)$ is a solution of the differential equation (I.98) and has at $t = t_1$ the same initial values as the function $x(t)$:

$$x(t_1) = x^*(t_1), \qquad \dot{x}(t_1) = \dot{x}^*(t_1) = 0$$

and therefore, by the theorem on the uniqueness of the solution of an initial-value problem it coincides with $x(t)$:

$$x^*(t) = x(t) \quad \text{for } t_1 \leq t \leq 2t_1 = t_2.$$

We have

$$x(2t_1) = x(t_2) = 0, \qquad \dot{x}(t_2) = -\,\dot{x}(0).$$

Now in the same way we can continue $x(t)$ by 'point-reflection' to a function $x^{**}(t)$

$$x^{**}(t) = x(2t_2 - t) \quad \text{for } t_2 \leq t \leq 2t_2 = t_4,$$

(Fig. I.49) and obtain

$$x(t_4) = 0 \quad \dot{x}(t_4) = \dot{x}(0).$$

$x(t)$ has at t_4 the same initial value as at $t = 0$, and by repeated continuation we obtain a solution of the differential equation with period t_4 and by the uniqueness theorem this solution is uniquely determined by its initial conditions.

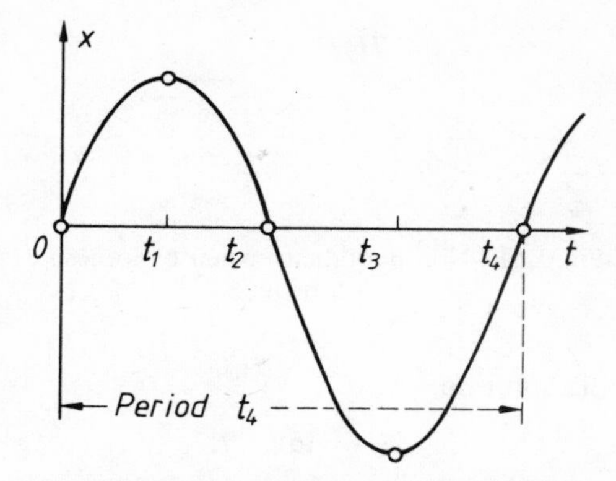

Fig. I.49. Reflection principle for proving periodicity

III. Now let $x(t)$ be an arbitrary solution, not identically zero, of the differential equation, which assumes at $t = \tilde{t}$ the initial values

$$x(\tilde{t}) = \tilde{x}, \qquad \dot{x}(\tilde{t}) = \tilde{v}$$

We calculate a value $\dot{x}_0$ from

$$\dot{x}_0^2 = \tilde{v}^2 + 2G(\tilde{x}).$$

The periodic solution $\hat{x}(t)$ of (I.98) determined by the initial conditions

$$x(0) = 0, \ \dot{x}(0) = \dot{x}_0$$

crosses the straight line $x = \tilde{x}$ with the slope $\dot{x} = \pm \tilde{v}$ and by a shift in the direction of the t-axis it can be brought into coincidence with the chosen solution $x(t)$; see Fig. I.50. $x(t)$ is therefore also periodic and has the same period as $\hat{x}(t)$.

Note. In this theorem the hypotheses (1) and (2) made to ensure periodicity can be somewhat weakened, but cannot be dispensed with entirely, as the following two counter-examples show.

1. The initial-value problem $x(0) = 0$, $\dot{x}(0) = 1$ for the differential equation (I.98) with

$$g(x) = 2(\tan x)(\cos x)^4 \operatorname{sgn} x$$

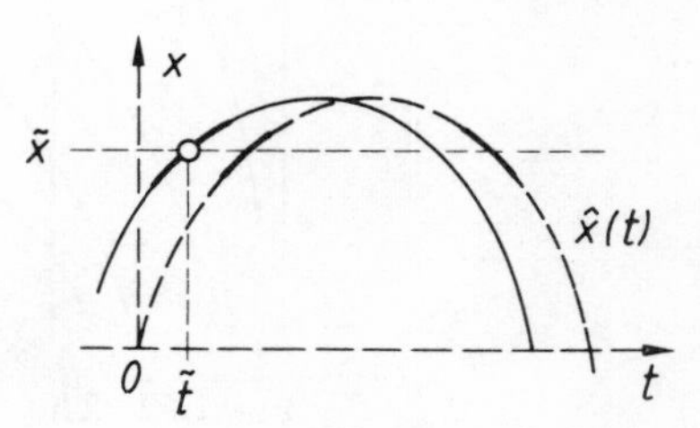

Fig. I.50. Periodicity with arbitrary initial values

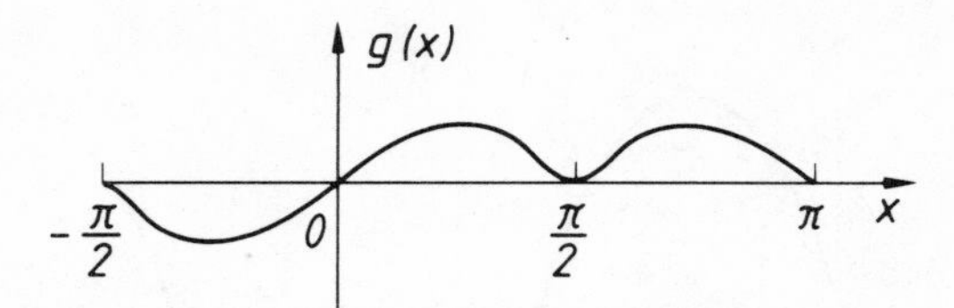

Fig. I.51. Non-periodicity when hypothesis (1) is infringed

has the non-periodic solution

$$x(t) = \tan^{-1} t;$$

here hypothesis (1) is infringed (see Fig. I.51) $g(\pi/2) = 0$ for example.

2. The differential equation (I.98) with

$$g(x) = \frac{\sinh x}{(\cosh x)^3}$$

and with the initial conditions $x(0) = 0$, $\dot{x}(0) = 1$ has the non-periodic solution

$$x(t) = \sinh^{-1} t.$$

Here hypothesis (2) is not satisfied.

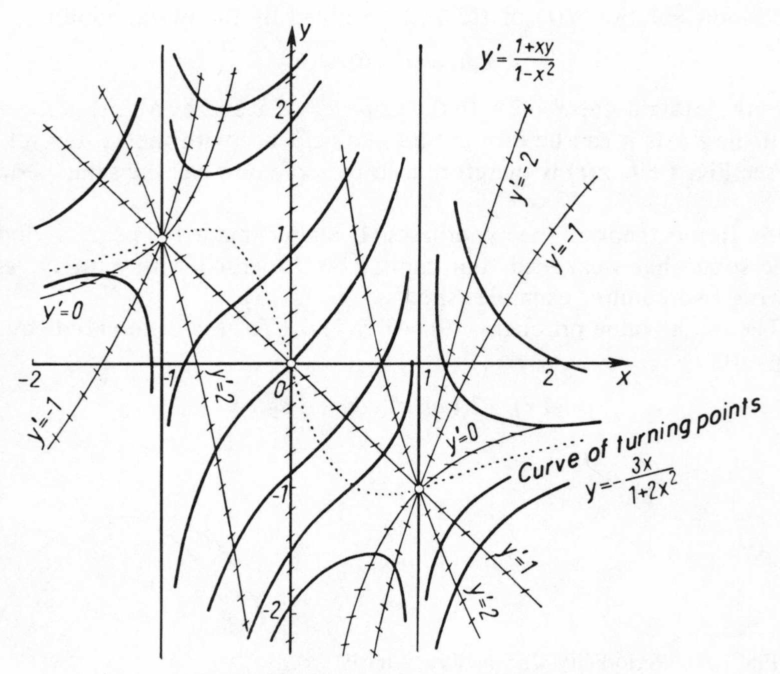

Fig. I.52. The direction-field in problem 1(d)

§8. Miscellaneous problems and solutions

30. PROBLEMS

1. For the following differential equations obtain the general solutions, any singular solutions which may exist, and any solutions noteworthy because of the simplicity of their analytical expression, and the locus of the points of inflexion of all solutions (differentiate the differential equation and put $y''=0$). In each case draw the direction-field (due to lack of space we show here the direction-field only for problem d; see Fig. I.52).

(a) $y' = y\tan x - 2\sin x$;

(b) $y' = \dfrac{x}{x^2 - y^2 - 1}$;

(c) $y' = \dfrac{y - xy^2}{x + x^2 y}$;

(d) $y' = \dfrac{1 + xy}{1 - x^2}$;

(e) $xy' = y + \sqrt{4x^2 - y^2}$;

(f) $y' = \dfrac{x + y}{x - y + 2}$;

(g) $y' = x(1 - y^2)$;

(h) the *orthogonal trajectories* for (g), i.e., the family of curves which cut the solution curves of (g) orthogonally. Their differential equation is obtained from that of (g) by replacing y' by $-1/y'$, so it is

$$y' = \frac{1}{x(y^2 - 1)};$$

(i) $y'(y^3 - xy) = 1$;

(k) $y' = \dfrac{x}{y}\dfrac{1 - y^2}{1 - x^2}$;

(j) the orthogonal trajectories for (k).

2. Solve the differential equation $y' = y^2 - x^2$ with the initial condition $y(0) = 1$ by expressing y as a power series

$$y = a_0 + a_1 x + a_2 x^2 + \cdots$$

and substituting series on both sides of the differential equation and comparing the coefficients of like powers of x.

3. Set up the differential equation for all circles which are symmetric relative to the x-axis and which touch the two straight lines $y = \pm x\tan\alpha$ (Draw a figure!)

4. A particle of mass m falls into a stream of air or liquid moving with constant velocity v_0 and is carried along with it; the particle is acted on by a force $k(v_0 - v)^2$, i.e., a force proportional to the square of the difference in

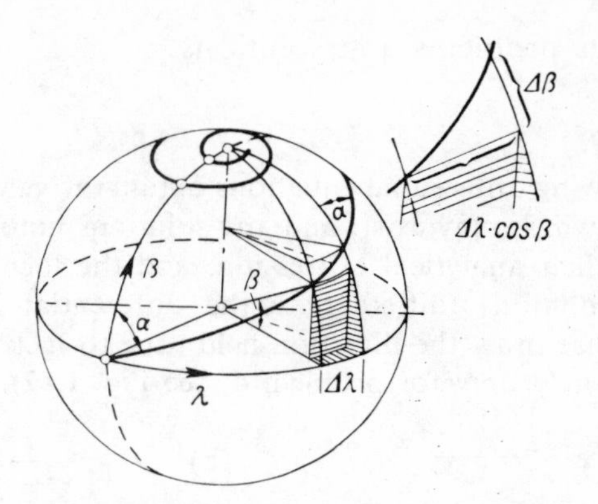

Fig. I.53. The loxodrome for the track of a ship steering a steady course N 60° E

velocities. Determine the distance x travelled and the velocity v as functions of the time t under the initial conditions $x = 0$ and $v = 0$ at $t = 0$.

5. A ship travels at a constant course of N 60° E on a 'loxodrome', starting from a point on the equator (Fig. I.53). What track does it describe on the surface of the earth, assumed to be spherical?

6. What shape is assumed by the water surface in the container on a rotating water-scooping wheel? See Fig. I.54.

7. The shape, in longitudinal section, of the surface of a stream of water flowing in a shallow bed of parabolic cross-section over a dam (see Fig. I.55) satisfies the differential equation

$$-\frac{\mathrm{d}h}{\mathrm{d}x} = \alpha\left[1 - \left(\frac{h_0}{h}\right)^4\right].$$

where α is the slope, assumed small, $h(x)$ is the height at a distance x from the dam, h_0 is the height at a great distance from the dam, and $H = h(0)$ is the height of the dam. Integrate the differential equation.

8. A cable 10 km long has a capacitance $C_B = 0.4\,\mu\text{F/km}$ (so a total capacitance $C = 4.10^{-36}\,\text{F}$) and an insulation resistance $R_B = 5 \cdot 10^5\,\Omega/\text{km}$. When a discharge occurs, how do the current i, the voltage u, and the charge

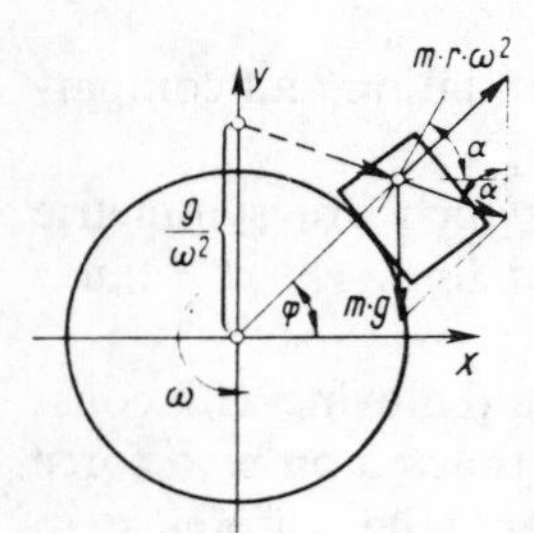

Fig. I.54. The water surface in the container on a rotating water-scooping wheel

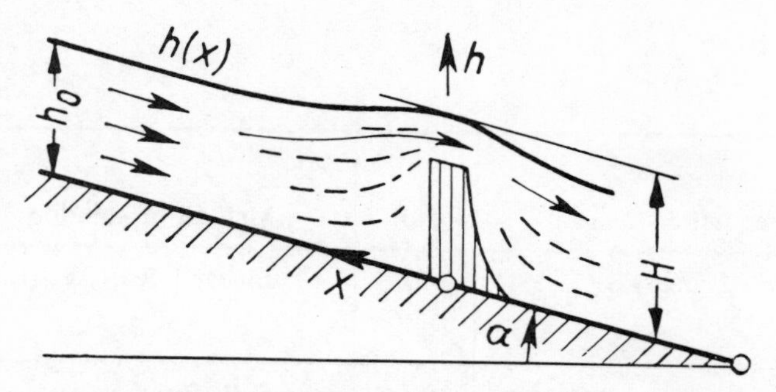

Fig. I.55. The shape, in longitudinal section, of the surface of a stream of water flowing over a dam

q vary from their initial values i_0, u_0, q_0, and after what times t_1, t_2 will they have decreased to 10% and 1% respectively?

31. Solutions

1. See table on pages 74, 75.

2. $y = 1 + x + x^2 + \frac{2}{3}x^3 + \frac{5}{6}x^4 + \frac{4}{5}x^5 + \cdots;$

for the coefficients a we obtain the recursion formula

$$a_{\nu+1}(\nu+1) = \sum_{\rho=0}^{\nu} a_\rho a_{\nu-\rho} + \begin{cases} 0 \text{ for } \nu = 0, 1, 3, \ldots \\ -1 \text{ for } \nu = 2. \end{cases}$$

We can hence prove by induction that $|a_\nu| \le 1$; the series therefore converges at least for $|x| < 1$.

3. From $(x-\xi)^2 + y^2 = \xi^2 \sin^2 \alpha$ we obtain by differention and elimination of ξ the differential equation

$$y'^2 y^2 \cos^2 \alpha - 2xyy' \sin^2 \alpha + y^2 - x^2 \sin^2 \alpha = 0.$$

4. The equation of motion reads $m\dot{v} = k(v_0 - v)^2$. Separation of the variables gives

$$v = v_0\left(1 - \frac{1}{1 + v_0(k/m)t}\right)$$

and an integration

$$x = v_0 t - \frac{m}{k} \ln (1 + v_0(k/m)t).$$

5. With the notation α, β, λ from Fig. I.53 we have

$$\tan \alpha = \frac{d\lambda \cos \beta}{d\beta}$$

Differential equations	No.	Method of solution
$y' = y \tan x - 2 \sin x$	(a)	Linear D.E. I, §2 (I.20)
$y'(x^2 - y^2 - 1) = x$	(b)	D.E. for $x(y)$ a Bernoulli equation
$y'(x + x^2 y) = y - xy^2$	(c)	Put $xy = z(x)$; then separate the variables
$y'(1 - x^2) = 1 + xy$	(d)	Linear D.E. I, §2
$xy' = y + \sqrt{4x^2 - y^2}$	(e)	Similarity D.E. I, §1
$y' = \dfrac{x + y}{x - y + 2}$	(f)	Put $x = \xi - 1$, $y = \eta + 1$; Similarity D.E. for $\eta(\xi)$
$y' = x(1 - y^2)$	(g)	Variables separate I, §1
$x(1 - y^2)y' = -1$	(h)	Variables separate I, §1
$y'(y^3 - xy) = 1$	(i)	For $x(y)$ a linear D.E. or integrating factor $e^{y^2/2}$
$y' = \dfrac{x}{y}\dfrac{1 - y^2}{1 - x^2}$	(k)	Variables separate I, §1
$y' = \dfrac{y}{x}\dfrac{1 - x^2}{y^2 - 1}$	(l)	Variables separate I, §1

General solution, c a parameter	Singular elements	Particularly noteworthy solutions	Curve of turning-points (inflexions)
$y = \cos x + \dfrac{c}{\cos x}$		$y = \cos x$	$y = \dfrac{2\cos x}{2 - \cos^2 x}$
$x^2 = y^2 + y + \frac{3}{2} + c\,e^{2y}$		$x^2 = y^2 + y + \frac{3}{2}$ Hyperbola	$2x^2(N - y) = N^2$ with $N = x^2 - y^2 - 1$
$x = cy\,e^{xy}$	$x = y = 0$ Node	y-axis ($x = 0$) x-axis ($y = 0$)	$xy = \pm\sqrt{3}$ Hyperbolae
$y = \dfrac{c + \arcsin x}{\sqrt{1 - x^2}}$ and $y = \dfrac{c - \ln(x + \sqrt{x^2 - 1})}{\sqrt{x^2 - 1}}$	$x = 1$, $y = -1$ and $x = -1$, $y = 1$ Saddle-points	$x = \pm 1$ Straight lines	$3x + y(2x^2 + 1 = 0$
$y = 2x\sin(\ln cx)$	$y = \pm 2x$	$y = \pm 2x$	$y = \pm\sqrt{2}x$
$\xi^2 + \eta^2 = c\,e^{2\arctan(\eta/\xi)}$ Logarithmic spirals	$x = -1$, $y = 1$ Focus		
$\dfrac{1 + y}{1 - y} = c\,e^{x^2}$		$y = \pm 1$	$2x^2 y = 1$
$x = c\,e^{(y^3 - 3y)/3}$		$x = 0$	$2y + (1 - y^2)^2 = 0$
$e^{x^2/2}(2 + x - y^2) = c$		$y^2 = x + 2$ Parabola	$y^4 - y^2(3 + x) + x = 0$
$y^2 + c(1 - x^2) = 1$ Pencil of circle sections	$x = \pm 1$; $y = \pm 1$; Four nodes	Circle $x^2 + y^2 = 2$ St. lines $x = \pm 1$, $y = \pm 1$, $y = \pm x$	
$xy = c\,e^{(x^2 + y^2)/2}$	$x = \pm 1$; $y = \pm 1$ Four centres	$x = 0$, $y = 0$	

and hence

$$\lambda = \tan\alpha \int_0^\beta \frac{d\varphi}{\cos\varphi} = \tan\alpha \ln\left[\tan\left(\frac{\beta}{2}+\frac{\pi}{4}\right)\right].$$

6. A particle of water of mass m is acted on by its weight mg vertically and the radially directed force of inertia (the 'centrifugal force') as shown in Fig. I.46. The water surface is perpendicular to the resultant of these two forces:

$$\tan\alpha = y' = \frac{mx\omega^2}{mg - my\omega^2}.$$

Separation of the variables yields $\left(\frac{g}{\omega^2} - y\right)^2 + x^2 = \text{const}$, so the surface is part of a circle about the point $x = 0$, $y = g/\omega^2$.

7. $-\alpha x = h - H + \frac{h_0}{4}\ln\left(\frac{h-h_0}{H-h_0}\,\frac{H+h_0}{h+h_0}\right) + \frac{h_0}{2}\left(\arctan\frac{H}{h_0} - \arctan\frac{h}{h_0}\right).$

8. $\frac{di}{dt} = -\frac{1}{RC}i$; $i = i_0 e^{-t/(RC)}$ and similarly $u = u_0\, e^{-t/(RC)}$; $q = q_0\, e^{-t/(RC)}$.

$$t_1 = RC \ln 10 = 46 \text{ s}, \quad t_2 = 2t_1 = 92 \text{ s}.$$

Further miscellaneous problems and their solutions are given in Chapter VI, §2, No. 7.

II

Ordinary differential equations of higher order

§1. SOME TYPES OF NON-LINEAR DIFFERENTIAL EQUATIONS

In some particular types of non-linear differential equations the order can be lowered.

1. THE DEPENDENT VARIABLE y DOES NOT APPEAR EXPLICITLY

If y does not appear explicitly but is only differentiated

$$F[x, y', y'', \ldots, y(n)] = 0, \tag{II.1}$$

then we can introduce a new function $z = y'$ and obtain a differential equation. of order $n-1$ for z.

Example. Motion with resistance depending on the velocity

Suppose a locomotive of mass m exerts a tractive force a and experiences a resistance proportional to the velocity, say bv, where the constant b depends on the friction and air resistance; let $x(t)$ be the distance travelled in time t. Then the equation of motion is

$$m\ddot{x} = a - \dot{x}b \tag{II.2}$$

(dots denote differentiation with respect to time). With $\dot{x} = v$, this becomes $mv = a - bv$, a first-order linear differential equation with the general solution (which can be obtained, e.g., from (I.20))

$$\dot{x} = v = \frac{a}{b} - C_1^* \exp\left(-\frac{b}{m}t\right), \tag{II.3}$$

where C_1^* is an integration constant. A second integration gives

$$x = \frac{a}{b}t + C_1 \exp\left(-\frac{b}{m}t\right) + C_2$$

as the general solution of (II.2). If the initial conditions

$$x=0,\ \dot{x}=0 \quad \text{for} \quad t=0,$$

are given, then it follows from $\dot{x}=0$ that $C_1 = am/b^2$ and from $x=0$ that $C_2 = -\ am/b^2$, and so the solution is

$$x=\frac{a}{b}\left\{t+\frac{m}{b}\left[\exp\left(-\frac{b}{m}t\right)-1\right]\right\}$$

2. THE EQUATION $y'' = f(y)$ AND THE ENERGY INTEGRAL

As a preparation for the next type we consider particularly

$$y''=f(y). \tag{II.4}$$

Its integration has already been carried out in Chapter I, 27, but because of the importance of this type another explanation of the method may be presented.

Explanation of the method in terms of the energy principle of mechanics. In a frictionless system suppose a mass m is accelerated by a force $K(x)$ which depends on the position of m. Then by Newton's law

$$m\ddot{x}=K(x).$$

Fig. II.1. shows a simple case where the force depends on the position of the mass.

During the motion, kinetic energy is converted into potential energy or *vice versa*. By multiplying the differential equation by $\dot{x}$ we obtain the power equation (force $\times$ velocity = power)

$$m\ddot{x}\dot{x}=\mathrm{K}(x)\dot{x}.$$

The left-hand side is the derivative of the kinetic energy T with respect to time:

$$T(t)=\tfrac{1}{2}mv^2=\tfrac{1}{2}m\dot{x}^2, \qquad \frac{\mathrm{d}T(t)}{\mathrm{d}t}=m\dot{x}\ddot{x}.$$

The work A which is expended in changing the kinetic energy of the mass, i.e., the difference in the two values of the potential energy, is the integral of the

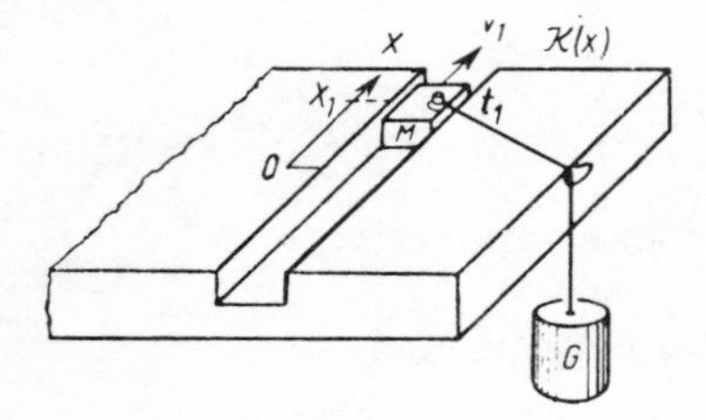

Fig. II.1. Example where the force depends on position

power with respect to time, say from t_1 to t_2 (let the corresponding positions be x_1 and x_2:

$$\underbrace{\int_{t_1}^{t_2} m\dot{x}\ddot{x}\,\mathrm{d}t}_{} = \underbrace{\int_{t_1}^{t_2} K(x)\frac{\mathrm{d}x}{\mathrm{d}t}\,\mathrm{d}t,}_{}$$

or

$$\int_{t_1}^{t_2} \frac{\mathrm{d}T(t)}{\mathrm{d}t}\,\mathrm{d}t = \int_{x_1}^{x_2} K(x)\,\mathrm{d}x$$

where, on the right-hand side, x has been introduced as a new variable. We thus obtain the energy equation

$$T(t_2) - T(t_1) = \tfrac{1}{2}mv_2{}^2 - \tfrac{1}{2}mv_1{}^2 = \int_{x_1}^{x_2} K(x)\,\mathrm{d}x = A; \tag{II.5}$$

on the left-hand side is the difference of the kinetic energy at two times t_1 and t_2, on the right the work done for the transformation. In the energy equation no accelerations appear, only velocities; but this means that the second-order differential equation has been reduced (by a multiplication by $\dot{x}$ and an integration) to one of first order.

3. THE GENERAL DIFFERENTIAL EQUATION IN WHICH THE INDEPENDENT VARIABLE x DOES NOT OCCUR EXPLICITLY

In a differential equation in which x does not appear explicitly

$$F[y, y', y'', \ldots, y(n)] = 0 \tag{II.6}$$

the order can be reduced by 1. To do this we regard y' as a new unknown function of y:

$$y' = p[y(x)]. \tag{II.7}$$

(This way of looking at the situation is possible in intervals $a \leq x \leq b$ in which $y' \neq 0$ so that y depends monotonely on x, and it can easily be illustrated graphically; see Figs. II.2 and II.3.). It follows that

$$y'' = \frac{\mathrm{d}p[y(x)]}{\mathrm{d}y}\frac{\mathrm{d}y}{\mathrm{d}x} = p\frac{\mathrm{d}p}{\mathrm{d}y},$$

$$y''' = \frac{\mathrm{d}\left(p\dfrac{\mathrm{d}p}{\mathrm{d}y}\right)}{\mathrm{d}y}\frac{\mathrm{d}y}{\mathrm{d}x} = \left(\frac{\mathrm{d}p}{\mathrm{d}y}\frac{\mathrm{d}p}{\mathrm{d}y} + p\frac{\mathrm{d}^2p}{\mathrm{d}y^2}\right)p = p\left(\frac{\mathrm{d}p}{\mathrm{d}y}\right)^2 + p^2\frac{\mathrm{d}^2p}{\mathrm{d}y^2}.$$

It is obvious (and it can be proved directly by induction) that, in general, the nth derivative $y(n)$ can be expressed in terms of $p(y)$ and its derivatives up to order $n-1$.

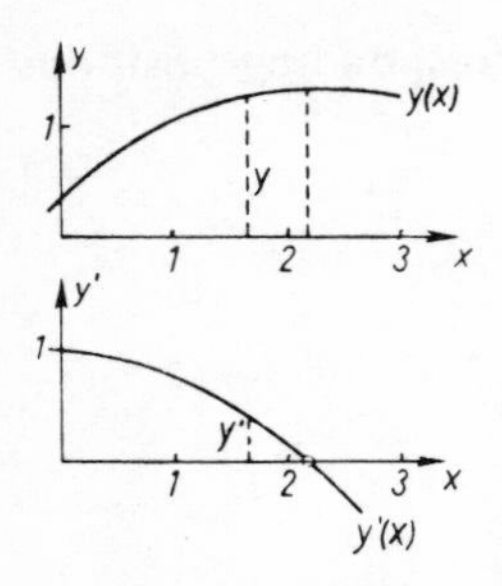

Fig. II.2. Introduction of y' as a function of y

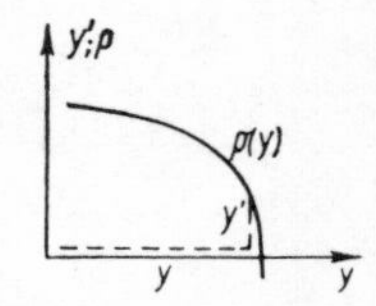

Fig. II.3. Graph of the function $y' = p(y)$

Example. An oscillator with a quadratic damping term

In an oscillatory system suppose there is a damping force of magnitude $k\dot{x}^2$ proportional to the square of the velocity $\dot{x}$. For the outward motion ($\dot{x} > 0$) and for the inward motion ($\dot{x} < 0$) we have two different differential equations

$$m\ddot{x} = -cx + \begin{cases} -k\dot{x}^2 & (\dot{x} > 0) \\ +k\dot{x}^2 & (\dot{x} < 0). \end{cases} \tag{II.8}$$

Each has to be solved separately. In the equation for the outward motion

$$m\ddot{x} + k\dot{x}^2 + cx = 0$$

a multiplication by $\dot{x}$ would no longer serve. (The energy principle (II.5) no longer holds, because part of the energy is lost as heat through the friction.) Putting $x = p(x)$ we obtain for p the first-order differential equation

$$mp\frac{dp}{dx} + kp^2 + cx = 0,$$

which can be solved in various ways, for example, as a Bernoulli differential equation. To do this, we put $p^2 = z$. The inhomogeneous linear differential equation

$$\frac{m}{2}\frac{dz}{dx} + kz + cx = 0$$

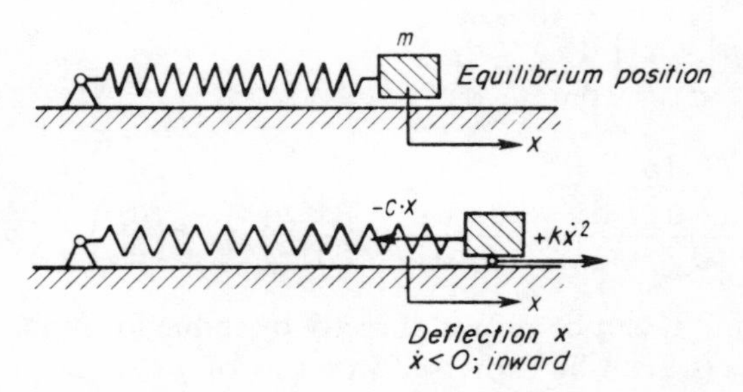

Fig. II.4. An oscillator with friction

with solution

$$z(x) = -\frac{c}{k}x + \frac{mc}{2k^2}\left[1 - \exp\left(-\frac{2k}{m}x\right)\right] + C^* \exp\left(-\frac{2k}{m}x\right)$$

yields

$$\dot{x} = p(x) = \pm\sqrt{z(x)} = \pm\sqrt{\frac{cm}{2k^2} - \frac{c}{k}x - C_1 \exp\left(-\frac{2k}{m}x\right)}$$

and

$$t = \int \frac{\mathrm{d}x}{\pm\sqrt{\frac{cm}{2k^2} - \frac{c}{k}x - C_1 \exp\left(-\frac{2k}{m}x\right)}} + C_2.$$

For evaluating this integral one is dependent on a graphical or other method of approximation.

We have to proceed similarly in solving the differential equation

$$m\ddot{x} - k\dot{x}^2 + cx = 0,$$

for the inward motion.

The first solution holds in each case for the 'half-period' with positive $\dot{x}$ until $\dot{x} = 0$, and then the second solution for negative $\dot{x}$ is fitted in. (Naturally the process is not periodic in the sense that there is a T such that $x(t+T) = x(t)$.)

In other words, for the whole oscillatory process the solutions must be 'patched together' half-period by half-period, naturally with appropriate boundary conditions such that the values of x and $\dot{x}$ fit in continuously with one another at the change-over points.

4. THE DIFFERENTIAL EQUATION CONTAINS ONLY THE RATIOS $y^{(\nu)}/y$

In the differential equation

$$F\left(x, \frac{y'}{y}, \frac{y''}{y}, \frac{y'''}{y}, \ldots, \frac{y^{(n)}}{y}\right) = 0$$

we make the substitution $u(x) = y'/y$. The derivatives of u can always be expressed in terms of quotients of the form $y^{(\nu)}/y$, as can easily be proved by induction:

$$u' = \frac{y''y - y'y'}{y^2} = \frac{y''}{y} - \left(\frac{y'}{y}\right)^2,$$

$$u'' = \frac{y'''}{y} - 3\frac{y'}{y}\frac{y''}{y} + 2\left(\frac{y'}{y}\right)^3, \quad \text{etc}$$

So we have

$$\frac{y'}{y} = u, \frac{y''}{y} = u' + u^2, \frac{y'''}{y} = u'' + 3u'u + u^3 \quad \text{etc.} \tag{II.10}$$

On substituting these we obtain a new differential equation in u of order $n - 1$.

A particular case is a homogeneous linear differential equation of the second order with coefficients which are functions of x:

$$p_2(x)y'' + p_1(x)y' + p_0(x)y = 0.$$

If we divide through by y, then as in (II.10) we obtain

$$u' + p(x)u + u^2 = r(x), \quad \text{where} \quad p(x) = \frac{p_1(x)}{p_2(x)} \quad \text{and} \quad r(x) = -\frac{p_0(x)}{p_2(x)},$$

which is a *Riccati* differential equation for u (cf. (I.22)).

The possibility of reducing a differential equation to one of the first order is particularly of value because then a direction-field can be drawn.

§2. FUNDAMENTAL THEOREMS ON LINEAR DIFFERENTIAL EQUATIONS

5. Notation

A very important type of differential equation of higher order is the *linear* differential equation. It has, cf. (I.15), the general from

$$p_n(x)y^{(n)} + p_{n-1}(x)y^{(n-1)} + \ldots + p_1(x)y' + p_0(x)y = r(x). \tag{II.11}$$

We also write the entire left-hand side briefly and symbolically as $L[y]$, thus

$$L[y] \equiv \sum_{\nu=0}^{n} p_\nu(x)y^{(\nu)} = r(x). \tag{II.12}$$

In this expression the zeroth derivative $y^{(0)}$ means the function y itself. The p_ν and r are given continuous functions of x in an interval $[a, b]$. As in Chapter 1, §2, we call the differential equation *homogeneous* if $r(x) \equiv 0$, otherwise *inhomogeneous*. If $p_n(x) \neq 0$ in (a, b), then we can divide (II.12) by $p_n(x)$, and we again obtain an equation with continuous coefficients, for which the theorem in Chapter I, 20 is applicable; according to this theorem, if the initial conditions $y^{(q)}(x_0) = y_0^{(q)}$ for $q = 0, 1, \ldots, n-1$, where x_0 must, of course, be in (a, b), are prescribed, then the solution $y(x)$ of the differential equation is uniquely determined. If nothing to the contrary is stated, we shall in the following assume $p_n(x) \neq 0$.

$L[y]$ is a *linear homogeneous differential expression*, which has these properties: if y, z are two n-times continuously differentiable functions and if

a is an arbitrary constant, then

$$\left.\begin{aligned} L[y+z] &= \sum_{\nu=0}^{n} p_\nu(x)\{y^{(\nu)}+z^{(\nu)}\} = L[y]+L[z], \\ L[ay] &= \sum_{\nu=0}^{n} p_\nu(x)\cdot ay^{(\nu)} = aL[y], \end{aligned}\right\} \tag{II.13}$$

In particular,

$$L[-y] = -L[y].$$

6. THE SUPERPOSITION THEOREM

The difference $z(x)=y_1(x)-y_2(x)$ of two arbitrary solutions $y_1(x)$ and $y_2(x)$ of an inhomogeneous differential equation is a solution of the corresponding homogeneous differential equation.

For, it follows immediately from (II.13) that

$$L[z] = L[y_1-y_2] = L[y_1]-L[y_2] = 0,$$

since y_1 and y_2 satisfy the inhomogeneous equation

$$L[y_1] = r(x),\ L[y_2] = r(x).$$

Hence z satisfies the homogeneous equations.

Now we let $y_1(x)$ traverse all the solutions of (II.12). If we subtract from the set of all these solutions—i.e., from the general solution $y_1(x)$—of an inhomogeneous linear differential equation an arbitrary particular solution $y_2(x)$ of the inhomogeneous equation, we obtain the general solution of the corresponding homogeneous equation.

For if z^* is any solution of the homogeneous equation, then x^*+y_2 is a solution of the inhomogeneous equation, and is therefore contained among the functions y_1; i.e., z^* is also contained among the functions z.

However, the theorem becomes of particular importance in its converse form:

The superposition theorem

The general solution of an inhomogeneous linear differential equation is the sum of the general solution of the corresponding homogeneous differential equation and any particular solution of the inhomogeneous equation.

Thus we have broken down the search for the general solution of the complete linear differential equation into two sub-problems, viz., firstly, the *determination of the general solution of the corresponding homogeneous or 'abbreviated' equation,* and secondly, the *search for a particular solution of the inhomogeneous equation.*

7. Reduction of the Order of a Linear Differential Equation

The order of the linear differential equation

$$L[y] \equiv \sum_{\nu=0}^{n} p_\nu(x) y^{(\nu)} = r(x)$$

can be reduced by unity if a particular solution $y_1(x)$ *of the corresponding homogeneous differential equation can be found* (this does not hold, of course, for the trivial solution $y \equiv 0$; we take $y_1 \not\equiv 0$), *and then making the product substitution*

$$y(x) = y_1(x) z(x). \qquad \text{(II.14)}$$

We form the individual expressions $p_\nu(x) y^{(\nu)}$ and add them, writing the result in abbreviated form by introducing new functions $P_\nu(x)$:

$$p_0 y = p_0 y_1 z$$

$$p_1 y' = p_1 y_1' z + p_1 y_1 z'$$

. . . .

$$p_n y^{(n)} = p_n y_1^{(n)} z + p_n \binom{n}{1} y_1^{(n-1)} z' + p_n \binom{n}{2} y_1^{(n-2)} z'' + \ldots + p_n \binom{n}{n-1} y_1' z^{(n-1)} + p_n y_1 z^{(n)}$$

$$\sum = L[y] = L[y_1] z + P_1(x) z' + \ldots + P_n(x) z^{(n)} = r(x).$$

Since by hypothesis $L[y_1] = 0$ and the $P_\nu(x)$ are known functions of x, we obtain for $z' = u$ a differential equation whose order has been reduced by unity:

$$L^*[u] = P_n(x) u^{(n-1)} + P_{n-1}(x) u^{(n-2)} + \cdots + P_2(x) u' + P_1(x) u = r(x).$$

If here a particular non-trivial solution $u_1(x)$ of the homogeneous equation can again by found, so that $L^*[u_1] = 0$, then we can repeat the process with the new substitution

$$u(x) = u_1(x) v(x)$$

and continue in this way until we reach a linear, first-order differential equation, which can be solved as in Chapter I, 9.

Example

$$y'' - y' = 1$$

The corresponding homogeneous equation $y'' - y' = 0$ has the solution $y_1(x) = e^x$, as may be verified immediately. The substitution

$$y = e^x \cdot z, \qquad y' = e^x(z + z'), \qquad y'' = e^x(z + 2z' + z'')$$

leads with $z' = u$ to $e^x(2z' + z'') = e^x(2u + u') = 1$, with the solution

$$u = c_1^* e^{-2x} + e^{-x} = z'.$$

So

$$z = c_1 + c_2 e^{-2x} - e^{-x}, \qquad y = e^x z = c_1 e^x + c_2 e^{-x} - 1.$$

In Chapter II, 11, a quicker way of finding this solution is given.

The procedure naturally works in the same way for the general linear second-order differential equation

$$p_2(x)y'' + p_1(x)y' + p_0(x)y = r(x).$$

If we know a particular solution $y_1(x) \not\equiv 0$ of the corresponding homogeneous equation $L''y_1] = 0$, then the general solution of the original inhomogeneous equation can be determined by quadrature alone.

The substitutions $y(x) = y_1(x)z(x)$ and $z'(x) = u(x)$ lead to

$$P_2(x)u' + P_1(x)u = r.$$

Its general solution reads

$$z' = u = \varphi_1(x) + C_2\varphi_2(x).$$

Here

$$\varphi_1(x) \equiv 0 \quad \text{in the case } r(x) \equiv 0.$$

Integration yields

$$z = \int \varphi_1(x)\,\mathrm{d}x + C_2 \int \varphi_2(x)\,\mathrm{d}x + C_1.$$

Hence,

$$y = \underbrace{y_1 \textstyle\int \varphi_1\,\mathrm{d}x}_{\psi(x)} + \underbrace{C_2 y_1 \textstyle\int \varphi_2\,\mathrm{d}x + C_1 y_1}_{y_2(x)}$$

or, in abbreviated form,

$$y = C_1 y_1(x) + C_2 y_2(x) + \psi(x).$$

In the case $r(x) \equiv 0$ we have $\psi(x) \equiv 0$, and therefore the function $y_2(x)$ must satisfy the homogeneous equation

Similarly, for the nth-order differential equation (II.12) we obtain, by induction, solutions of the form

$$y = C_1 y_1(x) + C_2 y_2(x) + \cdots + C_n y_n(x) + \psi(x). \qquad \text{(II.15)}$$

In the next section we shall obtain information about the form of the *general* solution, for we do not yet know whether all solutions of the differential equation are contained in this form (II.15).

§3. THE FUNDAMENTAL SYSTEMS OF A LINEAR DIFFERENTIAL EQUATION

8. Linear dependence of functions

If solutions of a linear differential equation (II.12) have been found in the form (II.15) with n free constants C_k, where the $y_k(x)$ represent different particular solutions of the homogeneous equation and $\psi(x)$ a particular solution of the inhomogeneous equation, then it is sensible to require that the y_k shall be 'linearly independent of one another'. For if, say, two of the y_k were proportional to one another, or if one function y_k were, apart from additive constants, an additive combination of several of the others, then the term in y_k could be dropped, and an expression with fewer free constants C_q would be obtained.

Definition

The n functions

$$u_1(x), \ldots, y_n(x)$$

defined in an interval J $(a < x < b)$ *are said to be 'linearly dependent' of one another if there are constants* $\alpha_1, \alpha_2, \ldots, \alpha_n$ *not all zero such that in J*

$$\alpha_1 y_1 + \cdots + \alpha_n y_n \equiv 0; \tag{II.16}$$

otherwise they are said to be 'linearly independent' of one another in J.

Expressed rather differently, the definition reads: *The functions* $y_\nu(x)$ *(for* $\nu = 1, \ldots, n$*) are said to be linearly independent of one another in the interval J if and only if it follows from*

$$\sum_{\nu=1}^{n} \alpha_\nu y_\nu(x) \equiv 0$$

in J that all the constants α_ν *vanish.*

Examples of linear dependence

1. If we have found as solutions of a third-order homogeneous differential equation the functions

$$y_1 = x, \; y_2 = x^2, \; y_3 = 4x - x^2,$$

then taking $\alpha_1 = -4$, $\alpha_2 = \alpha_3 = 1$ (as one of infinitely many possibilities):

$$-4 \cdot x + 1 \cdot x^2 + 1 \cdot (4x - x^2) \equiv 0.$$

The functions y_1, y_2, y_3 are therefore linearly dependent on one another (in any interval) and are not suitable to comprehend the general solution of the homogeneous differential equation. The functions in each pair are linearly independent of one another; only in a combination of the three do they become linearly dependent.

2. Linear dependence or independence can often be recognized immediately. Thus the powers x^m, x^n are linearly independent for $m \neq n$; so are $\sin x$ and $\sin 2x$, for example. It is advisable to be especially on the alert in combinations of exponential functions with trigonometric or hyperbolic functions.

3. The function x and $x - 1$ are not linearly dependent on one another in any interval.

4. Let the linear, homogeneous differential equation of third order

$$y'''(1-2x\cot 2x)-4xy''+4y'=0.$$

be given. Since y does not appear explicitly, $y=\text{const}$, say $y_1=1$, is a solution. Suppose further that by testing we have found

$$y_2=\cos 2x \quad \text{and} \quad y_2=\sin^2 x$$

to be further solutions; yet still

$$y=C_1\cdot 1+C_2\cos 2x+C_3\sin^2 x$$

is not the general solution, since y_1, y_2, and y_3 are linearly dependent of one another because

$$\cos 2x\equiv 1-2\sin^2 x.$$

So y_3 brings no 'new blood' into the solution compared with y_1 and y_2. For the sake of completeness we give the correct general solution, viz.,

$$y=C_1+C_2\cos 2x+C_3x^2.$$

We should further point out the significance of the specification of an interval J in the definition. J may extend in one or both directions to infinity; for the differential equations in the following sections J is often the whole real axis and we write $J=(-\infty,\ +\infty)$. Linear dependence may depend entirely on the choice of the interval. For example, the functions $y_1(x)=x$ and $y_2(x)=|x|$ are linearly dependent in the interval $(0, 1)$ (in which they are identical), and likewise in the interval $(-1, 0)$, but in the interval $(-1, 1)$ they are linearly independent since $\alpha_1 x+\alpha_2|x|\equiv 0$ for $-1\le x\le 1$ can be satisfied only by $\alpha_1=\alpha_2=0$ (see Fig. II.5).

Definition of a fundamental system. A system of *n particular solutions*

$$y_1, y_2, \ldots, y_n,$$

which are linearly independent of one another in an interval J, of a homogeneous differential equation of the nth order is called a *fundamental system* of the differential equation.

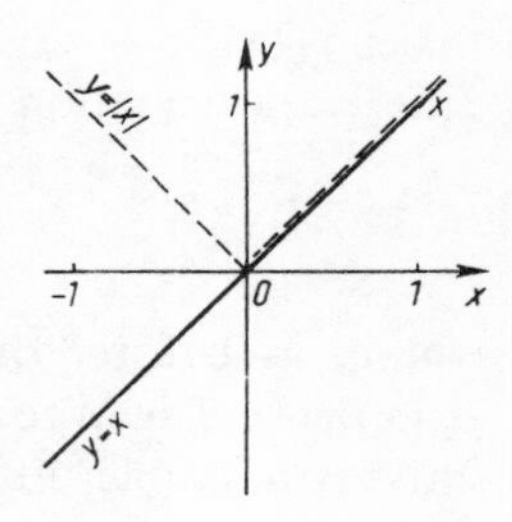

Fig. II.5. Effect of the interval on the linear dependence of functions

9. THE WRONSKIAN CRITERION FOR LINEAR INDEPENDENCE OF FUNCTIONS

There is a criterion for linear independence of n solutions $y_1, \ldots, y_n$ of an nth-order homogeneous differential equation.

First of all, let $y_1(x), \ldots, y_n(x)$ be any given functions which are $n-1$ times continuously differentiable in an interval (a, b). Now suppose that these functions are linearly dependent in (a, b); then there are constants C_k, not all zero, such that the combination

$$y = C_1 y_1 + C_2 y_2 + \cdots + C_n y_n \equiv 0 \tag{II.17}$$

vanishes identically in (a, b); but then the same is true for the derivatives

$$\begin{aligned} y' &= C_1 y_1' + C_2 y_2' + \cdots + C_n y_n' \equiv 0 \\ &\vdots \\ y^{(n-1)} &= C_1 y_1^{(n-1)} + C_2 y_2^{(n-1)} + \cdots + C_n y_n^{(n-1)} \equiv 0. \end{aligned} \tag{II.18}$$

This is a system of n linear homogeneous equations for the n quantities $C_1, C_2, \ldots, C_n$. It has a non-trivial solution, and so the determinant of the system of equations must have the value zero (at every point x of the interval); this determinant is called the *Wronskian determinant,* or simply the *Wronskian:*

$$D = \begin{vmatrix} y_1 & y_2 & \ldots y_n \\ y_1' & y_2' & \ldots y_n' \\ \cdots & \cdots & \cdots \\ \cdots & \cdots & \cdots \\ y_1^{(n-1)} & y_2^{(n-1)} & \ldots y_n^{(n-1)} \end{vmatrix} \tag{II.19}$$

Now the identical vanishing of the Wronskian is indeed a necessary, but by no means a sufficient, condition for linear dependence of the functions $y_1, \ldots, y_n$. For example the functions $y_1 = x^3, y_2 = |x|^3$ defined and continuously differentiable in the interval $(-1, +1)$ are linearly independent, but nevertheless their Wronskian vanishes identically:

$$\begin{vmatrix} x^3 & x^2 |x| \\ 3x^2 & 3x\,|x| \end{vmatrix} \equiv 0.$$

If, however, it is additionally assumed that the $y_1, \ldots, y_n$ are solutions of a homogeneous differential equation of order n, then the vanishing of the Wronskian is a criterion for linear dependence. This will now be made more precise.

Let $y_1(x_1, \ldots, y_n(x))$ be solutions in an interval (a, b) of the homogeneous differential equation

$$L[y] \equiv \sum_{\nu=0}^{n} p_\nu(x) y^{(\nu)}(x) = 0, \tag{II.20}$$

where, as before, the $p_\nu(x)$ are given continuous functions in (a, b) with $p_n(x) \neq 0$. The Wronskian D of the functions y_k, formed as in (II.19), then satisfies a simple differential equation. To set up this equation we form the derivative of D. A determinant is differentiated by differentiating each row in turn, keeping the other rows unchanged, and adding the n determinants so formed. Thus

$$\frac{\mathrm{d}D}{\mathrm{d}x}=\begin{vmatrix} y_1' & \dots y_n' \\ y_1' & \dots y_n' \\ y_1'' & \dots y_n'' \\ \dots & \dots \\ y_1^{(n-1)} & \dots y_n^{(n-1)} \end{vmatrix} + \begin{vmatrix} y_1 & \dots y_n \\ y_1'' & \dots y_n'' \\ y_1'' & \dots y_n'' \\ \dots & \dots \\ y_1^{(n-1)} & \dots y_n^{(n-1)} \end{vmatrix} + \dots + \begin{vmatrix} y_1 & \dots y_n \\ y_1' & \dots y_n' \\ y_1'' & \dots y_n'' \\ \dots & \dots \\ y_1^{(n)} & \dots y_n^{(n)} \end{vmatrix}$$

The first $(n-1)$ determinants on the right vanish, because in each there are two identical rows; so only the last determinant remains. In it we multiply the last row by $p_n(x)$, so that the value of the determinant is likewise multiplied by $p_n(x)$; then we multiply the first row by $p_0(x)$, the second row by $p_1(x)$, and so on, and finally the $(n-1)$th row by $p_{n-2}(x)$, and add the products to the nth row; the value of the determinant is unchanged by so doing. The terms in the last row are thus replaced by

$$p_n(x)y_k^{(n)}+\sum_{\nu=0}^{n-2} p_\nu(x)y_k^{(\nu)}. \qquad (k=1,\dots,n)$$

But from the differential equation (II.20) this is precisely equal to $-p_{n-1}(x)y_k^{(n-1)}$, and so we can take the factor $-p_{n-1}$ outside the determinant and thus obtain the original Wronskian; so

$$\frac{\mathrm{d}D}{\mathrm{d}x}=-\frac{p_{n-1}(x)}{p_n(x)}D. \tag{II.21}$$

This linear differential equation of the first order for D has the solution

$$D(x)=D(x_0)\exp\left(-\int_{x_0}^{x}\frac{p_{n-1}{}^{(s)}}{p_n(s)}\,\mathrm{d}s\right), \quad \text{(where } \exp(x)=\mathrm{e}^x\text{)} \tag{II.22}$$

Here x_0 is an arbitrarily chosen point in (a,b). If, therefore, $D=0$ (resp. $\neq 0$) at a point x_0, then $D=0$ (resp. $D\neq 0$) throughout the entire interval (a,b), since the exponential factor is always $\neq 0$; the Wronskian of the n solutions $y_1,\dots,y_n$ either vanishes identically or vanishes nowhere in the interval (a,b). It was proved earlier that if the y_k are linearly dependent in (a,b) then D vanishes identically. But the converse also holds: if D vanishes at a point in (a,b) (and therefore identically), then the y_k are linearly dependent in (a,b). Since $D(x_0)=0$, the system of equations (II.17), (II.18) written for the point x_0 has a non-trivial solution $C_1,\dots,C_n$, and the function

$$\varphi(x)=\sum_{k=1}^{n} C_k y_k(x)$$

formed using these constants is a solution of the homogeneous differential equation, with the initial values at the point x_0

$$\varphi(x_0)=\varphi'(x_0)=\cdots=\varphi(n-1)(x_0)=0.$$

But this initial-value problem has the solution $y\equiv 0$, and since the initial-value problem has a unique solution, we must have $\varphi\equiv 0$ in (a,b), i.e., the solutions $y_k(x)$ are linearly dependent. So the following theorem holds.

Theorem

n solutions $y_1(x), \ldots, y_n(x)$ of a homogeneous linear nth-order differential equation (II.20) *with continuous coefficients $p_\nu(x)$ and $p_n(x) \neq 0$ in an interval (a, b) form a fundamental system,* i.e., *are linearly independent, if and only if their Wronskian* (II.19) *is different from zero at an arbtrarily chosen point x_0 of the interval.*

10. THE GENERAL SOLUTION OF A LINEAR DIFFERENTIAL EQUATION

There is now the question whether all solutions of the inhomogeneous differential equation (II.12) are included in the form (II.15)

$$y = C_1 y_1(x) + \ldots + C_n y_n(x) + \psi(x),$$

where $\psi(x)$ is a particular solution of the inhomogeneous equation (II.12), if the y_k form a fundamental system.

By the general existence and uniqueness theorem (Chapter I, §6) the linear differential equation (II.12) has one and only one solution which satisfies the prescribed initial values

$$y(x_0) = y_0, \qquad y'(x_0) = y_0', \ldots, y^{(n-1)}(x_0) = y_0^{(n-1)} \tag{II.23}$$

If now by a suitable choice of the C_k in the form (II.15) arbitrarily prescribed initial conditions (II.23) can be satisfied, then *all* solutions of the differential equation must be included in this form (since every arbitrarily chosen solution of the functions y_k. It is not equal to 0 because the y_k form a fundamental uniquely determined by these initial values). We obtain for the C_k a system of linear equations

$$\sum_{k=1}^{n} C_k y_k^{(\nu)}(x_0) + \psi^{(\nu)}(x_0) = y_0^{(\nu)} \qquad (\nu = 0, 1, \ldots, n-1). \tag{II.24}$$

The determinant of this system of equations is precisely the Wronskian $D(x_0)$ of the functions y_k. If is not equal to 0 because the y_k form a fundamental system. The system of equations therefore has a uniquely determined solution C_k. Hence (II.15) is indeed the general solution of the differential equation (II.12).

The existence of a fundamental system is also shown if we construct a 'normalized fundamental system': we select first the initial condition

$$y = 1, \qquad y' = y'' = y''' = \cdots = y^{(n-1)} = 0$$

at the point $x = x_0$, Fig. II.6. By the general existence theorem there is a solution $\tilde{y}_1$ which satisfies these initial conditions. Similarly there is another particular solution $\tilde{y}_2$ which satisfies the initial conditions

$$y = 0, \qquad y' = 1, \qquad y'' = y''' = \cdots = y^{(n-1)} = 0$$

at the point $x = x_0$, Fig. II.6. And in general there is a $\tilde{y}_k$ with

$$y = y' = \cdots = y^{(k-2)} = 0, \qquad y^{(k-1)} = 1, \qquad y^{(k)} = y^{(k+1)} = \cdots = y^{(n-1)} = 0$$

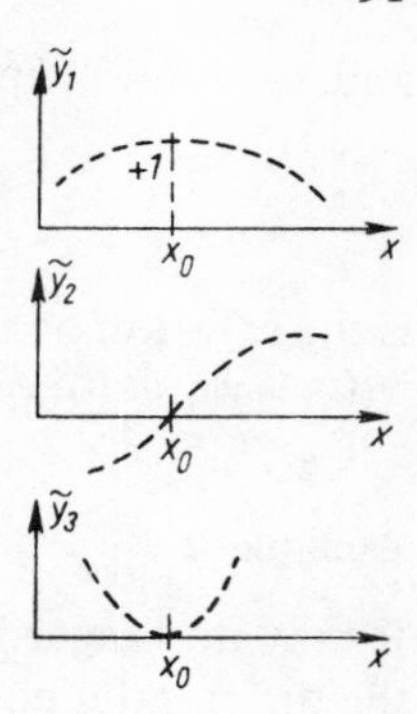

Fig. II.6. Determination of a particular fundamental system by definite initial conditions

at the point $x = x_0$. The corresponding Wronskian (for $x = x_0$)

$$D = \begin{vmatrix} 1 & 0 \ldots 0 \\ 0 & 1 \ldots 0 \\ \ldots & \ldots \\ 0 & 0 \ldots 1 \end{vmatrix} = 1$$

is not equal to 0, and so the $\tilde{y}_k(x)$ form a fundamental system. This normalized system enables the solution of the initial-value problem for the homogeneous differential equation with the intial values (II.23) to be given immediately:

$$y = y_0 \tilde{y}_1(x) + y_0' \tilde{y}_2(x) + y_0'' \tilde{y}_3(x) + \cdots + y_0^{(n-1)} \tilde{y}_n(x). \qquad \text{(II.25)}$$

Example 1

We can show immediately by substitution that (for $\omega \neq 0$)

$$y_1 = \cos \omega x \quad \text{and} \quad y_2 = \sin \omega x \qquad \text{(II.26)}$$

satisfy the differential equation

$$y'' + \omega^2 y = 0, \qquad \text{(II.27)}$$

Their Wronskian at, say, the point $x = 0$, is non-zero:

$$\begin{vmatrix} 1 & 0 \\ 0 & \omega \end{vmatrix} = \omega \neq 0,$$

so the general solution of (II.27) reads

$$y = C_1 \cos \omega x + C_2 \sin \omega x.$$

A normalized fundamental system for the point $x = 0$ is formed by the functions

$$\tilde{y}_1 = \cos \omega x, \; \tilde{y}_2 = \frac{1}{\omega} \sin \omega x; \qquad \text{(II.29)}$$

and so

$$y = y_0 \cos \omega x + \frac{y_0'}{\omega} \sin \omega x$$

is the solution of the initial-value problem for (II.27) with the initial conditions $y(0) = y_0$, $y'(0) = y_0'$.

Example 2

It may no longer be possible to construct a fundamental system in this way if the differential equation has a singularity at $x = x_0$. For example, suppose the general solution of a linear homogeneous second-order differential equation is

$$y = C_1 x + C_2 x^2. \tag{II.30}$$

Although both the particular solutions $y_1 = x$ and $y_2 = x^2$ are completely regular, the initial conditions $y = 1$ and $y' = 0$ for $x = 0$ cannot be satisfied by either of them, nor by a linear combination of them.

The reason for this is immediately apparent if we set up the relevant differential equation, which we easily obtain in determinantal form from the particular solutions and their derivatives

$$\begin{vmatrix} y & y' & y'' \\ x & 1 & 0 \\ x^2 & 2x & 2 \end{vmatrix} = x^2 y'' - 2xy' + 2y = 0. \tag{II.31}$$

When $x = 0$ the coefficient $p_n(x) = x^2$ of the highest derivative in the equation vanishes. Despite its regular solutions, the differential equation is singular at the point $x = 0$. This demonstrates the importance of the coefficient condition

$$p_n(x) \neq 0.$$

In contrast, the fundamental system for the point $x_0 = 1$ can be given immediately; a 'normalized fundamental system' at this point reads

$$\tilde{y}_1 = 2x - x^2, \qquad \tilde{y}_2 = -x + x^2.$$

§4. LINEAR DIFFERENTIAL EQUATIONS WITH CONSTANT COEFFICIENTS

11. A TRIAL SUBSTITUTION FOR THE SOLUTION; THE CHARACTERISTIC EQUATION

Among linear differential equations those with *constant* coefficients p_ν (and $p_n \neq 0$) are particularly important:

$$L[y] \equiv \sum_{\nu=0}^{n} p_\nu y^{(\nu)}(x) = r(x). \tag{II.32}$$

They turn up very often in applications and their general solution can always be determined by quadrature.

The earlier example of the differential equation $y' - ky = 0$ with its solution $y = Ce^{kx}$ suggests to us that for the homogeneous equation ($r(x) \equiv 0$)

$$L[y] \equiv \sum_{\nu=0}^{n} p_\nu y^{(\nu)}(x) = 0 \tag{II.33}$$

we should seek a substitution

$$y = e^{kx}, \qquad y' = ke^{kx}, \ldots, y^{(n)} = k^n e^{kx} \tag{II.34}$$

where the k is still to be determined.

Since e^{kx} is not zero for any finite exponent, we can divide through by e^{kx} and thus obtain the so-called *characteristic equation:*

$$P(k) \equiv p_n k^n + p_{n-1}k^{n-1} + \cdots + p_1 k + p_0 = 0. \tag{II.35}$$

This is an algebraic equation of the nth degree for k. Formally, we obtain it from the differential equation by replacing $y^{(\nu)}$ by k^ν.
Suppose its n roots are

$$k_1, k_2, \ldots, k_n.$$

We consider first the case where all the roots are distinct. Complex roots may occur. The further treatment when there are complex roots is given later (Example 2).

So now there are n different particular solutions

$$y_1 = e^{k_1 x}, \qquad y_2 = e^{k_2 x}, \ldots, y_n = e^{k_n x}.$$

These form a fundamental system, as can be established by evaluating the Wronskian D. We form D at the point $x = x_0$ say:

$$D = \begin{vmatrix} 1 & 1 & \ldots 1 \\ k_1 & k_2 & \ldots k_n \\ k_1^2 & k_2^2 & \ldots k_n^2 \\ \ldots & \ldots & \ldots \\ k_1^{n-1} & k_2^{n-1} & \ldots k_n^{n-1} \end{vmatrix} = \prod_{r>s} (k_r - k_s) = (k_2 - k_1)(k_3 - k_1)(k_3 - k_2) \ldots (k_n - k_1) \ldots (k_n - k_{n-1}).$$

This is a *Vandermonde* determinant; it is equal to the product of all possible differences of pairs of different values of k (cf., e.g., J. W. Archibald, *Algebra*, Pitman (London, 1961), p. 341, and so it is not equal to 0. Hence the y_ν are linearly independent. The general solution of (II.33) therefore reads:

$$y = C_1 e^{k_1 x} + C_2 e^{k_2 x} + \cdots + C_n e^{k_n x} \tag{II.36}$$

with the C_ν as free coefficients.

Example 1

For the differential equation

$$y''' - 7y' + 6y = 0$$

the characteristic equation is

$$k^3 - 7k + 0 = 0.$$

It has the roots $k_1 = 1, k_2 = 2$, and $k_3 = -3$. So the general solution of this differential equation is

$$y = C_1 \mathrm{e}^x + C_2 \mathrm{e}^{2x} + C_3 \mathrm{e}^{-3x}.$$

Example 2. The equation for oscillations; complex roots of the characteristic equation

For the differential equation

$$y'' + y = 0 \tag{II.37}$$

already solved by (II.26) the characteristic equation

$$k^2 + 1 = 0$$

has the roots

$$k_{1,2} = \pm \mathrm{i}.$$

Hence the general solution is, first of all,

$$y = C_1 \mathrm{e}^{\mathrm{i}x} + C_2 \mathrm{e}^{-\mathrm{i}x}.$$

Here complex solutions of the form $y(x) = u(x) + \mathrm{i}v(x)$ appear. If a linear homogeneous differential equation $L[y] = 0$ with real coefficients has such a complex-valued solution, then the real part $u(x)$ and the imaginary part $v(x)$ separately are each likewise solutions of the differential equation because

$$L[y] = L[u + \mathrm{i}v] = L[u] + \mathrm{i}L[v] = 0.$$

and so $L[u] = L[v] = 0$. Further, $Cy(x)$, where C is an arbitrary complex constant $C = A + \mathrm{i}B$, is also a solution of the differential equation, for in

$$Cy = (A + \mathrm{i}B)(u + \mathrm{i}v) = Au + \mathrm{i}Av + \mathrm{i}Bu - Bv$$

each one separately of the four terms on the right satisfies the differential equation.

Now in (II.35), since the given coefficients p_ν were assumed to be real, any complex roots occur, not singly, but always in pairs of *complex conjuagates*, of the form

$$k_1 = a + \mathrm{i}b, \qquad k_2 = a - \mathrm{i}b.$$

This is shown in the example above. The general solution is

$$y = C_1\,\mathrm{e}^{(a+\mathrm{i}b)x} + C_2\,\mathrm{e}^{(a-\mathrm{i}b)x} + \cdots$$

or

$$y = C_1\,\mathrm{e}^{ax}\mathrm{e}^{\mathrm{i}bx} + C_2\,\mathrm{e}^{ax}\,\mathrm{e}^{-\mathrm{i}bx} + \cdots.$$

Using the *Euler formula*

$$\mathrm{e}^{\pm\mathrm{i}z} = \cos z \pm i \sin z$$

we find that

$$y = \mathrm{e}^{nx}\{(C_1 + C_2)\cos bx + (C_1\mathrm{i} - C_2\mathrm{i})\sin bx\} + \cdots$$

C_1 and C_2 may be arbitrary complex constants; with new constants

$$C_1 + C_2 = C_1^* \quad \text{and} \quad C_1\mathrm{i} - C_2\mathrm{i} = C_2^*$$

we obtain

$$y = e^{ax}(C_1^* \cos bx + C_2^* \sin bx) + \cdots.$$

Here C_1^*, C_2^* may be complex; but if we are interested in solutions in real form, we can pick out the real part from this last expression, i.e., C_1^* and C_2^* will then be regarded as real constants. The linear independence of $\mathrm{e}^{ax} \cos bx$ and $\mathrm{e}^{ax} \sin bx$ can be shown immediately by means of the Wronskian.

In the initial example (II.37) we then arrive at the previously known solution

$$y = C_1^* \cos x + C_2^* \sin x.$$

12. Multiple Zeros of the Characteristic Equation

It can also happen that the algebraic equation (II.35) has a multiple zero; thus, at a double root, say,

$$k_1 = k_2, k_1 \neq k_m \quad \text{for } m = 3, \ldots, n.$$

Since in this case the particular solutions

$$y_1 = \mathrm{e}^{k_1 x} \quad \text{and} \quad y_2 = \mathrm{e}^{k_2 x}$$

are the same, the fundamental system lacks one solution. If all the n roots are equal,

$$k_1 = k_2 = \ldots = k_n,$$

then $n-1$ particular solutions are missing. In general, let k_1 be an r-fold root of (II.35) and let $y_1(x) = \mathrm{e}^{k_1 x}$. The 'missing' functions for a fundamental system can be determined by applying the substitution (II.14) used for reducing the order of a linear differential equation

$$y(x) = y_1(x)z(x).$$

Hence,

$$
\begin{aligned}
y &= e^{k_1x}z & \cdot p_0\\
y' &= e^{k_1x}(k_1z + z') & \cdot p_1\\
y'' &= e^{k_1x}(k_1^2z + 2k_1 + z'') & \cdot p_2\\
&\cdots\cdots\cdots\cdots\cdots\cdots\cdots\cdots\\
y^{(n)} &= e^{k_1x}\left[k_1^n z + \binom{n}{1} k_1^{n-1} z' + \cdots + z^{(n)}\right] & \cdot p_n
\end{aligned}
$$

The coefficients appearing in the square brackets are binomial coefficients. We multiply by $p_0, p_1, \ldots, p_n$ as indicated, and add. In (II.33), then, with $L[y] = L[y_1z]$ we obtain the following expression, multiplied by the common factor e^{k_1x}:

$$
\underbrace{z(p_0 + p_1k_1 + p_2k_1^2 + \cdots + p_nk_1^n)}_{\text{(I)}}
$$

$$
\underbrace{+ z'\left(p_1 + \binom{2}{1} p_2k_1 + \binom{3}{1} p_3k_1^2 + \cdots + \binom{n}{1} p_nk_1^{n-1}\right)}_{\text{(II)}}
$$

$$
\underbrace{+ z''\left(p_2 + \binom{3}{2} p_3k_1 + \binom{4}{2} p_4k_1^2 + \cdots + \binom{n}{2} p_nk_1^{n-2}\right)}_{\text{(III)}} + \cdots + \underbrace{z^{(n)}(p_n)}_{(N+1)}.
$$

The bracket marked (I) is the characteristic polynomial $P(k)$ once more of (II.35) at the point $k = k_1$; the brackets marked (II) is the first derivative of $P(k)$ with respect to k at the point $k = k_1$; the bracket (III) is its second derivative, divided by 2!, and so on up to the bracket marked $(N + 1)$, the nth derivative of $P(k)$ divided by $n!$:

$$
L[e^{k_1x}z] = e^{k_1x}\left\{zP(k_1) + \frac{z'}{1!}\frac{dP(k_1)}{dk} + \frac{z''}{2!}\frac{d^2P(k_1)}{dk^2} + \cdots + \frac{z^{(n)}}{n!}\frac{d^nP(k_1)}{dk^n}\right\}. \tag{II.38}
$$

If k_1 is a simple zero of $P(k)$, the bracket (II) is equal to 0; if k_1 is a double zero, then the first derivative of $P(k)$ also vanishes at k_1. In general, for an r-fold zero k_1, the derivatives up to and including the $(r - 1)$th order all vanish at k_1:

$$
P(k) = \frac{dP(k_1)}{dk} = \frac{d^2P(k_1)}{dk^2} = \cdots = \frac{d^{r-1}P(k_1)}{dk^{r-1}} = 0.
$$

The differential equation for z therefore begins with the rth derivative and has the form

$$q_r z^{(r)} + \cdots + q_n z^{(n)} = 0.$$

Hence any polynomial of at most the $(r-1)$th degree is a solution of the differential equation for z:

$$z = C_1 + C_2 x + \cdots + C_r x^{r-1},$$

and so the original differential equation (II.33) has the solution

$$y = e^{k_1 x} z = e^{k_1 x}(C_1 + C_2 x + \cdots + C_r x^{r-1}).$$

In general, if the characteristic polynomial (II.35) has the zeros $k_1, \ldots, k_s$ distinct from one another and of multiplicities $r_1, \ldots, r_s$ respectively with $r_1 + \cdots + r_s = n$, then the differential equation (II.33) has the n solutions

$$e^{k_1 x}, \quad x\, e^{k_1 x}, \ldots, \quad x^{r_1 - 1}\, e^{k_1 x}, \ldots, e^{k_s x}, \quad x\, e^{k_s x}, \ldots, x^{r_s - 1}\, e^{k_s x}. \tag{II.39}$$

If, for example, there are two double zeros $k_1 = k_2$, $k_3 = k_4$, then correspondingly the differential equation (II.33) has the solution

$$y = e^{k_1 x}(C_1 + C_2 x) + e^{k_3 x}(C_3 + C_4 x) + C_5\, e^{k_5 x} + \cdots + C_n\, e^{K_n x}.$$

We have still to show, however, that the newly adopted solutions of the form $x^s\, e^{kx}$ really do serve to complete the fundamental system as required, i.e., that the functions (II.39) are linearly independent of one another, so that no non-trivial combination of them can vanish identically. We show, more generally, that a relation

$$\sum_{j=1}^{s} P_j(x) e^{k_j x} \equiv 0, \tag{II.40}$$

where the $P_j(x)$ are polynomials, can hold only if all the polynomials $P_j(x)$ are identically zero. For if there were such a relation (II.40) in which not all the P_j vanish, we could assume, e.g., that P_1, say, is not identically zero. Division of (II.40) by $e^{k_s x}$ gives

$$\sum_{j=1}^{s-1} P_j(x)\, e^{(k_j - k_s)x} + P_s(x) \equiv 0. \tag{II.41}$$

Let P_j have degree g_j; we then differentiate (II.41) $(g_j + 1)$ times with respect to x. As a result, $P_s(x)$ disappears. A term $P_j(x)\, e^{\alpha x}$ gives on differentiation a term $Q_j(x) e^{\alpha x}$, where $Q_j(x)$ is again a polynomial, which, if $\alpha \neq 0$, is of the same degree as $P_j(x)$. Since $k_j \neq k_s$ for $j \neq s$, the condition $\alpha \neq 0$ here is always fulfilled. After differentiating $(g_s + 1)$ times an equation of the form

$$\sum_{j=1}^{s-1} P_j^*(x)\, e^{(k_j - k_s)x} \equiv 0 \tag{II.42}$$

arises, and the degree of P_j^* = degree of P_j.

This process is continued; we divide this equation by

$$e^{(k_{s-1} - k_s)x}$$

and differentiate $(g_{s-1}+1)$ times; so then P^*_{s-1} also disappears and we obtain an equation of the form (II.42), where now the summation is only from $j=1$ to $j=s-2$ and new polynomials appear as factors but these always have the same degree because the exponential functions have the form $e^{\alpha x}$ with $\alpha \neq 0$ (α is always the difference of two k_ν-values). On continuing the process we finally arrive at an equation

$$\tilde{P}_1(x)e^{(k_1-k_2)x} \equiv 0,$$

where $\tilde{P}_1(x)$ has the same degree as $P_1(x)$ and so cannot be identically zero. We have a contradiction, and therefore the functions (II.39) are indeed linearly independent.

Example

For the linear homogeneous differential equation

$$y^{\mathrm{IV}} + 4y'' + 4y = 0$$

which arises, for example, in studies of stability and oscillations, the characteristic equation is

$$k^4 + 4k^2 + 4 = (k^2+2)^2 = 0$$

with the two double roots $k_1 = k_2 = \sqrt{2}\mathrm{i}$ $k_3 = k_4 = -\sqrt{2}\mathrm{i}$. So we have the general solution

$$y = e^{\sqrt{2}\mathrm{i}x}(C_1 + C_2x) + e^{-\sqrt{2}\mathrm{i}x}(C_3 + C_4x)$$

or, using Euler's formula,

$$y = C_1^* \cos\sqrt{2}x + C_2^*\, sin\, \sqrt{2}x + C_3^* x \cos\sqrt{2}x + C_4^*\, x \sin\sqrt{2}x.$$

As the solution shows, the oscillatory process described by the differential equation is unstable, because the third and fourth terms do not remain

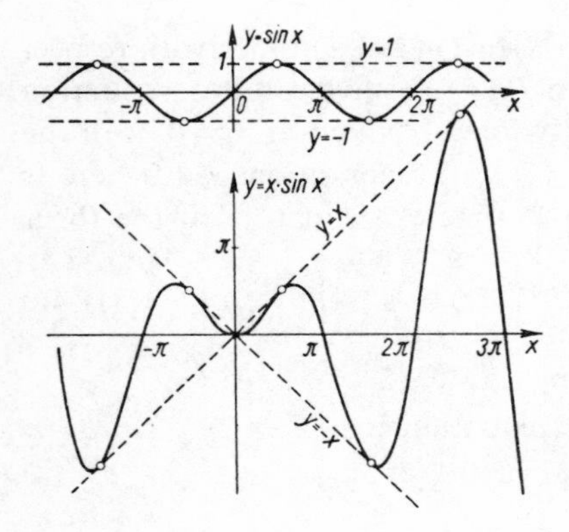

Fig. II.7. Solutions when the characteristic equation has a double root

bounded as $x \to \infty$. A comparison of the graphs of the simplified functions in Fig. II.7

$$y = \sin x \quad \text{and} \quad y = x \sin x$$

will illustrate this circumstance.

13. A CRITERION FOR STABILITY

The real parts of the zeros determine the stability.

The particular solutions have the general form

$$y_\nu = x^{q_\nu} e^{k_\nu x}.$$

1. If k_ν here is a real, positive root, then the exponential function increases monotonely without bound; as x increases, $y_\nu \to \infty$ for $x \to \infty$.
2. If k_ν is a real, negative root, then y_ν decreases to 0 as x increases, because the decrease in $e^{k_\nu x}$ outweighs the increase of the power x^{q_ν}.
3. If k_ν is complex and simple, i.e., if $k = a + ib$ (always there is a corresponding complex conjugate root $k_\nu = a - ib$), then oscillations occur (e^{ax} can always be factorized out here). Further
   (a) if a is positive, then the oscillations increase in amplitude without limit;
   (b) if a is negative, the oscillations are damped out.
   (c) if a is zero, the amplitude of oscillation remains constant.
4. If k_ν is complex and multiple, then the solutions decrease for negative a, increase for positive a and also for $a = 0$ without limit. Summing up, we can state the following

Theorem

The system discussed is stable, i.e., all the solutions of the differential equation (II.33) *remain bounded as x increases if and only if no zero k_ν of the characteristic polynomial* (II.35) *has a positive real part* ($a > 0$) *and so, in particular, if there are no positive real roots, and if any purely imaginary roots occurring $k = ib$* ($a = 0$) *are simple.*

14. THE EQUATION FOR FORCED OSCILLATIONS

A case often occurring in engineering is that of forced damped vibrations. It arises if a damped system capable of oscillation is excited by external influences, possibly periodic in time. A possible mechanical example is provided by a ship's propeller shaft, which is excited in one of its natural frequencies, generally in a decreasing cycle, due to the shaking and pitching motion of the vessel. This happens for transverse bending vibrations, longitudinal vibrations, and torsional vibrations of the shaft.

If $P(t)$ is the exciting force, and if the damping is proportional to the velocity, then the differential equation for the oscillatory variable x, which may be the deflection from the equilibrium position, is

$$m\ddot{x} + k\dot{x} + cx = P(t). \tag{II.43}$$

Here m is the vibrating mass, k the damping constant, and c the specific restoring force. If the excitation P, which may be harmonic, has amplitude P_0 and circular frequency ω

$$P(t) = P_0 \cos \omega t,$$

then (II.43) becomes

$$m\ddot{x} + k\dot{x} + cx = P_0 \cos \omega t. \tag{II.44}$$

Oscillations in an electrical circuit

There is a detailed analogy between mechanical and electrical oscillations; one can immediately associate a mechanical oscillatory system with an electrical one, and *vice versa.* (A table of equivalent quantities can be found, e.g., in J. P. Den Hartog, *Mechanical Vibrations,* McGraw-Hill, 1956, p. 28.) The calculation for the electrical series circuit shown in Fig. II.9 leads to a differential equation of exactly the same type as (II.44).

Let J = instantaneous value of the current as a function of the time t,
R = ohmic resistance (the 'damping')
L = inductance (the 'inertia')
C = capacitance (the reciprocal of the 'elasticity').

Let $U = U_m \sin \omega t$ be the applied electromotive force. Let any interior resistance be neglected or included in R. Let the instantaneous voltage at the capacitance be $U_C = U_C(t)$.

Then, by Kirchhoff's law,

$$U_C + L\frac{\mathrm{d}J}{\mathrm{d}t} + RJ = U(t)$$

which, combined with the capacitance equation

$$J = C\frac{\mathrm{d}U_c}{\mathrm{d}t},$$

leads to the differential equation for the voltage:

$$L\ddot{U}_C + R\dot{U}_C + \frac{1}{C}U_C = \frac{1}{C}U_m \sin \omega t, \tag{II.45}$$

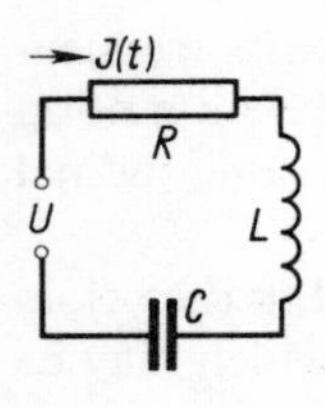

Fig. II.8. An electtrical oscillatory circuit

in complete analogy with (II.44). Whether sin or cos appears on the right-hand side depends merely on the instant chosen as the time origin. Correspondingly, for the current variation we obtain the same type of differential equation

$$L\ddot{J} + R\dot{J} + \frac{1}{C} J = \dot{E} = U_m \omega \cos \omega t.$$

15. Solution of the homogeneous equation for the oscillations

The substitution $x = e^{\lambda t}$ into the corresponding homogeneous differential equation leads to the quadratic equation $\lambda^2 + k\lambda + c = 0$ with the roots

$$\lambda_{1,2} = -\frac{k}{2m} \pm \frac{1}{2m}\sqrt{k^2 - 4\,cm}.$$

The roots are real or complex, depending on the value of k. In what follows, m, k, c are positive.

First case: weak damping. For

$$k^2 < 4\,cm$$

the square root is imaginary. Introducing the abbreviation

$$\frac{1}{2m}\sqrt{k^2 - 4cm} = \mathrm{i}\nu, \qquad \lambda_{1,2} = -\frac{k}{2m} \pm \mathrm{i}\nu$$

and using Euler's formula, we can write the general solution of the homogeneous differential equation as

$$x = \exp[-(k/2m)\,t]\,(C_1 \cos \nu t + C_2 \sin \nu t). \qquad \text{(II.46)}$$

The quantity ν, at first introduced merely formally, turns out to be the circular frequency of the damped oscillation. Using the addition formula for cosines, we can write this as

$$x = \exp[-k/2m)\,t]\;A \cos(\nu t - \varepsilon).$$

Here $A = \sqrt{C_1^2 + C_2^2}$ and $\varepsilon = \tan^{-1}(C_2/C_1)$ is the phase.

As the time increases, the factor $\exp[-(k/2m)\,t]$ decreases monotonely and tends to 0, i.e., the amplitude decreases exponentially (see Fig. II.9). The extreme values of the damped oscillation are obtained by differentiating and setting the derivative equal to 0

$$\dot{x} = A \exp\left(-\frac{k}{2m}t\right)\left\{-\frac{k}{2m}\cos(\nu t - \varepsilon) - \nu \sin(\theta t - \varepsilon)\right\} = 0.$$

Hence the time $t = t_n$ of an extremum value is given by

$$\tan(\nu t_n - \varepsilon) = -k/(2m\nu)$$

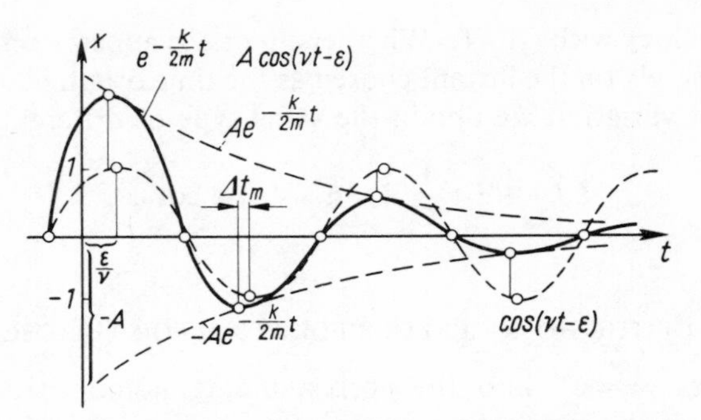

Fig. II.9. Solution of the oscillation equation with weak damping

or, taking the periodicity into account,

$$t_n = \frac{1}{\nu}\left\{n\pi + \varepsilon - \tan^{-1}\frac{k}{\nu\cdot 2m}\right\} \qquad (n = 0, 1, 2 \ldots).$$

The time difference Δt_n compared with the the extremum points of the pure cosine oscillation (Fig. II.9) is therefore constant:

$$\Delta t_n = \frac{1}{\nu}\tan^{-1}\frac{k}{\nu\cdot 2m}.$$

Second case: strong damping. For strong damping, $k^2 > 4cm$, the roots are negative reals, and the general solution is

$$x = C_1\,e^{\lambda_1 t} + C_2\,e^{\lambda_2 t}; \qquad \lambda_1 < 0, \qquad \lambda_2 < 0,$$

a combination of two decreasing functions (Fig. II.10).

Oscillations no longer take place. However, the constants can be so chosen that one passage through zero can take place (Fig. II.11).

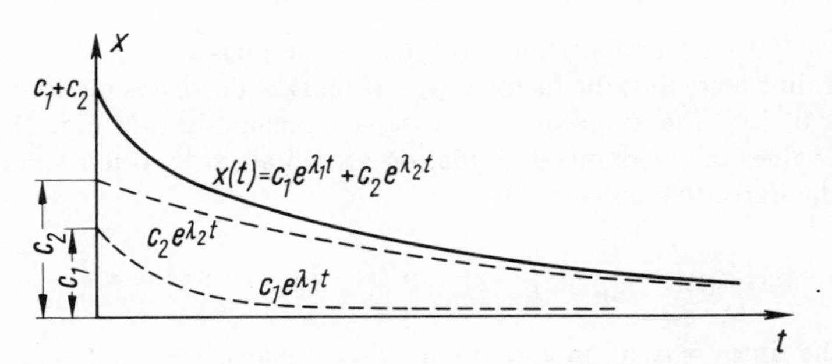

Fig. II.10. Solution of the equation for oscillations with strong damping

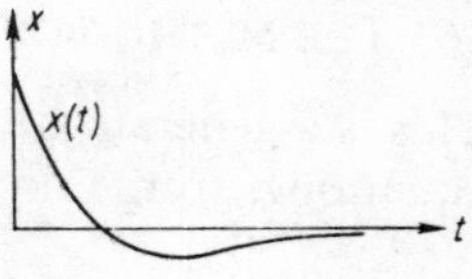

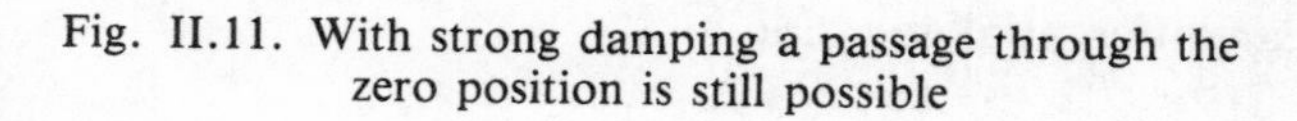
Fig. II.11. With strong damping a passage through the zero position is still possible

Third case: the aperiodic limiting case. Finally, if $k^2 = 4cm$, we obtain two equal zeros $\lambda_1 = \lambda_2 = \lambda$ and hence the general solution of the differential equation is

$$x = (C_1 + C_2 t)\, e^{\lambda t}.$$

Quantitatively, the behaviour is the same as the strong damping.

§5. DETERMINATION OF A PARTICULAR SOLUTION OF THE INHOMOGENEOUS LINEAR DIFFERENTIAL EQUATION

The cases just discussed are the transient, decaying processes of a damped system without external influences, or mathematically speaking, systems without a 'perturbing function' $P(t)$, as in (II.43).

In order to take into consideration the effect of the excitation $P(t)$, we need, in accordance with the superposition principle of Chapter II, 8, a particular solution of the inhomogeneous differential equation.

There are several ways of obtaining such a particular solution.

1. If the general solution of the homogeneous differential equation is known, we can, by the method described in Chapter II, 7, reduce the differential equation to a linear first-order equation, which can then be solved in closed form, as in Chapter II, 9.

2. If the general solution of the homogeneous differential equation is known, we can also use the method of variation of the constants, as described in Chapter II, 16, to carry out which, certain quadratures are necessary. So if a fundamental system of the homogeneous equation is available, we can always force out the solution of the inhomogeneous equation by quadratures.

3. For many cases occurring in applications, however, we can obtain very quickly a particular solution of the inhomogeneous equation by a rule-of-thumb method (to be described in II, 17). But this rule of thumb is not a general method, and it does not, by any means, always lead to the objective. In the cases where it does provide a solution, however, it is generally very much simpler and it leads to the result much more quickly than methods 1 and 2.

The canny calculator will, therefore, generally try to use the rule-of-thumb method, and only if this fails will he turn to one of the more cumbersome methods 1 or 2 which do, however, always leads to a result when a fundamental system is known.

16. THE METHOD OF VARIATION OF THE CONSTANTS

This is a generalization of the method descrbed in Chapter I, 9 for first-order equations. In the linear differential equation

$$L[y] = \sum_{\nu=0}^{n} p_\nu(x) y^{(\nu)}(x) = r(x) \tag{II.47}$$

suppose a fundamental system $y_1, y_2, \ldots, y_n$ is known for the homogeneous equation. Its general solution is therefore

$$y = C_1 y_1 + C_2 y_2 + \cdots + C_n y_n.$$

As before in Chapter I, 9 here too the consants C_μ are replaced by functions $C_\mu(x)$ which are to be determined, and so for a particular solution $u(x)$ of the inhomogeneous equation we make the substitution

$$u = C_1(x) y_1 + C_2(x) y_2 + \cdots + C_n(x) y_n.$$

By so doing we are really over-determining the solution substitution, since we actually need not n unknown functions $C_\mu(x)$ but only one solution function. We can therefore prescribe a further $n-1$ conditions, and we choose them so that certain expressions which appear when the derivatives $u, u', \ldots, u^{(n)}$ are formed shall vanish, and the substitution of $u, u', \ldots, u^{(n)}$ into the differential equation will then give a simple result. Thus we demand, for example, in forming the derivative

$$u' = C_1 y_1' + \cdots + C_n y_n' + C_1' y_1 + \cdots + C_n' y_n$$

that the second part $C_1' y_1 + \cdots + C_n' y_n$ shall vanish. There then remains simply

$$u' = C_1 y_1' + \cdots + C_n y_n',$$

and the next differentiation can easily be carried out. Proceeding in the same way with the successive derivatives we require that generally

$$C_1' y_1^{(\nu)} + \cdots + C_n' y_n^{(\nu)} = 0 \quad \text{for } \nu = 0, 1 \ldots, n-2. \tag{II.48}$$

We then have

$$u^{(\nu)} = C_1 y_1^{(\nu)} + \cdots + C_n y_n^{(\nu)}, \qquad \nu = 0, 1, \ldots, n-1,$$

$$u^{(n)} = C_1 y_1^{(n)} + \cdots + C_n y_n^{(n)} + C_1' y_1^{(n-1)} + \cdots + C_n' y_n^{(n-1)}.$$

Substituting all these derivatives into the inhomogeneous differential equation we obtain

$$L[u] = C_1\underbrace{L[y_1]}_{=0} + C_2\underbrace{L[y_2]}_{=0} + \cdots + C_n\underbrace{L[y_n]}_{=0} + \sum_{\mu=1}^{n} C_\mu{}' y_\mu^{(n-1)} p_n(x) = r(x);$$

since the y_μ are particular solutions.

Hence, we have the further condition for the C_μ':

$$C_1' y_1^{(n-1)} + \cdots + C_n' y_n^{(n-1)} = \frac{r(x)}{p_n(x)}. \tag{II.49}$$

(II.48) and (II.49) together form a system of n equations for the n unknowns $C_\mu'(x)$.

But the determinant of the coefficients of these equations is just the Wronskian (II.19) of the fundamental system y_ν and it is therefore non-zero. Hence the equations are uniquely soluble and enable $C_1'(x), C_2'(x), \ldots, C_n'(x)$ to be calculated; these functions are therefore continuous and consequently integrable. On integration we hence obtain the functions

$$C_\mu(x) = \int_{x_0}^{x} C_\mu'(x)\,\mathrm{d}x + C_\mu^*, \qquad (\mu = 1, 2, \ldots, n). \tag{II.50}$$

By including the integration constants C_μ^* we obtain after substitution the general solution of the complete differential equation.

17. The rule-of-thumb method

For finding a particular solution of an inhomogeneous linear differential equation there is a rule-of-thumb method (with the advantages and disadvantages mentioned at the beginning of this section), which is based on the fact that in numerous cases the functions in a particular solution are of the same types as the perturbing function $r(x)$. More precisely, the method works like this; for the differential equation (II.47) we differentiate $r(x)$ ν-times altogether, multiply the derivatives by the corresponding $p_\nu(x)$, and determine whether the results belong to one common class of function, e.g., whether only exponential functions appear, or only trigonometrical functions, or perhaps products of these two types of function, or whether only powers of x appear, etc.

We then choose as a substitution the linear combination

$$y(x) = \sum_\rho C_\rho \varphi_\rho(x),$$

in which the C_ρ are constant to be determined by identification in the differential equation, and the $\varphi_\rho(x)$ are functions of the observed type. For linear differential equations with *constant* coefficients a number of suitable substitutions are shown in the following table.

Perturbing function: $r(x) =$	Corresponding particular substitution $y(x) =$
a	A
x^n; $a_0 + a_1x + a_2x^2 + \cdots + a_nx^n$	$A_0 + A_1x + A_2x^2 + \cdots + A_nx^n$
$a\,e^{\lambda x}$	$A\,e^{\lambda x}$
$a \sin m$; $a \cos mx$; $a_0 \cos mx + a_1 \sin mx$	$A_0 \cos mx + A_1 \sin mx$
$a\,e^{\lambda x} \cos mx$; $a\,e^{\lambda x} \sin mx$; $a_0\,e^{\lambda x} \cos mx + a_1\,e^{\lambda x} \sin mx$	$A_0\,e^{\lambda x} \cos mx + A_1\,e^{\lambda x} \sin mx$

If linear combinations of these perturbing functions occur, then we have to choose the corresponding combination for the substitution. For example, if

$$r(x) = e^{3x} + 4 \sin 2x + x^2,$$

we try

$$y(x) = A_0\,e^{3x} + A_1 \cos 2x + A_2 \sin 2x + A_3 + A_4x + A_5x^2.$$

A really useful and common form is obtained by replacing z by x^s in (II.38). For the linear differential expression L as in (II.32) with constant coefficients p_ν and the characteristic polynomial $P(k)$ as in (II.35) we then get

$$L[e^{kx}x^s] = \left\{x^sP(k) + \binom{s}{1} x^{s-1} P'(k) + \binom{s}{2} x^{s-2}P''(k) + \cdots + \binom{s}{n} x^{s-n}P^{(n)}(k)\right\} e^{kx}. \tag{II.51}$$

If $k = k_1$ is an r-fold root of $P(k) = 0$, i.e., if

$$P(k) = P'(k) = \cdots = P^{(r-1)}(k) = 0, \quad \text{but } P^{(r)}(k) \neq 0,$$

then on the right-hand side there will be e^{kx} multiplied by a polynomial of degree $s - r$, if s is a natural number $\geq r$. Equation (II.51) also holds even if s is not a natural number. An example of the use of this formula is given in No. 19.

If we apply the rule-of-thumb method to the equation for forced vibrations (II.44)

$$m\ddot{x} + k\dot{x} + cx = P_0 \cos \omega t,$$

then we try as a particular solution

$$x = a \cos \omega t + b \sin \omega t,$$

since on differentiating the perturbing function twice, only functions of this sort appear.

18. Introduction of a complex differential equation

By working with complex quantities we can, in the last example, reach our objective even quicker. The perturbing function

$$r(x) = P_0 \cos \omega t$$

is, in fact, the real part of the complex-valued function

$$P_0\,\mathrm{e}^{\mathrm{i}\omega t} \equiv \underbrace{P_0 \cos \omega t}_{\text{real part}} + \mathrm{i}\underbrace{P_0 \sin \omega t}_{\substack{\text{imaginary}\\ \text{part}}}. \quad \text{(Euler's identity).}$$

One writes, in abbreviated form,

$$P_0 \cos \omega t = \mathrm{Re}(P_0\,\mathrm{e}^{\mathrm{i}\omega t}),\; P_0 \sin \omega t = \mathrm{Im}(P_0\,\mathrm{e}^{\mathrm{i}\omega t}).$$

So the differential equation now reads:

$$m\ddot{x} + k\dot{x} + cx = \mathrm{Re}(P_0\,\mathrm{e}^{\mathrm{i}\omega t}).$$

If we solve instead of this differential equation, the at first seemingly more complicated equation with the perturbing function extended by the imaginary term

$$m\ddot{x} + k\dot{x} + cx = P_0 \cos \omega t + \mathrm{i}P_0 \sin \omega t = P_0\,\mathrm{e}^{\mathrm{i}\omega t},$$

we also obtain a complex-valued particular solution of the form

$$x(t) = u(t) + \mathrm{i}v(t).$$

On substituting for $x(t)$ in the differential equation, whose coefficients are assumed to be real, the $u(t)$ provides the real part, and $v(t)$ the imaginary part of the perturbing function.

If, therefore, we pick out from the solution of the differential equation, extended to have a complex-valued perturbing function, the real part of that solution, we shall have the required particular solution of the original differential equation.

The saving of work thus achieved the reader can check for himself by working out the result with a purely real substitution.

So we put

$$x = C\,\mathrm{e}^{\mathrm{i}\omega t}$$

into the differential equation:

$$-mC\,\omega^2 + kC\,\mathrm{i}\omega + Cc = P_0$$

giving

$$C = \frac{P_0}{c - m\omega^2 + ik\omega}.$$

The extension with the complex conjugate as denominator enables us to separate the real and imaginary parts, since by this trick the denominator becomes real:

$$C = \frac{P_0}{(c - m\omega^2) + ik\omega} \cdot \frac{c - m\omega^2 - ik\omega}{(c - m\omega^2) - ik\omega} = P_0 \frac{c - m\omega^2 - ik\omega}{(c - m\omega^2)^2 + k^2\omega^2}.$$

Using the abbreviations α, β (the denominator is at first assumed to be non-zero)

$$C = P_0 \Bigg[\underbrace{\frac{c - m\omega^2}{(c - m\omega^2)^2 + k^2\omega^2}}_{\alpha} + \mathrm{i} \underbrace{\frac{-k\omega}{(c - m\omega^2)^2 + k^2\omega^2}}_{\beta}\Bigg], \tag{II.52}$$

$$C = P_0(\alpha + \mathrm{i}\beta)$$

we can write the particular solution for the complex-valued disturbing function as

$$x = P_0(\alpha + \mathrm{i}\beta)\, \mathrm{e}^{\mathrm{i}\omega t}.$$

From this we now seek the real part:

$$x = P_0(\alpha + \mathrm{i}\beta)(\cos \omega t + \mathrm{i} \sin \omega t),$$
$$x = P_0(\alpha \cos \omega t - \beta \sin \omega t) + \underbrace{\mathrm{i}(\ldots)}_{\text{of no interest!}}$$

To obtain the general solution, we now add, in accordance with the principle of superposition, Chapter II, 9, the general solution (II.46) of the homogeneous differential equation and the particular solution of the complete inhomogeneous differential equation:

$$x = \exp\left(-\frac{k}{2m}t\right)(C_1 \cos \nu t + C_2 \sin \nu t) + P_0(\alpha \cos \omega t - \beta \sin \omega t). \tag{II.53}$$

The transient damped vibration, which is of practical significance only for the initial process of setting up the vibrations, is superimposed on the undamped forced vibration (see Fig. II.12). These two have, in general, different circular

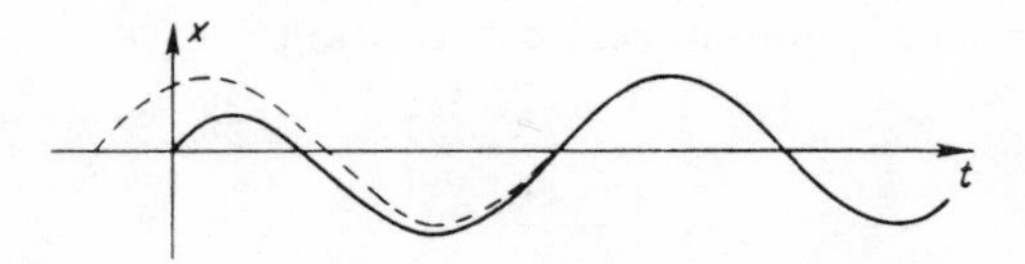

Fig. II.12. The process of initiating the oscillations

frequencies and different amplitudes. The constants C_1 and C_2 are chosen to fit the initial conditions.

19. The case of resonance

We now consider the case where the denominator in (II.52) vanishes:

$$(c - m\omega^2)^2 + k^2\omega^2 = 0.$$

This happens if simultaneously $c - m\omega^2 = 0$ and $k\omega = 0$; since $\omega \neq 0$ it must be that $k = 0$, i.e., the system is undamped; and $\omega^2 = c/m$ means that the exciting frequency is the same as the natural frequency of the system itself. This case is known as the *resonance case.* The differential equation is therefore

$$m\ddot{x} + cx = P_0 \cos \omega t \quad \text{with } \omega^2 = c/m. \tag{II.54}$$

Here again extending the right-hand side of the equation to be complex-valued is useful:

$$m\ddot{x} + cx = P_0\, \mathrm{e}^{\mathrm{i}\omega t}. \tag{II.55}$$

The rule-of-thumb method fails here, as we have already established, but formula (II.51) is applicable; in it t has now to be written instead of x. $k = \mathrm{i}\omega$ is a root of the characteristic polynomial; in order that e^{kt} multiplied by a constant shall appear on the right-hand side we have to choose $s = 1$. The substitution $x = Ct\,\mathrm{e}^{\mathrm{i}\omega t}$ into (II.55) yields

$$2iC\omega m\, \mathrm{e}^{\mathrm{i}\omega t} = P_0\, \mathrm{e}^{\mathrm{i}\omega t}$$

or

$$x = -\mathrm{i}\frac{P_0}{2m\omega} t\, \mathrm{e}^{\mathrm{i}\omega t} = -\mathrm{i}\frac{P_0}{2m^{\omega}} t \cos \omega t + \frac{P_0}{2m\omega} t \sin \omega t.$$

The real part is the required particular solution of the real inhomogeneous original equation:

$$x = \frac{P_0}{2m\omega} t \sin \omega t \tag{II.56}$$

and represents an oscillation whose amplitude increases linearly with time and hence increases without bound, as in Fig. II.7.

§6. THE EULER DIFFERENTIAL EQUATION

20. The substitution for the solution; the characteristic equation

A further type of integrable linear differential equations, with non-constant coefficients, is the *Euler* differential equation

$$p_0 y + p_1 x y' + p_2 x^2 y'' + \cdots p_n x^n y^{(n)} = r(x).$$

The p_ν are given constants with $p_n \neq 0$.

As in Chapter II, 16 it suffices to discuss the homogeneous equation

$$\sum_{\nu=0}^{n} p_\nu x^\nu y^{(\nu)} = 0, \tag{II.57}$$

The obvious substitution

$$y = x^r, \qquad y' = rx^{r-1}, \qquad y'' = r(r-1)\,x^{r-2}, \ldots,$$
$$y^{(n)} = r(r-1)(r-2)\cdots(r-n+1)x^{r-n} \tag{II.58}$$

leads, after dividing through by x^r, to the characteristic equation

$$p_0 + p_1 r + p_2 r(r-1) + \cdots + p_n r(r-1)(r-2)\cdots(r-n+1) = 0. \tag{II.59}$$

This is an algebraic equation of the nth degree; let the roots be

$$r_1, r_2, r_3, \ldots, r_n.$$

If all the roots are distinct, then the general solution of the differential equation reads

$$y = C_1 x^{r_1} + C_2 x^{r_2} + \cdots + C_n x^{r_n}.$$

It now has to be shown that the powers $x^{r_1}, \ldots, x^{r_n}$ actually are linearly independent. However, we can save ourselves the trouble of proving this (not difficult), and of investigating what happens when the characteristic equation has multiple roots, if we reduce the differential equation (II.57) to a linear equation with constant coefficients, for which the circumstances have already been completely discussed in §4. This reduction is performed by the transformation, for $x > 0$ (for $x < 0$ replace x by $-x$)

$$z = \ln x, \; x = e^z, \qquad \frac{dx}{dz} = Dx = x, \tag{II.60}$$

where the symbol D has been introduced to denote differentiation with respect to z. So then

$$Dy = \frac{dy}{dz} = \frac{dy}{dx}\frac{dx}{dz} = xy'$$
$$D^2 y = D(xy') = xy' + xy'' \cdot x$$

or

$$x^2 y'' = D^2 y - xy' = D^2 y - Dy = D(D-1)y.$$

In general, for the qth derivative we have

$$x^q y^{(q)} = D(D-1)\ldots(D-q+1)y. \tag{II.61}$$

This rule can be proved immediately by induction; the step from q to $q+1$ is obtained by multiplying by $D-q$:

$$(D-q)x^q y^{(q)} = D(x^q y^{(q)} - qx^q y^{(q)} = qx^q y^{(q)} + x^q y^{(q+1)} \cdot x - qx^q y^{(q)}$$
$$= x^{q+1} y^{(q+1)}.$$

On the other hand,

$$(D-q)\, x^q y^{(q)} = D(D-1) \ldots (D-q)\, y.$$

The transformation thus gives the new differential equation with constant coefficients for the function $v(z) = y(e^z)$

$$\sum_{\nu=0}^{n} p_\nu D(D-1) \ldots (D-\nu+1) v = 0. \tag{II.62}$$

The characteristic equation (II.35) for this is identical with the characteristic equation (II.59), but now the results of the theory in §4 are available: if all the roots $r_1, \ldots, r_n$ are distinct, then $x^{r_1}, \ldots, x^{r_n}$ form a fundamental system; if r_1 is a p-fold root, then

$$e^{r_1 z}, \qquad z\, e^{r_1 z}, \ldots, z^{p-1}\, e^{r_1 z}$$

or, transformed,

$$x^{r_1}, \qquad (\ln x)\, x^{r_1}, \ldots, (\ln x)^{p-1}\, x^{r_1}$$

are solutions of (II.57).

21. Examples

1. $$x^3 y''' - 3x^2 y'' + 6xy' - 6y = 0.$$

The substitution

$$y = x^r$$

leads to

$$r^3 - 6r^2 + 11r - 6 = 0$$

with the roots $r_1 = 1$, $r_2 = 2$, $r_3 = 3$. The general solution is therefore

$$y = C_1 x + C_2 x^2 + C_3 x^3.$$

2. If conjugate complex roots occur

$$r_1 = a + ib, \qquad r_2 = a - ib,$$

one converts the particular solutions

$$x^{r_1} = x^{a+ib} = x^a\, x^{ib}, \qquad x^{r_2} = x^{a-ib} = x^a x^{-ib}$$

by Euler's formula (Chapter II, 11) into

$$x^{\pm ib} = e^{\pm (\ln x) ib} = \cos(b \ln x) \pm i \sin(b \ln x).$$

Then, with $C_1^* = C_1 + C_2$ and $C_2^* = iC_1 - iC_2$ the solution is obtained as

$$y = x^a [C_1^* \cos(b \ln x) + C_2^* \sin(b \ln x)] + \cdots. \tag{II.63}$$

For example, for

$$x^2 y'' - xy' + 2y = 0$$

the substitution $y = x^r$ leads to $r^2 - 2r + 2 = 0$, and so $r_{1,2} = 1 \pm \mathrm{i}$.
From

$$y = C_1 x^{1+\mathrm{i}} + C_2 x^{1-\mathrm{i}}$$

we get the general solution in real form

$$y = x[C_1^* \cos(\ln x) + C_2^* \sin(\ln x)].$$

3. $$x^2 y'' + 3xy' + y = 0.$$

The substitution $y = x^r$ leads to $r^2 + 2r + 1 = (r+1)^2 = 0$, therefore $r_1 = r_2 = -1$. There are equal roots. Therefore the general solution reads

$$y(x) = \frac{C_1}{x} + C_2 \frac{\ln x}{x}.$$

§7. SYSTEMS OF LINEAR DIFFERENTIAL EQUATIONS

In applications, systems of linear ordinary differential equations are often encountered. Even in those cases where it is possible by eliminating some of the unknowns to convert such a system into a single equivalent differential equation of higher order, it is often more sensible to attack the system of equations directly. Therefore we shall present here briefly some of the properties of such systems.

22. Example: Vibrations of a motor-car; (types of coupling)

Let us consider vertical vibrations of a car by idealizing the car as a rigid beam on two elastic supports (see Fig. II.13); thus the front-wheel springs and the rear-wheel springs are each combined into a single spring, with the spring constants c_1 and c_2 respectively. We consider small vibrations, with small deflections x_1, x_2 at the points of action P_1, P_2 of the springs and small angles of inclination φ, so that, as a first approximation, each point of the beam can be assumed to move along a vertical straight line. Let the centre of gravity S of the beam be at distances l_1, l_2 from the points P_1, P_2, and have the vertical

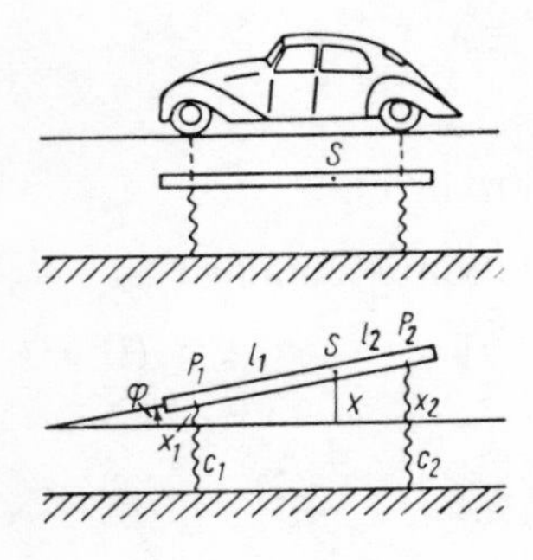

Fig. II.13. Vibrations of a motor-car

deflection x from its equilibrium position. It is a well-known result in mechanics that the centre of gravity moves as if the total mass M of the beam were concentrated at it and the forces, here the spring forces $-c_1x_1$ and $-c_2x_2$, acted directly on it:

$$M\ddot{x} = -c_1x_1 - c_3x_2. \tag{II.64}$$

If the beam has the moment of inertia Θ relative to the centre of gravity, then for φ we have the equation (see (III.52))

$$\Theta\ddot{\varphi} = c_1x_1l_1 - c_2x_2l_2. \tag{II.65}$$

For small deflections we may put

$$x_1 = x - l_1\varphi, \qquad x_2 = x + l_2\varphi,$$

and so

$$x = \frac{l_2x_1 + l_1x_2}{l_1 + l_2}, \qquad \varphi = \frac{x_2 - x_1}{l_1 + l_2}.$$

Substituting these expressions for x_1 and x_2 in (II.64), (II.65), we obtain the system of two second-order differential equations for x and φ

$$\begin{aligned} M\ddot{x} &= -a_{11}x - a_{12}\varphi \\ \Theta\ddot{\varphi} &= -a_{21}x - a_{22}\varphi \end{aligned} \tag{II.66}$$

where

$$a_{11} = c_1 + c_2, \ a_{12} = a_{21} = c_2l_2 - c_1l_1, \ a_{22} = c_1l_1^2 + c_2l_2^2.$$

If by chance the dimensions of the system are such that $a_{12} = 0$, then the system (II.66) is 'decoupled', and we have one differential equation for x only and one for φ only. But in the general case, with $a_{12} \neq 0$, the system is sometimes said to be written as 'deflection-coupled'. But there is little sense in calling the system 'deflection-coupled'. For if the equations (II.66) are solved for x and φ, we can write

$$\begin{aligned} x &= \alpha_{11}\ddot{x} + \alpha_{12}\ddot{\varphi} \\ \varphi &= \alpha_{21}\ddot{x} + \alpha_{22}\ddot{\varphi} \end{aligned}$$

which are 'acceleration-coupled'. We can also, as equation (II.72) will show, describe the system as entirely 'decoupled'. For the following, it will be assumed that the equations (II.66) are not yet decoupled, $a_{12} \neq 0$, $a_{21} \neq 0$ and that a_{12} and a_{21} have the same sign, $a_{12}a_{21} > 0$ (It can now happen that $a_{12} \neq a_{21}$).

To solve the differential equations (II.66) we make the substitution for 'eigen-oscillations', in which x and φ oscillate in the same rhythm, i.e., with the same frequency and same phase:

$$x = A_1 \cos \omega t, \qquad \varphi = A_2 \cos \omega t;$$

the equations (II.66) then become, dropping the factor $\cos \omega t$,

$$\begin{aligned}(a_{11} - M\omega^2)\, A_1 + a_{12}A_2 \qquad\qquad &= 0 \\ a_{21}A_1 + (a_{22} - \Theta\omega^2)A_2 &= 0.\end{aligned} \tag{II.67}$$

These two linear homogeneous equations for A_1, A_2 have just one non-trivial solution if the determinant vanishes:

$$\begin{vmatrix} a_{11} - M\omega^2 & a_{12} \\ a_{21} & a_{22} - \Theta\omega^2 \end{vmatrix} = 0. \tag{II.68}$$

This is the 'frequency-equation'

$$M\Theta\omega^4 - (Ma_{22} + \Theta a_{11})\omega^2 + D = 0 \quad \text{with } D = a_{11}a_{22} - a_{12}\alpha_{21}, \tag{II.69}$$

which has the squares ω'^2 and ω''^2 of the two natural frequencies ω' and ω'' as its roots. We shall see that this equation has two different roots, and if $D > 0$, $a_{11} > 0$, $a_{22} > 0$ they are both positive; suppose $0 < \omega' < \omega''$. The vibration with frequency ω' is called the *lower-frequency vibration* or *fundamental,* that with ω'' is called the *upper-frequency vibration* or *harmonic.* Further, for each of these vibrations the ratio of the amplitudes

$$\frac{A_1}{A_2} = \frac{x}{\varphi} = s \tag{II.70}$$

has a definite, fixed value. The fact that $x = \varphi s$ implies that the beam (or its extension) always passes through a fixed point, a *node*; let s', s'' be the distances from S of the nodes K', K'' corresponding to ω', ω''; see Fig. II.14. From (II.70), (II.67)

$$\frac{a_{21}A_1}{A_2} = a_{21}s = \Theta\omega^2 - a_{22} \tag{II.70a}$$

and using this to replace ω^2 by s in (II.69) we find

$$Ma_{21}s^2 + (Ma_{22} - \Theta a_{11})s - \Theta a_{12} = 0. \tag{II.71}$$

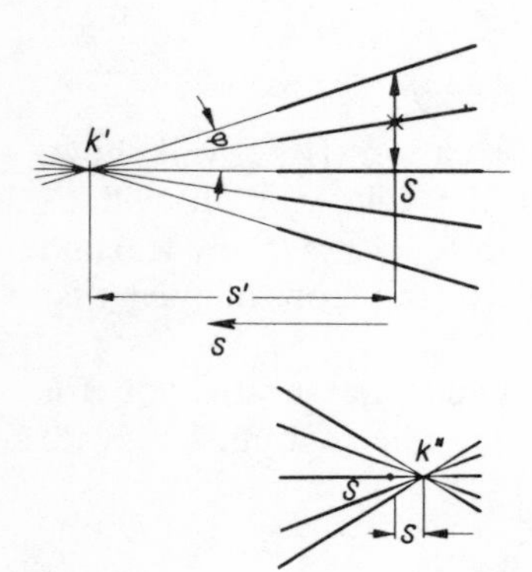

Fig. II.14. The nodes for the lower and higher frequency vibrations

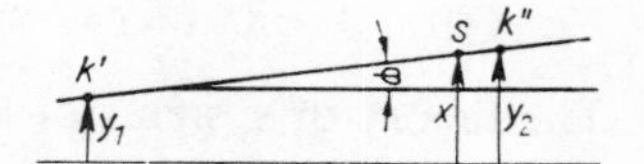

Fig. II.15. Introduction of normal co-ordinates

From this quadrant equation in s it follows that $s's'' = -\Theta a_{12}/(Ma_{21}) < 0$, so s' and s'' are real and have different signs, i.e., the nodes K' and K'' lie on different sides of the centre of gravity S. Hence $\omega'^2 \neq \omega''^2$ and ω'^2 and ω''^2 are real. In the case $D > 0$, ω'^2 and ω''^2 have the same sign by (II, 69), and by (II.70a) they cannot both be negative if $a_{22} > 0$.

Now we introduce as new co-ordinates for describing the system the deflections y_1, y_2 at the nodes (Fig. II.15):

$$y_1 = x - s'\varphi, \qquad y_2 = x - s''\varphi, \qquad y_2 - y_1 = \hat{s}\varphi,$$

where $\hat{s} = s' - s''$ is the distance between the two nodes. Hence,

$$x = \frac{-s''y_1 + s'y_2}{\hat{s}}, \qquad \varphi = \frac{y_2 - y_1}{\hat{s}},$$

and the equations (II.66) become

$$-Ms''\ddot{y}_1 + Ms'\ddot{y}_2 = (a_{11}s'' + a_{12})\, y_1 - (a_{11}s' + a_{12})\, y_2$$
$$-\Theta\ddot{y}_1 + \Theta\ddot{y}_2 = (a_{21}s'' + a_{22})\, y_1 - (a_{21}s' + a_{22})\, y_2.$$

If we eliminate $\ddot{y}_2$ from these equations (by multiplying the upper equation by Θ, the lower equation by $-Ms'$, and adding), then by (II.71) the y_2 also drops out, and there is left an equation of the form

$$M\Theta\hat{s}\ddot{y}_1 = Cy_1, \tag{II.72}$$

where C is a constant. Similarly, by eliminating $\ddot{y}_1$ an equation is obtained in which only $\ddot{y}_2$ and y_2 appear; i.e., the equations are 'decoupled'; y_1 and y_2 are called the *principal* or *normal co-ordinates* of the system.

A numerical example

In a motor-car let the weight be 15 kN, the moment of inertia about the centre of gravity be given by the radius of gryration $i_s = 1.2$ m, so that $\Theta = Mi_s^2$; let $l_1 = 1.2$ m, $l_2 = 1.8$ m, $c_1 = 70$ kN/m, $c_2 = 60$ kN/m. Here equations (II.66) become

$$-M\ddot{x} = 130x + 24\varphi, \quad -\Theta\ddot{\varphi} = 24x + 295, .2\varphi.$$

The frequency equation (II.69) gives $M\omega'^2 = 125$ kN/m for the fundamental, and $M\omega''^2 = 210$ kN/m for the harmonic vibration. The nodes are then calculated from (II.70a) to be at $s' = -4.8$ m, $s'' = 0.3$ m. If the deflections y_1, y_2 at the nodes are introduced, the decoupled equations

$$-M\ddot{y}_1 = 210y_1, \qquad -M\ddot{y}_2 = 125y_2$$

are obtained.

23. THE FUNDAMENTAL SYSTEM OF SOLUTIONS

The theory of systems of linear differential equations has acquired increasing significance in recent times for its practical applications. For its rational presentation matrices are the most suitable mathematical tool. Due to lack of space, however, we can here neither provide an introduction to the matrix calculus nor present the theory of systems in the same detail as in other sections of the book. Since, however, matrix theory is fundamental for many purposes, for physical processes too—for example, the theory of small oscillations (see No. 27) is best presented using matrices—we shall here briefly present the relevant mathematical lines of thought. Readers who are unfamiliar with the matrix calculus or who find the presentation in these sections too concise may skip Chapter II, Nos. 23 to 28; we shall not subsequently assume them to be known.

Consider a linear system of n first-order differential equations for n unknown functions $y_1(x), \ldots, y_n(x)$:

$$y_j'(x) = \sum_{k=1}^{n} a_{jk}(x)\, y_k(x) + r_j(x), \qquad (j = 1. \ldots, n). \tag{II.73}$$

The $a_{jk}(x)$ and $r_j(x)$ are continuous functions defined in an interval $a \le x \le b$; if the $r_j(x) \equiv 0$ in (a, b), then the system is said to be *homogeneous,* otherwise it is *inhomogeneous.*

With the matrix notation

$$y(x) = \begin{pmatrix} y_1(x) \\ \vdots \\ y_n(x) \end{pmatrix}, \qquad r(x) = \begin{pmatrix} r_1(x) \\ \vdots \\ r_n(x) \end{pmatrix}, \qquad A(x) = \begin{pmatrix} a_{11}(x) \ldots a_{1n}(x) \\ \ldots\ldots\ldots\ldots \\ a_{n_1}(x) \ldots a_{nn}(x) \end{pmatrix} \tag{II.47}$$

the system (II.73) is written more shortly as

$$y'(x) = A(x)y(x) + r(x). \tag{II.75}$$

By the general existence and uniqueness theorem in Chapter I, 20, when the initial values $y(x_0)$ at a point $x = x_0$ in (a, b) are prescribed, there is one and only one solution of (II.75) which assumes these initial values (for brevity we say 'solution' instead of 'system of solutions'). First we consider a homogeneous system with $r(x) \equiv 0$, and several, say n, different solutions

$$y_{(1)}, y_{(2)}, \ldots, y_{(n)}.$$

Each of these solutions has n components, $y_{(j)} = \begin{pmatrix} y_{1j} \\ \vdots \\ y_{nj} \end{pmatrix}$.

By writing these column-vectors one beside another, we obtain a matrix T:

$$Y(x) = (y_{(1)}(x), \ldots, y_{(n)}(x)) = \begin{pmatrix} y_{11}(x) & \cdots & y_{1n}(x) \\ \cdots & \cdots & \cdots \\ y_{n1}(x) & \cdots & y_{nn}(x) \end{pmatrix}. \qquad \text{(II.76)}$$

Since each column is a solution

$$Y'(x) = A(x) \cdot Y(x). \qquad \text{(II.77)}$$

also holds.

Definition

The solutions $y_{(j)}$ are said to form a *fundamental system* or *principal system* if the determinant D of $Y(x)$ for any x_0 in (a, b) does not vanish:

$$D(x_0) = \det Y(x_0) \neq 0. \qquad \text{(II.78)}$$

If now $C = (c_{jk})$ is a non-singular matrix (i.e, a matrix with a non-vanishing determinant, $\det C \neq 0$) whose elements c_{jk} are constant numbers, then if Y is a matrix whose columns are solutions, so also is $Y(x)C$, because solutions can be superimposed and multiplied by constant factors, and so the columns of $Y(x)C$ are still solutions. If Y is a fundamental system, then so is YC, since

$$\det(Y(x_0)\, C) = \det\, Y(x_0) \det\, C \neq 0. \qquad \text{(II.79)}$$

Now $D(x)$, exactly like the Wronskian in Chapter II, 9 satisfies a homogeneous, linear, first-order differential equation. For, regarding D as a function of its elements y_{jk}, we have

$$D'(x) = \frac{dD(x)}{dx} = \sum_{j,k=1}^{n} \frac{\partial D(x)}{\partial y_{jk}} \frac{dy_{jk}}{dx}.$$

$\partial D / \partial y_{jk}$ is the minor Y_{jk} of the element y_{jk} in D, and, from (II.77),

$$\frac{dy_{jk}}{dx} = \sum_{l=1}^{n} a_{jl} y_{lk}.$$

So, using the Kronecker symbol

$$\delta_{jl} = \begin{cases} 0 & \text{for } j \neq l \\ 1 & \text{for } j = l \end{cases}$$

we have

$$D'(x) = \sum_{j,l=1}^{n} a_{jl} \sum_{k=1}^{n} Y_{jk} y_{lk} = \sum_{j,l=1}^{n} a_{jl} \delta_{jl} D;$$

since the inner sum, by the rules for expanding determinants by rows, represents D itself when $j = l$ and vanishes for $j \neq l$; so there remain from the summation only the terms with $j = l$:

$$D'(x) = S_A(x) \cdot D(x) \quad \text{where } S_A(x) = \sum_{j=1}^{n} a_{jj}(x). \qquad \text{(II.80)}$$

S_A is the sum of the elements in the principal diagonal, i.e., it is the *trace* of the matrix A.

As in (II.22) it follows that

$$D(x) = D(x_0) \cdot \exp\left(\int_{x_0}^{x} s_A(t)\,\mathrm{d}t\right). \tag{II.81}$$

This form includes (II.22) as a special case. As before, it is also the case here that:

$D(x_0) \neq 0$ implies that $D(x)$ does not vanish anywhere in (a, b);
$D(x_0) = 0$ implies that $D(x)$ vanishes identically in (a, b).

24. Solution of the inhomogeneous system by the variation-of-the-constants method

This is a generalization of chapter II, 16. As in that section, we suppose a fundamental system $Y(x)$ of solutions of the homogeneous equations (II.77) is known, and we seek a solution of the inhomogeneous equations (II.75) in the form

$$y(x) = Y(x)\,z(x), \tag{II.82}$$

where $z(x)$ again is a vector.

Substitution into (II.75) leads to

$$Y'(x)\,z(x) + Y(x)z'(x) = A(x)\,Y(x)\,z(x) + r(x).$$

Subtracting (II.77), there remains

$$z'(x) = Y^{-1}(x)\,r(x); \tag{II.83}$$

(the inverse matrix Y^{-1} exists because $D(x) \neq 0$ for all x in (a, b)). Hence we have a particular solution of the inhomogeneous equations:

$$y(x) = Y(x) \cdot \left[z(x_0) + \int_{x0}^{x} Y^{-1}(t)\,r(t)\,\mathrm{d}t\right], \tag{II.84}$$

in which we can at the same time choose the constant vector $z(x_0)$ so that prescribed initial values $y(x_0)$ are assumed; we have

$$z(x_0) = Y^{-1}(x_0) \cdot y(x_0). \tag{II.85}$$

25. Matrix A constant; characteristic roots of a matrix

For applications, and particularly for the theory of small oscillations of mechanical systems about their equilibrium position, but also for other mechanical and electrical oscillations, systems with constant coefficients, $a_{jk} = \text{const}$, are of particular importance.

As in Chapter II, 11 we first enquire about solutions of homogeneous equations

$$y' = Ay \tag{II.86}$$

which are of the form (the 'eigen-motions' of the system)

$$y(x) = v \cdot e^{k(x-x_0)}, \qquad y'(x) = vk\, e^{k(x-x_0)} \tag{II.87}$$

where k is a fixed number still to be determined and v is a constant column-vector. Substitution in (II.86) yields, after removal of an exponential factor,

$$(A - kE)\, v = 0, \tag{II.88}$$

where $E = (\delta_{jk})$ is the n-rowed unit matrix.

These n homogeneous equations for the n components of v have a non-trivial solution (which alone is of interest) if and only if the 'frequency determinant' vanishes:

$$\varphi(k) = \det(A - kE) = \begin{vmatrix} a_{11} - k & a_{12} & \dots & a_{1n} \\ a_{21} & a_{22} - k & \dots & a_{2n} \\ \dots & \dots & \dots & \dots \\ a_{n1} & a_{n2} & \dots & a_{nn} - k \end{vmatrix} = 0 \tag{II.89}$$

This *characteristic equation* has n roots $k_1, \dots, k_n$, the *characteristic roots of the matrix A,* for each of which a corresponding *eigen-vector* $v_{(1)}, \dots, v_{(n)}$ (not identically zero) of the matrix A can be calculated. Provided the k_j are distinct from one another, we certainly obtain n different eigen-vectors; but if some of the k_j coincide, then certain complications mentioned below occur.

26. THE THREE MAIN CLASSES IN THE THEORY OF SQUARE MATRICES

It is assumed in the following that the concept of 'linear dependence of vectors' is familiar. The results of the theory hold also in the case where the elements a_{jk} of the matrix A are not real but complex; the theorems are formulated first as they hold for reals, and the small modifications necessary for the complex case are indicated in square brackets, for example, 'unitary' instead of 'orthogonal'. Let us recall the concept of orthogonality: two vectors v, w with components v_j, w_j are said to be *orthogonal* [*unitary*] if

$$\sum_{j=1}^{n} v_j w_j = 0 \qquad \left[\text{resp. } \sum_{j=1}^{n} v_j \bar{w}_j = 0\right] \tag{II.90}$$

where $\bar{w}$ denotes the complex conjugate of w. The *transpose* of a matrix, i.e., the result of reflecting the elements in the principal diagonal, is denoted by a prime $'$: $A = (a_{jk})$, $A' = (a_{kj})$.

Matrices can be classified as follows:

1. A is called a *normal* matrix if A has a system of n linearly independent eigen-vectors which are pairwise orthogonal [*resp. unitary*].

It can be shown that normal matrices have various other characteristics. Thus, A is normal if and only if A commutes with A' [resp. $\bar{A}'$], i.e., if

$AA' = A'A$ [resp. $A\bar{A}' = \bar{A}'A$]. Normal matrices are precisely those matrices which can be brought into diagonal form by an orthogonal [resp. unitary] transformation by means of an orthogonal [resp. unitary] matrix U, i.e., if $U^{-1}AU = D$ is a diagonal matrix (a diagonal matrix has elements which are different from 0 only in the main diagonal). Here a matrix U is said to be orthogonal [resp. unitary] if $U'U = E$[resp. $\bar{U}'U = E$], where E is the unit matrix. In this connection, see II, 29, or, e.g., MacLane and Birkhoff, *Algebra,* Macmillan, 1979, p. 360.

Normal matrices include one particular case which is extremely important for applications, viz., real, symmetric [resp. Hermitian] matrices. A matrix A is said to be symmetric [resp. Hermitian] if $A = A'$ [resp. $A = \bar{A}'$] or $a_{jk} = a_{jk}$ [resp. $a_{jk} = \bar{a}_{kj}$], i.e. if the matrix on reflection in the main diagonal becomes itself [resp. its complex conjugate]. That these matrices are normal, i.e., that there are n eigen-vectors with the desired properties, is proved in the 'principal axis theorem' of quadratic forms in analytical geometry (see II, 29 or e.g., MacLane and Birkhoff, *Algebra,* Macmillan, 1979, p. 364.

2. If a matrix A has n linearly independent eigen-vectors $v_{(j)}$ anyway (not necessarily pairwise orthogonal), then it is said to be *normalizable.* Such matrices include normal matrices as a special case, and so they constitute a more comprehensive class. We mention without proof at the moment (proof in II, 29) that they are also precisely those matrices which can be brought into diagonal form $P^{-1}AP = D$ by a non-singular transformation by means of a matrix P with det $P \neq 0$. The $v_{(j)}$ can serve as the basis vectors of an, in general, oblique-angled co-ordinate system, so that an arbitary vector z can be 'expanded' in terms of the $v_{(j)}$, i.e., can be expressed as a linear combination of the $v_{(j)}$. In the older classical literature, normalizable matrices are precisely those with 'linear elementary divisors'.

3. If a matrix A no longer has n linearly independent eigen-vectors, then it is said to be *non-normalizable.* Here it is no longer possible, as we merely mention without proof, to transform the matrix A to diagonal form by a non-singular transformation with matrix P, but only to the *Jordan canonical form* (in which in certain places in the diagonal immediately to the right of the main diagonal a 1 occurs instead of 0 and all other elements except those in the main diagonal are 0). These matrices can appear only if (II.89) has multiple roots k_j.

One can ascertain in the following systematic way to which class a given matrix A belongs: (one can often see this immediately from the matrix, for example, real symmetric matrices belong, as already mentoned, to the class 1, the normal matrices). The n roots k_j of the characteristic equation (II.89) are calculated. If they are all different, then A belongs to class 2 or its sub-class 1. Next, for each of the k_j values obtained, one solves the system of homogeneous equations (II.88) and determines how many linearly independent vectors are obtained, and whether these vectors are pairwise orthogonal [resp. unitary] or not. Thereby the class into which the given matrix falls is determined.

Examples

1. $A = \begin{pmatrix} a & b \\ b & c \end{pmatrix}$, with real a, b, c, is real, symmetric, and so certainly normal.

2. $A = \begin{pmatrix} 1 & 1 \\ 0 & 2 \end{pmatrix}$ is normalizable, since A has two different characteristic numbers, $k_1 = 1$, $k_2 = 2$, but A is not normal, because it is easily seen that $A' = \begin{pmatrix} 1 & 0 \\ 1 & 2 \end{pmatrix}$ and $AA' = \begin{pmatrix} 2 & 2 \\ 2 & 4 \end{pmatrix}$ is different from $A'A = \begin{pmatrix} 1 & 1 \\ 1 & 5 \end{pmatrix}$.

3. $A = \begin{pmatrix} a & 1 \\ 0 & a \end{pmatrix}$ is not normalizable for any real or complex a, because $k = a$ is a double root of (II.89) and from

$$0 \cdot v_1 + 1 \cdot v_2 = 0, \qquad 0 \cdot v_1 + 0 \cdot v_2 = 0$$

only the one-parameter family of eigen-vectors $v = c\begin{pmatrix} 1 \\ 0 \end{pmatrix}$ with c arbitrary is obtained.

27. Application to the Theory of Oscillations

If A is normalizable, there are n linearly independent eigen-vectors $v_{(j)}$, where $v_{(j)}$ belongs to the characteristic number k_j. An arbitrary initial vector $y(x_0)$ can therefore be expressed as a linear combination of the $v_{(j)}$:

$$y(x_0) = \sum_{j=1}^{n} b_j v_{(j)}. \tag{II.91}$$

A solution, uniquely determined by these initial values, of the homogeneous equation reads

$$y(x) = \sum_{j=1}^{n} b_j v_{(j)} \, e^{k_j(x-x_0)}. \tag{II.92}$$

Since the $v_{(j)}$ are linearly independent, their determinant does not vanish:

$$D = \det \, (v_{(1)}, \ldots, v_{(n)}) \neq 0;$$

it is also the determinant of the solutions $v_{(j)} \, e^{k_j(x-x_0)}$ for $x = x_0$, and so these solutions constitute a principal system. We have therefore the fundamental distinction between cases:

Theorem

Suppose a physical system leads to a system of differential equations of the first order with constant coefficients: $y' = Ay$. If the matrix $A = (a_{jk})$ is *normalizable (this is the case, for example, if A is real, symmetric), then the general solution can be obtained by the superposition of the natural modes of vibration and it has the form (II.92). If, on the other hand, A is not normalizable, there are not sufficient natural modes of vibration for the state of the system to be described by a superposition of them.*

The rest we shall have to give without proof. One naturally asks what appears in place of the missing eigen-vectors. The answer is: 'principal vectors' of higher degree, which lead to 'secular' motions; (a detailed presentation can be found, e.g., in L. Collatz: *Eigenwertaufgaben mit technischen Anwendungen,* Leipzig 1949, p. 312ff. A vector h, which does not vanish identically, is called a *principal vector* of A if there is an integer $r \geq 1$ and a (real or complex) number k such that

$$(A - kE)^r h = 0, \tag{II.93}$$

and h is called a *principal vector of degree r* if r is the smallest integer having such a property. The eigen-vectors are principal vectors of degree 1. Only the characteristic numbers of A can appear as possible values of k. Then the following representation theorem holds: *an arbitrary vector z can always be expressed as a sum of principal vectors. If h_r is a principal vector of degree r corresponding to the characteristic number k, then*

$$\left(h_r + \frac{t}{1!} h_{r-1} + \frac{t^2}{2!} h_{r-2} + \cdots + \frac{t^{r-1}}{(r-1)!} h_1\right) e^{kt}$$

where

$$(A - kE)h_s = h_{s-1} \quad (\textit{for } s = r, r-1, \ldots, 2)$$

is a solution of the homogeneous differential equation; the solution corresponding to an arbitrarily prescribed initial state $y(x_0)$ is obtained thus: $y(x_0)$ can be written as a sum of principal vectors

$$y(x_0) = h_{r_1} + h_{r_2} + \cdots + h_{r_p}$$

and $y(x)$ is then the sum of the solutions (II.94) *which correspond to the individual $h_{r_1}, h_{r_2}, \ldots, h_{r_p}$.*

28. Example of a Physical System Which Leads to a Non-Normalizable Matrix

Consider a simple mechanical system such as is shown in Fig. II.16. For small deflections x_1, x_2 the masses m_1, m_2 move in a straight line. They are connected by a spring, and damping forces proportional to velocity are also present. Using the notation shown in Fig. II.16, we can write the equations of motion as

$$m_1\ddot{x}_1 = d(x_2 - x_1) - k_1\dot{x}_1 + k_2(\dot{x}_2 - \dot{x}_1)$$
$$m_2\ddot{x}_2 = d(x_1 - x_2) + k_2(\dot{x}_1 - \dot{x}_2),$$

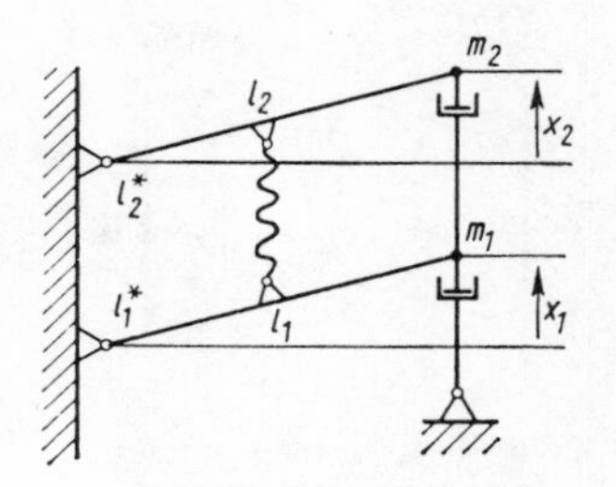

Fig. II.16. An oscillatory system with a non-normalizable matrix

where d is proportional to the spring constant. For simplicity we take

$$m_1 = m_2, \qquad k_1 = k_2, \qquad \frac{d}{m_1} = 1, \qquad \frac{k_1}{m_1} = 1.$$

Then the vector y with components $(x_1, x_2, \dot{x}_1, \dot{x}_2)$ satisfies the system of first-order differential equations $\dot{y} = Ay$ with the matrix

$$A = \begin{pmatrix} 0 & 0 & 1 & 0 \\ 0 & 0 & 0 & 1 \\ -1 & 1 & -2 & 1 \\ 1 & -1 & 1 & -1 \end{pmatrix}.$$

Here the characteristic equation (II.89) reads $k(k+1)^3 = 0$. For $k = 0$ and for $k = -1$ there are in each case only the eigen-vectors

$$C\begin{pmatrix} 1 \\ 1 \\ 0 \\ 0 \end{pmatrix} \quad \text{and } C^* \begin{pmatrix} 1 \\ 0 \\ -1 \\ 0 \end{pmatrix}$$

where C and C^* are free constants. To these eigen-vectors correspond the 'natural modes of vibration' indicated in Fig. II.17: $k = 0$ implies deflection without velocity. Both masses are deflected, the spring is not stretched or compressed, and there is no return to the equilibrium position. To $k = -1$ corresponds an aperiodic process for m_1, while m_2 remains in its equilibrium position. At m_2 the damping and spring forces balance. These two natural nodes are insufficient to obtain the general state of motion by their superposition.

Here there are principal vectors of degree 2 and 3 corresponding to $k = -1$. h_3 for example is obtained thus: we find

$$(A - (-1)E)^3 = \begin{pmatrix} 1 & 0 & 1 & 0 \\ 0 & 1 & 0 & 1 \\ -1 & 1 & -1 & 1 \\ 1 & -1 & 1 & 0 \end{pmatrix}^3 = \begin{pmatrix} 1 & 0 & 1 & 1 \\ 1 & 0 & 1 & 1 \\ 0 & 0 & 0 & 0 \\ 0 & 0 & 0 & 0 \end{pmatrix}.$$

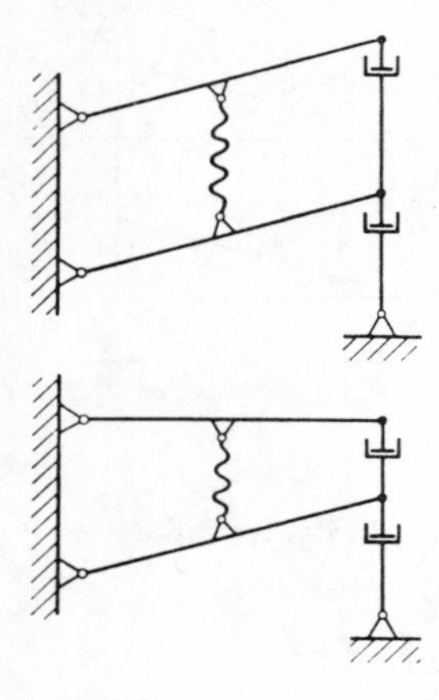

Fig. II.17. Natural modes of vibration of the system in Fig. II.16

The vectors

$$h_3 = \begin{pmatrix} a \\ b \\ c \\ -a-c \end{pmatrix},$$

where a, b, c are free constants, are solutions of $(A+E)^3 h_3 = 0$ but if $f = a - b + c = 0$ then h_3 reduces to a principal vector of lower degree. To h_3 corresponds the solution (II.94) with three free parameters, and a fourth free parameter corresponds to the eigen-vector for $k = 0$, and so, with the 4 free parameters a, b, f, C we obtain the solution

$$y = \begin{pmatrix} a + t(b+f) - \frac{t^2}{2} f \\ b - tf \\ b + f - a + t(-2f - b) + \frac{t^2}{2} f \\ -b - f + tf \end{pmatrix} e^{-t} + \begin{pmatrix} C \\ C \\ 0 \\ 0 \end{pmatrix},$$

which can be adjusted to fit arbitrary initial conditions and which therefore represents the general solution of the system considered.

29. Transformation of a normal or normalizable matrix to diagonal form

We use the same notation as in No. 26, and consider the system of differential equations $y' = Ay$. We now try, by introducing new co-ordinates (a linear transformation of the co-ordinates y into new co-ordinates z by means of a constant matrix T)

$$y = Tz, \tag{II.95}$$

to decouple the system of equations, i.e., to bring them into a new form

$$z' = \hat{A}z, \tag{II.96}$$

where $\hat{A}$ is a diagonal matrix. The transformation T is, of course, to be non-singular, i.e., det $T \neq 0$, and so, since

$$y' = Tz', \qquad z' = T^{-1}AT_z$$

we obtain immediately, for the matrix $\hat{A}$ in the new co-ordinate system,

$$\hat{A} = T^{-1}AT. \tag{II.97}$$

If we denote the column-vectors of the matrix T by $t_1, \ldots, t_n$, then we can write T in symbolic form as

$$T = (t_1 \mid \ldots \mid t_n).$$

If in particular we choose an eigen-vector y_j of the matrix A (so that

$$Ay_j = k_j y_j, \tag{II.98}$$

as the jth column-vector of the matrix T, i.e, put $t_j = y_j$, then by the definition of the product of a matrix with a vector,

$$Te_j = t_j,$$

where $e_1, \ldots, e_n$ denote the unit vectors. Hence $e_j = T^{-1}t_j$, and by formula (II.97) we have

$$\hat{A}e_j = T^{-1}ATe_j = T^{-1}At_j = k_j T^{-1}t_j = k_j e_j. \tag{II.99}$$

But now $\hat{A}e_j$ denotes just the jth column of $\hat{A}$, and by the last equation this contains in the main diagonal position the number k_j and zeros elsewhere.

If we now assume the matrix A to be normalizable, i.e., that it has n linearly independent eigen-vectors $y_1, \ldots, y_n$, we can choose all the $t_j = y_j$ and so obtain a transformation matrix T which is non-singular and is such that in each column of $\hat{A}$ a k_j occurs in the principal diagonal position and all the other elements in the column are zero, i.e., A has the desired diagonal form

$$\hat{A} = \begin{pmatrix} k_1 & 0 & \ldots 0 \\ 0 & k_2 & \ldots 0 \\ \ldots & \ldots & \ldots \\ 0 & 0 & \ldots k_n \end{pmatrix} = (k_j \delta_{j,l}). \tag{II.100}$$

In the following proof it will be assumed that A is normal, i.e., by definition, that A commutes with $\bar{A}'$

$$A\bar{A}' = \bar{A}'A \tag{II.101}$$

If A is a Hermitian matrix, i.e., if $A = \bar{A}'$, we have a special case of a normal matrix.

If U is an arbitrary unitary matrix, i.e., if $\bar{U}' = U^{-1}$ (as in No. 26) and so also $|\det \bar{U}'| = |\det U| = 1$, then U is not singular, and under the unitary transformation U the matrix A goes into $B = U^{-1}AU$. Using the rule for the transpose of a product, $(CD)' = D'C'$, we have immediately $\bar{B}' = U^{-1}\bar{A}'U$. The equations can, of course, be solved for A and $\bar{A}'$, e.g., $A = UBU^{-1}$. Hence we see that $B = \bar{B}'$ if and only if $A = \bar{A}'$, so the property of a matrix of being Hermitian is invariant under a unitary transformation. It also immediately follows that

$$\begin{cases} B\bar{B}' = U^{-1}A\bar{A}'U \\ \bar{B}'B = U^{-1}\bar{A}'AU, \end{cases}$$

i.e., $B\bar{B}' = \bar{B}'B$ if and only if $A\bar{A}' = \bar{A}'A$, or in other words, the property of a matrix of being normal is invariant under unitary transformations.

We next prove the property of a normal matrix that is can be transformed into diagonal form under a unitary transformation, by induction from $n-1$ to n. The assertion is true for a one-rowed matrix, since we can take the unit matrix as the unitary matrix. A has at least one characteristic number k_1 and a corresponding eigen-vector y_1 which is already normalized (i.e., the sum of the squares of the components equals 1). Further vectors $z_2, \ldots, z_n$ can be chosen so that the matrix $T = (y_1 | z_2 | \ldots | z_n)$ is unitary. We have merely to complete y_1 by means of the vectors $z_2, \ldots, z_n$ into an orthonormal system (ONS) of vectors. If under the transformation T the matrix A is transformed into a new matrix $F = T^{-1}AT$, then F has the unit vector e_1 as an eigenvector corresponding to the characteristic number k_1, because $Fe_1 = k_1e_1$ by (II.99), and by the considerations at the beginning of this section F therefore has the form

$$F = \left(\begin{array}{c|c} k_1 & f_2 \ldots f_n \\ \hline 0 & \\ \vdots & C \\ 0 & \end{array}\right), \tag{II.102}$$

where the other elements in the first row have been denoted by $f_2, \ldots, f_n$, and the remaining $(n-1)$-rowed square matrix has been denoted by C. Then F' has the form

$$\bar{F}' = \left(\begin{array}{c|c} k_1 & 0 \ldots 0 \\ \hline \bar{f_2} & \\ \vdots & \bar{C}' \\ \bar{f_n} & \end{array}\right),$$

and since normality is preserved under a unitary transformation,

$$F\bar{F}' = \bar{F}'F.$$

Of the matrices standing here on both sides only the first main-diagonal element is used; we calculate immediately

$$FF' = \left(\begin{array}{c|c} |k_1|^2 + \sum_{\nu=2}^{n} |f_\nu|^2 & \dots \\ \hline \dots & \dots \end{array}\right), \qquad \bar{F}'F = \left(\begin{array}{c|c} |k_1|^2 & \dots \\ \hline \dots & \dots \end{array}\right).$$

and so

$$|k_1|^2 + \sum_{\nu=2}^{n} |f_\nu|^2 = |k_1|^2$$

or $f_2 = \dots = f_n = 0$. The transformed matrix is therefore of the form

$$F = \left(\begin{array}{c|c} k_1 & 0 \dots 0 \\ \hline 0 & \\ \vdots & C \\ 0 & \end{array}\right),$$

and since in forming the products $F\bar{F}'$, $\bar{F}'F$ the first row and the other rows do not become mixed up with one another, it also follows that $C\bar{C}' = \bar{C}'C$, i.e., C is an $(n-1)$-rowed normal matrix. But by the induction hypothesis for $(n-1)$-rowed normal matrices it is already true that they can be transformed into diagonal form by means of unitary matrices. Therefore there is a unitary matrix R with $R^{-1}CR = \tilde{D}$, where $\tilde{D}$ is a diagonal matrix

$$\tilde{D} = \begin{pmatrix} d_2 & & 0 \\ & \ddots & \\ 0 & & d_n \end{pmatrix}.$$

The matrix R is now extended into an n-rowed matrix S (likewise unitary) by bordering it with the elements shown below:

$$S = \left(\begin{array}{c|c} 1 & 0 \dots 0 \\ \hline 0 & \\ \vdots & R \\ 0 & \end{array}\right).$$

The matrix F is then transformed into diagonal form by the unitary transformation S: $S^{-1}FS = D$, where the diagonal matrix D has $k_1, d_2, \dots, d_n$ as its principal-diagonal elements.

Now the matrix $U = TS$ is formed. Unitary matrices form a group relative to multiplication; the product of two unitary matrices TS is again unitary, as we can check immediately:

$$\bar{U}'U = (\overline{TS})'TS = \bar{S}'\bar{T}'TS = \bar{S}'S = E.$$

The original matrix A is transformed by U into the diagonal form D:

$$U^{-1}AU = S^{-1}T^{-1}ATS = S^{-1}FS = D.$$

If we introduce new co-ordinates z by $y = Uz$, where the matrix U has the column vectors $u_1, \ldots, u_n$, $U = (u_1 \mid \ldots \mid u_n)$, then for the transformed matrix D the eigenvalue problem reads $Dz_j = d_j z_j$, with $d_1 = k_1$. This has the unit vectors e_j as the eigen-vectors. If we now transform back to the co-ordinate system y, the e_j transform into the eigen-vectors $y_j = u_j$, and since U is a unitary matrix, the original matrix A therefore has an orthonormal system of pairwise unitary eigen-vectors. Since the characteristic numbers are unaltered in this process, it follows that $d_j = k_j$ also for $j = 2, \ldots, n$.

The converse can now be proved similarly. Let A now be a matrix which has an orthonormal system of pairwise unitary eigen-vectors y_j. Then $U = (y_1 \mid \ldots \mid y_n)$ is thus a unitary matrix which transforms A into the diagonal form

$$D = \begin{pmatrix} k_1 & & & 0 \\ & k_2 & & \\ & & \ddots & \\ 0 & & & k_n \end{pmatrix} = U^{-1}AU \tag{II.103}$$

But a diagonal matrix is always a normal matrix (since by definition of a matrix product we have immediately $D\bar{D}' = \bar{D}'D$), and since normality is preserved under unitary transformations, A too is a normal matrix. For a matrix A, therefore, the properties that $A\bar{A}' = \bar{A}'A$ and that there is an orthonormal system of pairwise unitary eigen-vectors are equivalent.

Now let A be Hermitian, i.e., $A = \bar{A}'$, and let y_j be an eigen-vector of A corresponding to the characteristic number k_j, $Ay_j = k_j y_j$. We can then form the number $s = \bar{y}_j' A y_j$, which represents a matrix with just one element, and which therefore is equal to the reflected matrix. So $s = \bar{y}_j' A y_j = s' = (\bar{y}_j' A y_j)'$ $= y_j' A' \bar{y}_j = \bar{s}$ and $s = \bar{s}$ means that s is real. On the other hand, $s = \bar{y}_j' k_j y_j = k_j \mid y_j \mid^2$. Since not all the components of an eigen-vector can vanish, $\mid y_j \mid^2$ being the square of the length of the eigen-vector must be positive; hence k_j is real.

Now we also prove the converse. Let A be a normal matrix which has only real characteristic numbers k_j. It can be brought unitarily into diagonal form according to (II.103). A diagonal matrix D with real elements in the principal diagonal is, however, Hermitian: $D = \bar{D}'$, and since the property of a matrix of being Hermitian is invariant under unitary transformations, A too is Hermitian. Hermitian matrices are therefore characterized by the fact that among the normal matrices they are precisely those which have real characteristic numbers.

If A is an arbitrary real matrix, i.e., a matrix with real elements a_{jk}, and if A has a real characteristic number k_j, there is no reason at all why a

Matrix A	Characterization by means of A', $\bar{A}$, $\bar{A}'$	A is transformable by means of a matrix T to $\hat{A} = T^{-1}AT$		ONS = orthonormal system EV = eigen-vectors A has	
Real symmetric	$A = A' = \bar{A}$	T real orthogonal	$\hat{A}$ real diagonal matrix	A real ONS of EV	Only real characteristic numbers k_j
Hermitian	$A = \bar{A}'$	T unitary		An ONS of EV	
Normal	$A\bar{A}' = \bar{A}'A$		$\hat{A}$ diagonal matrix		
Normalizable		T non-singular		n linearly independent EV	
Not normalizalbe			$\hat{A}$ Jordan normal form	Not n linearly independent EV, but principal vectors of higher than the first degree	

corresponding eigen-vector y_j should be real. But the vector w_j which has as its components the real parts of the individual components of y_j, $w_j = \text{Re}\, y_j$ (or $w_j = \text{Im}\, y_j$ if $\text{Re}\, y_j$ should be zero) is a real eigen-vector of A corresponding to the characteristic number k_j. For a real matrix, therefore, corresponding to a real characteristic number k_j there is always at least one real eigen-vector. However, we cannot obtain the principal-axes transformation for real symmetric matrices A from the theorem for Hermitian matrices simply by taking the real part of all the numbers; for, if two complex eigen-vectors are unitary to one another, their real parts are not in general orthogonal to one another. However, we can validate the assertions that a real symmetric matrix has an orthonormal system of real eigen-vectors, that its characteristic numbers are all real, and that the matrix can be brought into diagonal form by a real, orthogonal transformation, by repeating the proof given in this No. in the reals, i.e., by simply leaving out the bars everywhere, replacing 'unitary' by 'orthogonal', 'normal' by 'real-normal', etc.

Altogether, we arrive at the classification of the separate types of matrices shown on page 129.

§8. LINEAR DIFFERENTIAL EQUATIONS WITH PERIODIC COEFFICIENTS

30. ENGINEERING EXAMPLES LEADING TO DIFFERENTIAL EQUATIONS WITH PERIODIC COEFFICIENTS

In recent times differential equations with periodic coefficients have acquired great importance. Here we shall consider only linear systems of such equations; in matrix notation as in equation (II.75), let all the given functions have the same period, say 2π:

$$y'(x) = A(x)\, y(x) + p(x) \quad \text{with } A(x+2\pi) = A(x),\ p(x+2\pi) = p(x). \tag{II.104}$$

Example 1. Metrology

By periodic shaking of a measuring instrument a false setting of the pointer can be caused. If we consider the pointer of the instrument to be a pendulum, then any arbitrary position of the pointer can be stable if the point of suspension is agitated in a suitable direction periodically with a suitable frequency. The system is idealized as a compound pendulum, with moment of inertia θ, in a gravitational field; its weight is G, mass m, and s is the distance from the point of rotation O to the centre of gravity S, see Fig. II.18. The point of rotation O, is made to execute a harmonic motion $u \cos \Omega t$ along a line making an angle δ with the vertical. (This example is taken from K. Klotter, *Technische Scwingungslehre,* vol. 1, 2nd edition, Berlin–Göttingen, Heidelberg 1951, p. 355.)

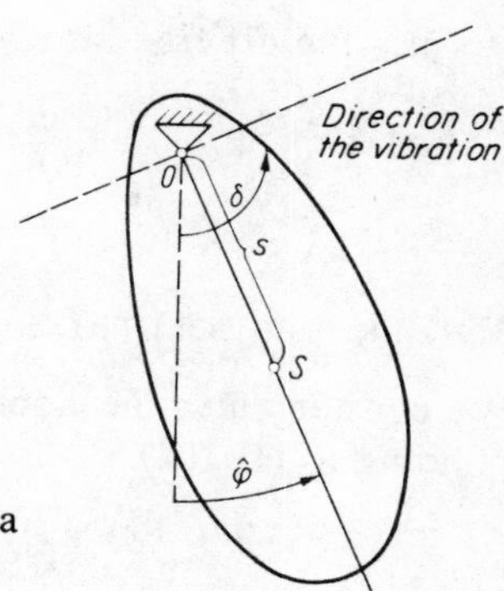

Fig. II.18. Periodic vibration of the support of a compound pendulum

If $\hat{\varphi}$ is the angle which the straight line OS makes with the vertical at time t, the equation of motion in the moving co-ordinate system is

$$\theta\ddot{\hat{\varphi}} + Gs \sin \hat{\varphi} - mus\Omega^2 \sin(\delta - \hat{\varphi}) \cos \Omega t = 0; \tag{II.105}$$

For small motions $\varphi(t)$ in the neighbourhood of a fixed mean position α with $\hat{\varphi} = \alpha + \varphi$, $\varphi \ll 1$ we can linearize (II.105):

$$\begin{aligned}\theta\ddot{\varphi} + \varphi(Gs \cos \alpha + mus\Omega^2 \cos(\delta - \alpha) \cos \Omega t)\\ = -Gs \sin \alpha + mus\Omega^2 \sin(\delta - \alpha) \cos \Omega t.\end{aligned} \tag{II.106}$$

In the particular case where the mean position and the direction of vibration lie in the vertical, $\delta = 0$ and $\alpha = 0$ or $\alpha = \pi$, this equation reduces to a simple Mathieu differential equation (cf. (II.116))

$$\theta\ddot{\varphi} \pm \varphi(Gs + mus\Omega^2 \cos \Omega t) = 0. \tag{II.107}$$

Example 2

Klotter (loc. cit.) has discussed various other examples, such as a strut under pulsating longitudinal forces, cables under varying tension, etc.

Example 3

A microphone circuit includes a current source U, a microphone of resistance $R_M = R_0 + R_1 \cos \omega t$ when it is excited at frequency ω, a further ohmic resistance R_2 and a self-inductance L, see Fig. II.19. The current intensity i

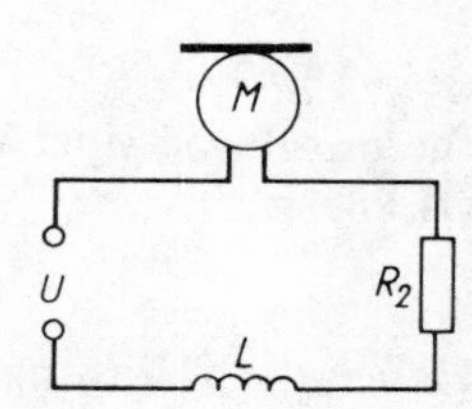

Fig. II.19. Microphone circuit

satisfies the differential equation

$$L\frac{di}{dt} + (R_0 + R_1 \cos \omega t + R_2)\, i = U.$$

31. Periodic solutions of the homogeneous system

We consider first the homogeneous system of differential equations corresponding to (II.104)

$$y'(x) = A(x)y(x) \quad \text{where } A(x+2\pi) = A(x). \tag{II.108}$$

Just as in (II.76), let

$$Y(x) = (y_{(1)}(x), \ldots, y_{(n)}(x)) \quad \text{with det } Y(x) \neq 0 \tag{II.109}$$

be a matrix whose column vectors $y_{(\nu)}(x)$ form a fundamental system.

By the results of No. 23, $Y(x+2\pi)$ is then also a matrix with a fundamental system of column vectors, and every $y_{(\nu)}(x+2\pi)$ is a linear combination of the $y_{(1)}(x), \ldots, y_{(n)}(x)$, i.e., there is a 'transition matrix' $B = (b_{jk})$ with constant elements b_{jk} and det $B \neq 0$ such that

$$Y(x+2\pi) = Y(x)B. \tag{II.110}$$

Every solution $y(x)$ of (II.108) can be expressed in terms of $y_{(\mu)}(x)$ by means of a constant vector w:

$$y(x) = Y(x) \cdot w,$$

and we then have

$$y(x+2\pi) = Y(x+2\pi)\, w = Y(x) \cdot B \cdot w.$$

We are now interested in the question whether there are solutions $y(x)$ which reproduce themselves (up to a constant factor k) after the expiry of one period:

$$y(x+2\pi) = ky(x). \tag{II.111}$$

If $k = 1$ such solutions are called *periodic solutions of the first kind;* if $k \neq 1$ (and also for complex k) they are called *periodic solutions of the second kind,* although these periodic solutions of the second kind are periodic only if there is a natural number r such that $k^r = 1$ (k is then an rth root of unity). For $k = -1$, the solution is said to be *semi-periodic.* From

$$Y(x)\, Bw = kY(x)\, w$$

there follows, since det $Y(x) \neq 0$, the well-known eigenvalue equation, as in (II.88)

$$Bw = kw.$$

Hence the following theorem holds.

Theorem. *The homogeneous system of differential equations* (II.108) *with periodic coefficients has at least one periodic solution of the second kind* $y(x)$ *with*

$$y(x+2\pi)=\mathrm{e}^{2\pi\mu}y(x), \tag{II.113}$$

where the 'characteristic exponent' μ *is calculated from a characteristic number* k *of the transition matrix* B *by means of* $k=\mathrm{e}^{2\pi\mu}$. *A periodic solution of the first kind exists if and only if* $k=1$ *is a characteristic number of the transition matrix* B.

This theorem includes the Floquet theorem, which G. Floquet proved in 1884 for a single differential equation of higher order.

32. Stability

By means of the Jordan normal form we can easily describe the structure of the solutions more precisely. As in (II.97) let T be a constant, non-singular matrix which transforms B into the Jordan normal form $\hat{B}=J$ (cf. No. 26, case 3):

$$\hat{B}=T^{-1}BT=J. \tag{II.114}$$

Here J consists of a number of 'Jordan blocks', and has the form shown below.

If the matrix B is normalizable, then J is a diagonal matrix and contains in the principal diagonal the characteristic numbers of B, and zeros elsewhere.

$Y(x)T$ is now again a matrix whose columns form a fundamental system:

$$Z(x)=Y(x)T=(z_{(1)}(x),\ldots,z_{(r_1)}(x),\ldots),$$

and we have

$$Z(x+2\pi)=Y(x+2\pi)T=Y(x)BT=Z(x)T^{-1}BT=Z(x)J \tag{II.115}$$

or, written out for a Jordan block,

$$\begin{aligned}
z_{(1)}(x+2\pi)&=k_1z_{(1)}(x)\\
z_{(2)}(x+2\pi)&=k_1z_{(2)}(x)+z_{(1)}(x)\\
&\cdots\cdots\cdots\cdots\cdots\cdots\cdots\cdots\\
z_{(r_1)}(x+2\pi)&=k_1z_{(r_1)}(x)+z_{(r-1)}(x).
\end{aligned}$$

$$J=\left|\begin{array}{c|c|c}
\begin{matrix}k_1 & 1 & 0\\ & \ddots\ \ddots & 1\\ 0 & & k_1\end{matrix} & 0 & 0\\
\hline
0 & \begin{matrix}k_2 & 1 & 0\\ & \ddots\ \ddots & 1\\ 0 & & k_2\end{matrix} & 0\\
\hline
0 & 0 & \ldots
\end{array}\right|
\begin{matrix}r_1\text{ rows}\\[3em] r_2\text{ rows}\\[3em] \ \end{matrix}$$

If the matrix B is normalizable, there are n linearly independent solutions $z_{(\upsilon)}(x)$ (for $\upsilon = 1, 2, \ldots, n$), which are all periodic solutions of the first or second kind.

Definition

The system (II.108) is said to be *stable* if every solution remains bounded as $x \to +\infty$ and *unstable* if there is at least one solution which is unbounded as $x \to +\infty$.

Now, by (II.115),

$$Z(x + 2m\pi) = Z(x) J^m \qquad (m = 1, 2, 3, \ldots).$$

When J^m is formed, the individual Jordan blocks are raised to the mth power separately and do not mix with one another. For the mth power of an individual block, for $m > r$, we have, writing k, r instead of k_1, r_1,

$$\begin{bmatrix} k^m & \binom{m}{1} k^{m-1} & \binom{m}{2} k^{m<2} & \ldots & \binom{m}{r-1} k^{m-r+1} \\ 0 & k^m & \binom{m}{1} k^{m-1} & \ldots & \\ \cdots & \cdots & \cdots & \cdots & \cdots \\ 0 & 0 & 0 & \ldots & k^m \end{bmatrix}.$$

Hence, it follows for the asymptotic behaviour of the components $z_{(p)}(x)$ $\quad (p = 1, 2, \ldots, r)$ corresponding to this Jordan block that

for $|k| < 1$ $\quad z_{(p)}(x) \to 0$ as $x \to +\infty$

for $|k| = 1$ $\quad$ and $r = 1$, $z_{(1)}(x)$ remains bounded, but for $r > 1$, $z_{(r)}(x)$ is not bounded, and

for $|k| > 1$, $\quad z_{(p)}(x)$ is not bounded as $x \to +\infty$.

If the matrix B is normalizable, the result is easily formulated: there is stability if and only if all the characteristic numbers k of B have modulus $|k| \leq 1$.

The stability circumstances have been particularly studied for the Mathieu differential equation

$$y''(x) + (\lambda + \gamma \cos x)\, y(x) = 0, \tag{II.116}$$

which is a particular case of Hill's differential equation

$$y'' + (\lambda + \varphi(x))y = 0 \quad \text{with } \varphi(x + 2\pi) = \varphi(x), \tag{II.117}$$

How complicated the circumstances become is shown by the Strutt diagram, Fig. II.20, in which in the λ, γ-plane the regions with λ, γ pairs for which (II.116) has stable solutions are shown shaded (Klotter, loc. cit., p. 366).

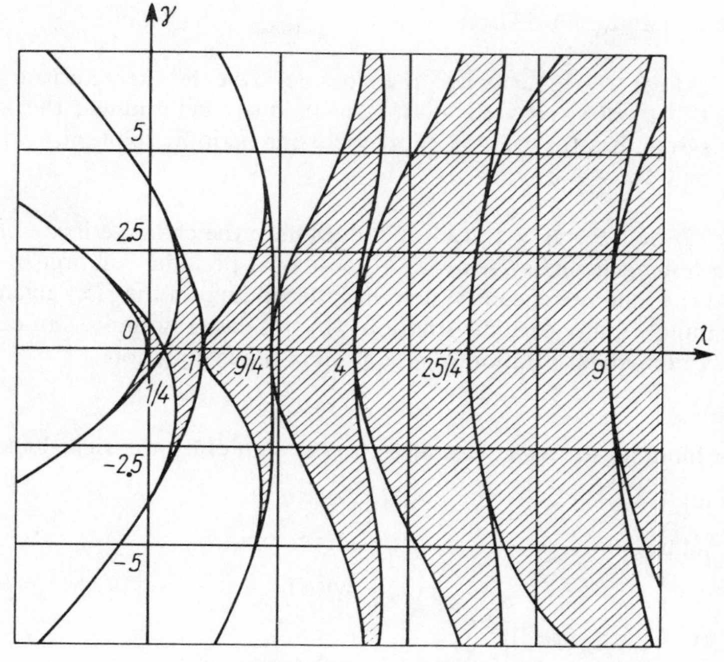

Fig. II.20. The Strutt diagram

33. Periodic solutions for the inhomogeneous system

The considerations now link up directly with No. 31 again, and they do not require the Jordan normal form. The solution of the initial-value problem for the inhomogeneous system (II.104)

$$y'(x) = A(x)y(x) + r(x), \qquad y = y(0) \quad \text{for } t = 0$$

reads, according to (II.84), (II.85)

$$y(x) = Y(x)[Y^{-1}(0)y(0) + \int_0^x Y^{-1}(s)r(s)\,ds]. \tag{II.118}$$

here we are interested in the question whether there are periodic solutions; this is equivalent to $y(2\pi) = y(0)$; with (II.110) this means

$$y(0) \equiv Y(0)\,Y^{-1}(0)\,y(0) = Y(0)\,B[Y^{-1}(0)y(0) + q]$$

with

$$q = \int_0^2 Y^{-1}(s)r(s)\,ds,$$

Since $\det Y(0) \neq 0$ it follows from this that

$$(B - E)Y^{-}(0)\,y(0) = -\,Bq. \tag{II.119}$$

We thus obtain two different cases:

Case 1. If $\det(B-E) \neq 0$, i.e., if B does not have the characteristic number 1, then $Y^{-1}(0)y(0)$ and with it $y(0)$ are uniquely determined; the inhomogeneous system has for every $r(x)$ precisely one periodic solution, and in particular $y \equiv 0$ for $r \equiv 0$.

Case 2. If $\det(B-E) = 0$, i.e., if B does have the characteristic number 1, then the homogeneous system has a non-trivial periodic solution of the 1st kind and the inhomogeneous system has, depending on the $r(x)$ given, either no or infinitely many periodic solutions. In this case resonance can occur: let $p(t)$ be a non-trivial solution of the homogeneous equation

$$p' = A(x)p;$$

then the inhomogeneous equation with $r(x) = p(x)$ on the right-hand side

$$y' = A(x)y + p(x)$$

has the solution

$$y(x) = xp(x), \tag{II.120}$$

as may be verified directly.

34. An Example for the Stability Theory

As a very simple example which can be worked out in closed form we consider an undamped oscillator and suppose that the restoring force is switched off for periodic time intervals; we choose 1 as the period of the whole time cycle, and so

$$\left.\begin{aligned} u'' + a^2u = 0 \quad &\text{for } 0 < x < b \text{ and generally for } m < x < m+b \\ u'' = 0 \quad &\text{for } b < x < 1, \text{ and generally for } m+b < x < m+1 \\ m = 0, 1, 2, \ldots. & \end{aligned}\right\},$$

We can, of course, also write this as a system for a vector

$$y = \begin{pmatrix} u \\ u' \end{pmatrix}.$$

With prescribed initial values $u(0) = u_0$, $u'(0) = u_0'$ we have, in the first interval $0 < x < b$,

$$y = \begin{pmatrix} u_0 \cos(ax) + \dfrac{u_0'}{a} \sin(ax) \\ -au_0 \sin(ax) + u_0' \cos(ax) \end{pmatrix}$$

with final end-values

$$y(b) = \begin{pmatrix} u_0 \cos(ab) + \dfrac{u_0'}{a} \sin(ab) \\ -au_0 \sin(ab) + u_0' \cos(ab) \end{pmatrix} = \begin{pmatrix} u_b \\ u_b' \end{pmatrix};$$

with these values as the new initial values in the interval $b < x < 1$ we integrate $u'' = 0$ and obtain in that interval

$$y(x) = \begin{pmatrix} u_b + u_b'(x-b) \\ u_b' \end{pmatrix}$$

and so, at the end-point $x = 1$, of the period

$$y(1) = \begin{pmatrix} u_0[\cos(ab) - a(1-b)\sin(ab)] + u_0'\left[\dfrac{1}{a}\sin(ab) + (1-b)\cos(ab)\right] \\ -u_0 a \sin(ab) + u_0' \cos(ab) \end{pmatrix}$$
$$= B\begin{pmatrix} u_0 \\ u_0' \end{pmatrix}.$$

The expressions multiplying u_0 and u_0' respectively are the elements of the transition matrix B.

The characteristic numbers k of the matrix B now have to be calculated, i.e., the zeros of

$$\varphi(k) = \det(B - kE)$$
$$= \begin{vmatrix} \cos(ab) - a(1-b)\sin(ab) - k & \dfrac{1}{a}\sin(ab) + (1-b)\cos(ab) \\ -a\sin(ab) & \cos(ab) - k \end{vmatrix} = 0;$$

we find

$$\varphi(k) = k^2 - 2\rho k + 1 = 0$$

with

$$2\rho = 2\cos(ab) - a(1-n)\sin(ab).$$

The roots $k = \rho \pm (\rho^2 - 1)^{1/2}$ are complex for $|\rho| < 1$ and lie on the unit circle, and the system is then stable. For $|\rho| > 1$ a root k is greater than 1 in absolute value, and the system is then unstable. In Fig. II.21 the shaded areas of stability are separated from the areas of instability by the curves

$$2\rho = 2\cos(ab) - a(1-b)\sin(ab) = \pm 2\,ab,$$

the boundary line $b = 0$ belongs to the domain of instability; the curves $\rho = \pm 1$ consist of the hyperbolas $ab = p\pi\,(p = 1, 2, 3 \ldots)$ and of further branches with the parametric representation (p is the parameter)

$$a = p + \frac{2(\cos p \pm 1)}{\sin p}, \qquad b = \frac{p}{a}.$$

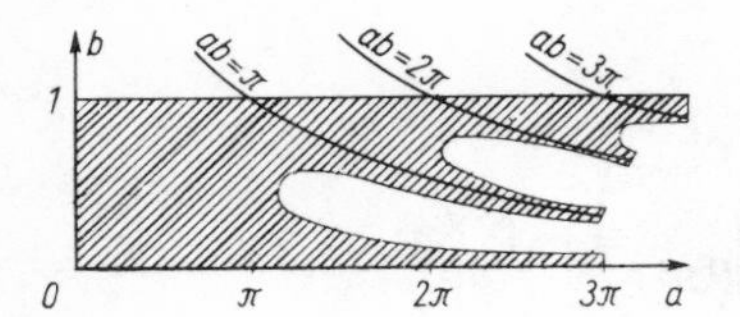

Fig. II.21. The stability regions

§9. THE LAPLACE TRANSFORMATION

The Laplace transformation is frequently used to determine the solutions of certain classes of initial-value problems, for both ordinary and partial differential equations, and also for certain types of integral equations and other functional equations. With ordinary differential equations it is used principally with inhomogeneous linear differential equations with constant coefficients and with given initial conditions. Among the many spheres of application of Laplace transformations we may mention particularly electric circuitry, automatic control technology, stability questions, etc.

The following treatment may serve as an introduction for the user. The underlying mathematical theory is gone into only so far as seems necessary for a proper understanding of the process and for the simplest applications; the difficult general theory of integral transformations cannot be dealt with here.

35. Definition of the Laplace transformation

The Laplace transformation associates with a given function $f(x)$ of a real variable x a function $F(s)$ according to the transformation rule

$$F(s) = \int_0^\infty f(x)\,\mathrm{e}^{-sx}\,\mathrm{d}x = \mathscr{L}[f(x)]. \tag{II.121}$$

$F(s)$ is called the *Laplace transform* of $f(x)$, and the operator $\mathscr{L}$ which associates with $f(x)$ the function $F(s)$ is called the *Laplace operator*. $f(x)$ is called the *original function*, and $F(s)$ the *image function*. The inverse transformation is denoted by $\mathscr{L}^{-1}$, provided it exists and is uniquely determined (see No. 40 on this question):

$$f(x) = \mathscr{L}^{-1}[F(s)].$$

As an illustration let us consider a simple example:

$$f(x) = a\,\mathrm{e}^{cx}, \quad \text{where } a \text{ and } c \text{ are constants.}$$

Then we obtain, for $s > c$,

$$F(s) = \int_0^\infty a\,\mathrm{e}^{cx}\,\mathrm{e}^{-sx}\,\mathrm{d}x = \int_0^\infty a\,\mathrm{e}^{-(s-c)x}\,\mathrm{d}x = \left[\frac{a\,\mathrm{e}^{-(s-c)\,x}}{c-s}\right]_{x=0}^{x=\infty}$$

$$= \frac{a}{s-c}. \tag{II.122}$$

Here we have used the fact that

$$\lim_{x \to \infty} e^{-(s-c)x} = 0;$$

this still remains true in the complex domain, provided Re $s >$ Re c.

For $c = k + i\omega$ with real k, ω (and $a = 1$) (II.122) gives

$$\mathscr{L}[e^{(k+i\omega)x_1}] = \mathscr{L}[e^{kx}\cos(\omega x) + i\, e^{kx}\sin(\omega x)] = \frac{1}{s-k-i\omega}$$

$$= \frac{s-k+i\omega}{(s-k-i\omega)(x-k+i\omega)} = \frac{s-k+i\omega}{(s-k)^2+\omega^2}.$$

Separating the real and imaginary parts gives the much-used formulae

$$\mathscr{L}[e^{kx}\cos(\omega x)] = \frac{s-k}{(s-k)^2+\omega^2}, \qquad \mathscr{L}[e^{kx}\sin(\omega x)] = \frac{\omega}{(s-k)^2+\omega^2}. \tag{II.123}$$

By a linear combination of these two formulae we hence obtain the following useful formula for the inversion of the Laplace transformation:

$$\mathscr{L}^{-1}\left[\frac{As+B}{(s-k)^2+\omega^2}\right] = e^{kx}[A\cos(\omega x) + (Ak+B)\sin(\omega x)]. \tag{II.124}$$

The operator $\mathscr{L}$ is a linear operator; this means that, if $f_\nu(x)$, $\nu = 1, 2, \ldots, p$ are functions for which the Laplace transforms $F_\nu(s)$ exist

$$\mathscr{L}[f_\nu(x)] = F_\nu(s)$$

then, for any (real or complex) numbers a_ν

$$\mathscr{L}\left[\sum_{\nu=1}^{p} a_\nu f_\nu(x)\right] = \sum_{\nu=1}^{p} a_\nu F_\nu(s).$$

In particular, we have

$$\mathscr{L}\left[\sum_{\nu=1}^{p} a_\nu e^{c_\nu x}\right] = \sum_{\nu=1}^{p} \frac{a_\nu}{s-c_\nu}$$

provided Re $s >$ Re c_ν for $\nu = 1, \ldots, p$.

For the question of the existence of the Laplace transform we use the following

Definition

A function defined for real, non-negative x is said to be *admissible* if

(1) $f(x)$ is *piecewise smooth* in every finite interval of the positive x-axis, and

(2) there are constants M, k (with $M > 0$, k real) such that

$$|f(x)| \le M e^{kx} \quad \text{for } 0 \le x < \infty. \tag{II.125}$$

A function $f(x)$ *is pieceswise smooth in an interval I* if I can be split into a finite number of sub-intervals and if $f(x)$ is continuously differentiable in the interior of each sub-interval and has finite limit-values as x approaches the boundary points of each sub-interval.

A sufficient criterion. Every admissible function $f(x)$ has a Laplace transform $F(s)$ for $s > k$. Because, for every $w > 0$, we have

$$\int_0^w |f(x)|\, \mathrm{e}^{-sx}\, dx \leq \mathrm{M} \int_0^w \mathrm{e}^{kx}\, \mathrm{e}^{-sx}\, dx \leq M \int_0^w \mathrm{e}^{(k-s)x}\, dx = \frac{M}{s-k}.$$

It follows that, as $w \to \infty$, $F(s) \leq M/(s-k)$.

At the same time the proof shows that the improper integral (II.121) is absolutely convergent for every admissible function $f(x)$, and that it converges uniformly for $s \geq \hat{s} > k$, where $\hat{s}$ is a fixed number $> k$. If complex quantities are brought in, it follows that the integral is absolutely and uniformly convergent for $\mathrm{Re}\, s \geq \hat{s} > k$.

The definition given here of admissible functions suffices for most applications. Thus, for example, all polynomials are admissible, and so are all periodic functions which are piecewise continuous in the period interval.

Lemma. If $f(x)$ is admissible, then so is

$$g(x) = \int_a^x f(u)\, du, \quad \textit{where } a \textit{ is a fixed number.}$$

Proof. Let $f(x)$ satisfy the estimate (II.125), where we can take $k > 0$. We can then make the estimate, for $x \geq a$,

$$|g(x)| = \left| \int_a^x f(u)\, du \right| \leq \int_a^x |f(u)|\, du \leq M \int_a^x \mathrm{e}^{ku}\, du = M \frac{\mathrm{e}^{kx} - \mathrm{e}^{ka}}{k}$$

$$\leq M \frac{\mathrm{e}^{kx} + \mathrm{e}^{kx}}{k} = \frac{2M}{k} \mathrm{e}^{kx}.$$

$g(x)$ therefore satisifies the estimate (II.125). The other condition(1) is clearly satisfied.

If $f_1(x)$ and $f_2(x)$ are admissible, then it follows immediately from the definition that $f_1(x) + f_2(x)$ and $f_1(x) \cdot f_2(x)$ are also admissible.
Hence

$$f(x) = \sum_{\nu=1}^{p} P_\nu(x)\, \mathrm{e}^{c_\nu x}, \qquad \text{(II.126)}$$

where the c_ν are (real or complex) constants, and the $P_\nu(x)$ are polynomials in x, is also admissible. Hence also every solution $y(x)$ of an ordinary linear inhomogeneous differential equation

$$\sum_{\nu=1}^{n} p_\nu y^{(\nu)}(x) = r(x)$$

with constant coefficients p_ν ($p_n \neq 0$) and $r(x)$ a continuous admissible function is admissible, because every solution of the homogeneous equation ($r(x) \equiv 0$) is of the form (II.126), and in solving the inhomogeneous equation ($r(x) \not\equiv 0$) by the method of variation of the constants (§5, No. 16), in carrying out the integrations we remain in the domain of admissible functions.

As a further example, we shall calculate the Laplace transform of x^ν for a fixed x and with $\operatorname{Re} \nu > 0$. For this purpose we require the gamma function $\Gamma(\nu)$, which also crops up in various applications. This function can be defined when $\operatorname{Re} \nu > 0$ by the then convergent integral

$$\Gamma(\nu) = \int_0^\infty e^{-t} t^{\nu-1} \, dt$$

and for $\operatorname{Re} \nu < 0$ (but with $\sin(\pi\nu) \neq 0$) by $\Gamma(\nu) \cdot \Gamma(1-\nu) = \dfrac{\pi}{\sin(\pi\nu)}$.

For integral values $\nu = 1, 2, \ldots$ we obtain by repeated integration by parts

$$\Gamma(\nu) = (\nu - 1)!.$$

For $\operatorname{Re} \nu > 0$ the substitution $t = s \cdot x$ with x as the new variable of integration gives

$$\Gamma(\nu) = \int_0^\infty e^{-sx} s^\nu x^{\nu-1} \, dx = s^\nu \cdot \mathscr{L}[x^{\nu-1}]$$

or

$$\mathscr{L}[x^{\nu-1}] = \frac{\Gamma(\nu)}{s^\nu} \quad \text{for } \operatorname{Re} s > 0 \text{ and } \operatorname{Re} \nu > 0.$$

36. Differentiation and Integration of the Original Element

Theorem. *If $f(x)$ is continuous and if its derivative $f'(x)$ is an admissible function, the following relation holds:*

$$\mathscr{L}[f'(x)] = s \cdot \mathscr{L}[f(x)] - f(0). \tag{II.127}$$

Proof. Since $f'(x)$ is assumed to be an admissible function, $f(x)$ as the integral of $f'(x)$ is also, by the lemma, an admissible function. For every $w > 0$ we then have by integration by parts

$$\int_0^w f'(x) e^{-sx} \, dx = \Big[f(x) e^{-sx} \Big]_{x=0}^{x=w} + s \int_0^w f(x) e^{-sx} \, dx.$$

Now letting $w \to \infty$ we obtain

$$\int_0^\infty f'(x) e^{-sx} \, dx = -f(0) + s \int_0^\infty f(x) e^{-sx} \, dx,$$

which is precisely the formula (II.127) we had to prove.

We shall make several applications of this important formula.

(I) If $g(x)$ is the integral of the function $f(x)$

$$g(x) = \int_0^x f(u)\,du \quad \text{with } g(0) = 0,$$

then (II.127) gives, when we replace $f(x)$ by $g(x)$

$$\mathscr{L}[g'(x)] = \mathscr{L}[f(x)] = s \cdot \mathscr{L}[g(x)] - g(0) = s \cdot \mathscr{L}\left[\int_0^x f(u)\,du\right]$$

or

$$\mathscr{L}\left[\int_0^x f(u)\,du\right] = \frac{1}{s} \cdot \mathscr{L}[f(x)]. \qquad \text{(II.128)}$$

(II) For $f(x) = x^n\,e^{cx}$ where n is an integer ≥ 1 (II.127) gives

$$\mathscr{L}[nx^{n-1}\,e^{cx} + cx^n\,e^{cx}] = n \cdot \mathscr{L}[x^{n-1}\,e^{cx}] + c \cdot \mathscr{L}[x^n\,e^{cx}] = s \cdot \mathscr{L}[x^n\,e^{cx}].$$

Hence we have the recursion formula

$$\mathscr{L}[x^n\,e^{cx}] = \frac{n}{s-c}\,\mathscr{L}[x^{n-1}\,e^{cx}] \quad \text{for Re } s > 0.$$

It follows that

$$\mathscr{L}[x^n\,e^{cx}] = \frac{n}{s-c}\,\frac{n-1}{s-c}\,\mathscr{L}[x^{n-2}\,e^{cx}] = \cdots = \frac{n(n-1)\ldots 1}{(s-c)^n}\,\mathscr{L}[e^{cx}],$$

or, by (II.122),

$$\mathscr{L}[x^n\,e^{cx}] = \frac{n!}{(s-c)^{n+1}} \quad (\text{Re } s > 0). \qquad \text{(II.129)}$$

(III) Now suppose $f''(x)$ exists and is admissible; then (II.127) gives, on replacing f by f',

$$\mathscr{L}[f''(x)] = s \cdot \mathscr{L}[f'(x)] - f'(0) = s^2 \cdot \mathscr{L}[f(x)] - s \cdot f(0) - f'(0),$$

Continuing this process we have the following

Theorem. Let the nth derivative $f^{(n)}$ of a function $f(x)$ be admissible ($n = 1, 2, \ldots$), and let $f^{(n-1)}(x)$ be continuous for $x > 0$ (and so $f(x)$, $f'(x), \ldots, f^{(n-2)}(x)$) are also continuous for $x > 0$). Then

$$\mathscr{L}[f^{(n)}(x)] = s^n \cdot \mathscr{L}[f(x)] - s^{n-1}f(0) - s^{n-2}f'(0) - \cdots - sf^{(n-2)}(0) - f^{(n-1)}(0). \qquad \text{(II.130)}$$

37. USING THE LAPLACE TRANSFORMATION TO SOLVE INITIAL-VALUE PROBLEMS FOR ORDINARY DIFFERENTIAL EQUATIONS

Using the auxiliary results already obtained we can now handle initial-value problems for ordinary linear differential equations (as in II.32, though there L had a different meaning)

$$Ly = \sum_{\nu=0}^{n} p_\nu y^{(\nu)}(x) = r(x), \qquad y^{(j)}(0) = y_{j,0} \qquad (j = 0, 1, \ldots, n-1) \qquad \text{(II.131)}$$

with constant coefficients p_ν (with $p_n \neq 0$). Here it is assumed that $r(x)$ is an admissible function, and that the initial values $y_{j,0}$ are given.

By No. 35 the solution $y(x)$ of the initial-value problem (II.131) is also admissible. We now apply the Laplace transformation to both sides of (II.131) and introduce the characteristic polynomial

$$P(k) = \sum_{\nu=0}^{n} p_\nu k^\nu$$

as in (II.35). By (II.130),

$$\mathscr{L}[p_\nu y^{(\nu)}] = p_\nu s^\nu \mathscr{L}[y] + q_\nu(s, y_{j,0}) \qquad (\nu = 0, 1, \ldots, n)$$

where $q_\nu(s, y_{j,0})$ is a polynomial in s of degree $\nu - 1$ with the given initial-values $y_{j,0}$ as coefficients:

$$q_\nu(s, y_{j,0}) = -p_\nu \sum_{\rho=0}^{\nu-1} s^{\nu-1-\rho} y_{\rho,0},$$

Let the Laplace transform of the given right-hand side $r(x)$ be

$$\mathscr{L}[r(x)] = R(s).$$

Under the Laplace transformation, equation (II.131) becomes

$$\mathscr{L}\left[\sum_{\nu=0}^{n} p_\nu y^{(\nu)}\right] = \sum_{\nu=0}^{n} p_\nu \mathscr{L}[y^{(\nu)}] = \sum_{\nu=0}^{n} (p_\nu s^\nu) \cdot \mathscr{L}[y] + Q(s)$$

$$= P(s)\,\mathscr{L}[y] + Q(s) = \mathscr{L}[r(x)] = R(s),$$

where $Q(s)$ arises as a combination of all the polynomials $q_\nu(s, y_{j,0})$ and it is therefore itself a polynomial in s. If all the initial values $y_{j,0}$ were zero, then $Q(s)$ would also be zero. Formally we have thus obtained the representation

$$Y = \mathscr{L}[y] = \frac{R(s)}{P(s)} - \frac{Q(s)}{P(s)} = M(s), \quad \text{say.} \qquad \text{(II.132)}$$

We thus know the Laplace transform Y of the required function y, and we have obtained the solution y formally in the form

$$y(x) = \mathscr{L}^{-1}[M(s)].$$

Example (I)

Consider the initial-value problem

$$y'' - 4y = e^{-3x}, \qquad y(0) = 1, y'(0) = -1.$$

We introduce the image function $Y(s) = \mathscr{L}[y]$ as the unknown. (II.130) here gives, with the use of (II.122),

$$\mathscr{L}[y'] = sY - y(0)$$

$$\mathscr{L}[y''] = s^2Y - sy(0) - y'(0)$$

$$\mathscr{L}[e^{-3}x] = R(s) = \frac{1}{s+3}$$

$$\mathscr{L}[y'' - 4y] = s^2Y - sy(0) - y'(0) - 4Y = R(s) = \frac{1}{s+3}.$$

Hence,

$$Y = \left(\frac{1}{s+3} + s \cdot -1\right)\frac{1}{s^2-4} = M(s) = \frac{s^2+2s-2}{(s+3)(s^2-4)}.$$

$M(s)$ has to be split into partial fractions; this can be done by well-known methods, and we find

$$M(s) = \frac{1}{5}\,\frac{1}{s+3} + \frac{1}{2}\,\frac{1}{s+2} + \frac{3}{10}\,\frac{1}{s-2}.$$

Since by (II, 122) the inverse transformation can be carried out on the individual fractions, we obtain immediately

$$y(x) = \mathscr{L}^{-1}[M(s)] = \frac{1}{5}\,e^{-3x} + \frac{1}{2}\,e^{-2x} + \frac{3}{10}\,e^{2x}.$$

(If the denominator of any of the partial fractions had had a multiple zero, we should have had to use (II.129).)

In this process (II.122) has been used in the form

$$\mathscr{L}^{-1}\left[\frac{a}{s-c}\right] = a\,e^{cx} \quad \text{for } s > \operatorname{Re} c,$$

while it is not yet certain whether the inverse transformation is single-valued. But in the present case it is already known from the theory of differential equations that the solution of the initial-value problem exists and is unique; so the conclusion we have drawn is justified.

Example (II)

If in $M(s)$ a denominator has a pair of conjugate complex zeros, but the coefficients of the differential equations are real, then the calculation can be carried out entirely in the reals by using formula (II.124). For the initial-value problem

$$y'' + 4y' + 5y = e^{-3x} \cos x, \qquad y(0) = 2, \qquad y'(0) = 1,$$

writing

$$\mathscr{L}[y] = Y, \qquad \mathscr{L}[y'] = sY - 2, \qquad \mathscr{L}[y''] = s^2 Y - s \cdot 2 - 1,$$

$$\mathscr{L}[e^{-3x} \cos x] = \frac{s+3}{(s+3)^2 + 1}$$

we obtain for Y the equation

$$s^2 Y - 2s - 1 + 4sY - 8 + 5Y = \frac{s+3}{(s+3)^2 + 1}$$

or

$$Y = \frac{1}{s^2 + 4s + 5}\left[2s + 9 + \frac{s+3}{(s+3)^2 + 1}\right]$$

$$= \frac{1}{5}\frac{9s + 46}{s^2 + 4s + 5} + \frac{1}{5}\frac{s+1}{(s+3)^2 + 1}$$

(II.124) gives immediately

$$y(x) = \frac{1}{5} e^{-2x}(9 \cos x + 28 \sin x) + \frac{1}{5} e^{-3x}(\cos x - 2 \sin x).$$

Example (III)

A system of two coupled oscillatory circuits (Fig. II.22).

For two coupled circuits with inductances L_k, resistances R_k, capacitances C_k, and currents $j_k(t)$, ($k = 1, 2$ in each case), and mutual inductance G, with an E.M.F. u applied as shown in Fig. II.22, the equations (corresponding to the last equation in No. 14) are

$$L_1 \ddot{j}_1 + R_1 \dot{j}_1 + C_1^{-1} j_1 + G \ddot{j}_2 = u,$$
$$G \ddot{j}_1 + L_2 \ddot{j}_2 + R_2 \dot{j}_2 + C_2^{-1} j_2 = 0.$$

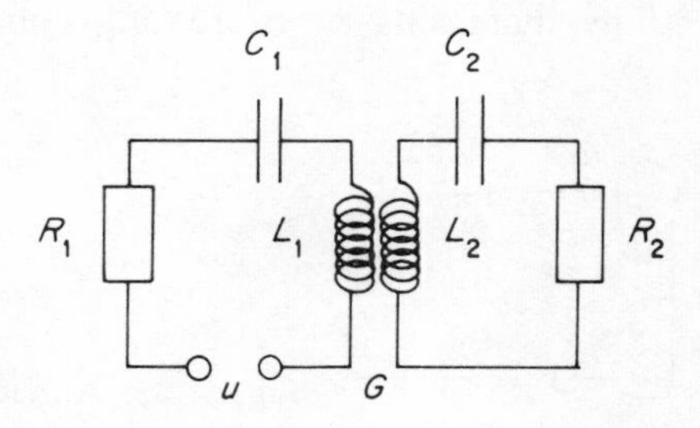

Fig. II.22. Coupled electrical circuits

let $\mathscr{L}[j_k] = J_k$, $\mathscr{L}[u] = U$, and let the initial values at $t = 0$ be $j_{ko}, \dot{j}_{ko}$. Then the Laplace transformation gives

$$L_1(s^2J_1 - sj_{10} - \dot{j}_{10}) + R_1(sJ_1 - j_{10}) + C_1^{-1}J_1 + G(s^2J_2 - sj_{20} - \dot{j}_{20}) = U,$$

$$G(s^2J_1 - sj_{10} - \dot{j}_{10}) + L_2(s^2J_2 - sj_{20} - \dot{j}_{20}) + R_2(sJ_2 - j_{20}) + C_2^{-1}J_2 = 0.$$

J_1 and J_2 can be found from these 2 linear equations and then the inverse transformation gives $j_1(t)$ and $j_2(t)$.

38. Transients and periodic solutions

I. Transients in the build-up of oscillations

Often in stable oscillatory processes only the steady state is of interest. This may be illustrated by the example of build-up of current in an electrical circuit having inductance L and resistance R (Fig. II.23); $u(t)$ is the E.M.F. and $j(t)$ the current strength. The equation now reads (in contrast to (I.7), u is not constant but sinusoidal):

$$L\frac{\mathrm{d}j}{\mathrm{d}t} + Rj = u = u_0 \sin(\omega t).$$

At time $t = 0$ let $j(0) = 0$. Writing $\mathscr{L}[j] = J$, we have from the Laplace transformation

$$LsJ(s) - Lj(0) + RJ(s) = \frac{u_0\omega}{s^2 + \omega^2},$$

or

$$J(s) = \frac{1}{Ls + R}\,\frac{u_0\omega}{s^2 + \omega^2}.$$

We express $J(s)$ in partial fractions with the constants a, b, c still to be determined:

$$J(s) = \frac{a}{Ls + R} + \frac{bs + c}{s^2 + \omega^2}.$$

The process described by $j(t)$ consists of the superposition of an exponentially increasing or decreasing component $j_e(t)$ and a steady (undamped)

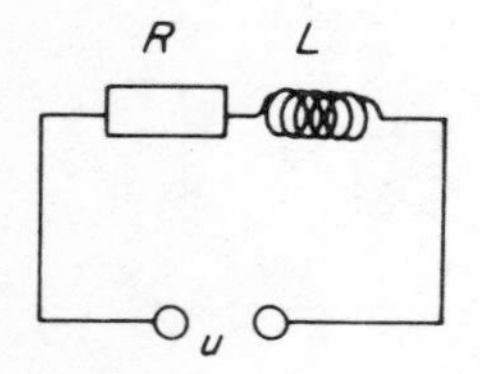

Fig. II.23. An electrical circuit

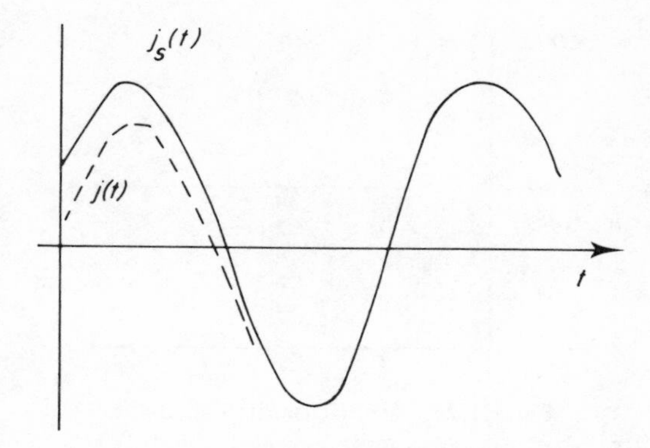

Fig. II.24. A build-up process

oscillatory component $j_s(t)$, which are calculated from

$$J_e = \frac{a}{Ls + R}, \qquad J_s = \frac{bs + c}{s^2 + \omega^2};$$

see Fig. II.24.

Since the denominator of J_e has a negative zero, $j_e(t)$ represents a transient component. The values of a, b, c are obtained immediately from

$$u_0\omega = a(s^2 + \omega^2) + (bs + c)(Ls + R).$$

Putting $s = -R/L$ we have

$$a = \frac{u_0\omega L}{R^2 + \omega^2 L^2};$$

and comparison of the coefficients of s^2 and s gives

$$b = -\frac{u_0\omega L}{R^2 + \omega^2 L^2}, \qquad c = -\frac{bR}{L} = \frac{u_0\omega R}{R^2 + \omega^2 L^2}.$$

The inverse transformation gives, by (II.124),

$$j(t) = \frac{u_0\omega L}{R^2 + \omega^2 L^2}\frac{1}{L}\mathrm{e}^{-Rt/L} + \frac{u_0\omega}{R^2 + \omega^2 L^2}(-L\cos\omega t + R\sin\omega t).$$

II. Periodic processes

Another useful formula is provided by the

Shift theorem. *Let $f(x)$ be an admissible function, with $t(x) = 0$ for $x < 0$, and let $F(s)$ be its Laplace transform. The function f_1 obtained by a shift through a positive amount $T, f_1(x) = f(x - T)$ (see Fig. II.25), has the Laplace transform*

$$\mathscr{L}[f(x - T)] = \mathrm{e}^{-sT}\mathscr{L}[f(x)]. \qquad \text{(II.133)}$$

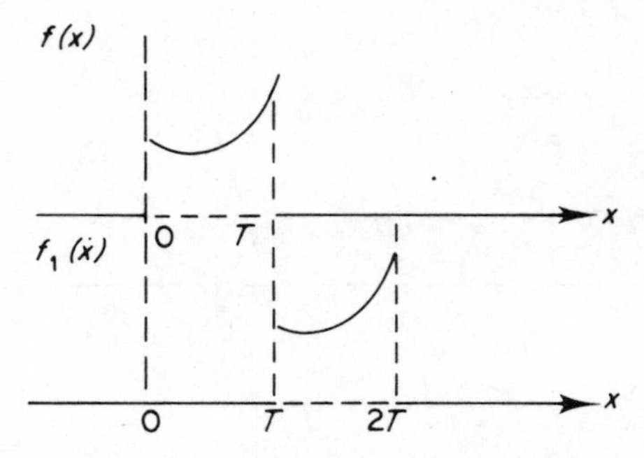

Fig. II.25. A shift or displacement

Proof. Writing $x - T = v$, we have

$$\mathscr{L}[f_1(x)] = \int_0^\infty f(x-T)\,\mathrm{e}^{-sx}\,\mathrm{d}x = \int_T^\infty f(x-T)\,\mathrm{e}^{-sx}\,\mathrm{d}x$$

$$= \int_0^\infty f(v)\,\mathrm{e}^{-s(T+v)}\,\mathrm{d}v = \mathrm{e}^{-sT}\int_0^\infty f(v)\,\mathrm{e}^{-sv}\,\mathrm{d}v = \mathrm{e}^{-sT}\mathscr{L}[f(x)].$$

Hence, we have immediately the following

Lemma

Let $f(x)$ be an admissible function with

$$f(x) = 0 \ \textit{for}\ x < 0 \ \textit{and for}\ x > T.$$

From $f(x)$ we form, for $x > 0$, the periodic function

$$f_T(x) = \sum_{k=0}^{\infty} f(x - kT), \qquad \textit{Fig. II.26.}$$

The corresponding Laplace transform is

$$\mathscr{L}[f_T(x)] = \frac{1}{1-\mathrm{e}^{-sT}}\mathscr{L}[f(x)]. \tag{II.134}$$

Proof. Let $f_{1T}(x) = f_T(x - T)$ be the function obtained by a shift from the periodic function $f_T(x)$, and put

$$\psi = \mathscr{L}[f(x)] = F_T(s).$$

Then, see Fig. II.26, we have

$$f_T(x) = f(x) + f_{1T}(x), \quad \text{and by (II.133)}$$
$$F_T(s) = \mathscr{L}[f(x)] + \mathrm{e}^{-sT}\mathscr{L}[f_T(x)] = \mathscr{L}[f(x)] + \mathrm{e}^{-sT}F_T(s),$$

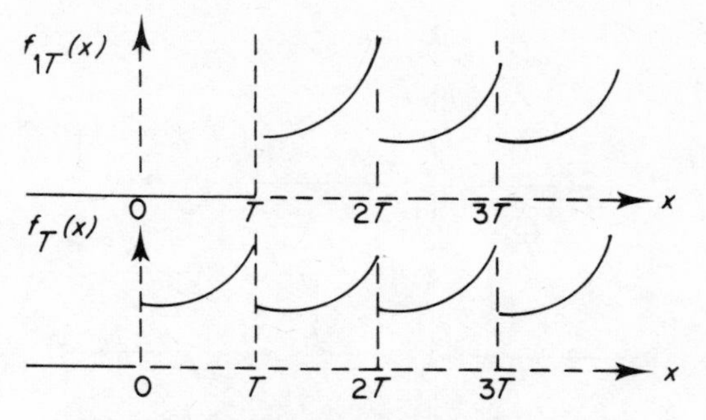

Fig. II.26. Laplace transformation of a periodic function

So

$$F_T(s) = \frac{1}{1 - e^{-sT}} \mathscr{L}[f(x)], \quad \text{as asserted.}$$

We can also put this result into the following form.

Lemma

Let $f(x)$ be admissible, $f(x) = 0$ for $x < 0$, and let f have the period T for $x > 0$. Then

$$\mathscr{L}[f(x)] = \frac{1}{1 - e^{-sT}} \int_0^T f(x)\, e^{-sx}\, dx. \qquad \text{(II.135)}$$

Example I

$$f(x) = \begin{cases} h = \text{const. for } 0 < a < T \\ 0 \text{ otherwise} \end{cases} \qquad \text{(see Fig. II.27).}$$

$$F(s) = \int_0^\infty f(x)\, e^{-sx}\, dx = h \int_0^a e^{-sx}\, dx = \frac{h}{s}(1 - e^{-sa}).$$

For the corresponding periodic function $f_T(x)$ we have, using (II.134),

$$\mathscr{L}[f_T(x)] = \frac{1 - e^{-sa}}{1 - e^{-sT}} \cdot \frac{h}{s}. \qquad \text{(II.136)}$$

If $a = T$, we obtain the earlier formula

$$\mathscr{L}[f_T(x)] = \mathscr{L}[h] = h/s.$$

Example II

In Example I, let $a \to 0$ and $h \to \infty$ keeping $ah = 1$. Then $f(x)$ becomes the

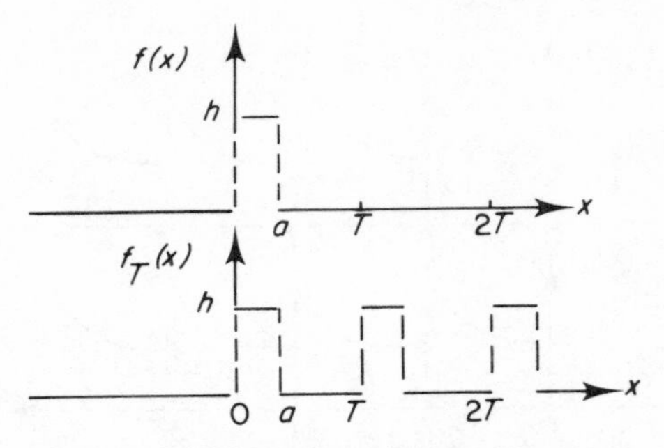

Fig. II.27. A periodic piecewise constant function

'δ-distribution', $\delta(x)$, also known as Dirac's delta function or the unit impulse. In this passage to the limit we obtain

$$\mathscr{L}[\delta(x)] = 1.$$

$\delta(x)$ is not a function. There is a far-reaching theory of distributions, but we cannot go into it here. We mention only the much-used relation

$$\int_{-\infty}^{\infty} \delta(x) f(x)\, \mathrm{d}x = f(0)$$

which holds for every function f continuous at $x = 0$.

Suppose now that unit impulses act at the times kT ($k = 0, 1, 2, \ldots$). The corresponding periodic (for positive t) distribution δ_T is represented symbolically in Fig. II.28; it has the Laplace transform

$$\mathscr{L}[\delta_T] = \frac{1}{1 - \mathrm{e}^{-sT}}.$$

Example III

$$f(x) = \begin{cases} \sin(\omega x) & \text{for } 0 \le x \le \pi/\omega \le T \\ 0 & \text{otherwise.} \end{cases} \qquad \text{(Fig. II.29)}$$

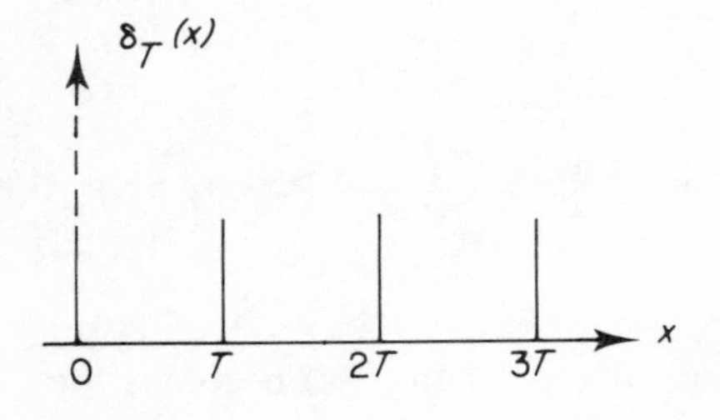

Fig. II.28. A periodic impulse

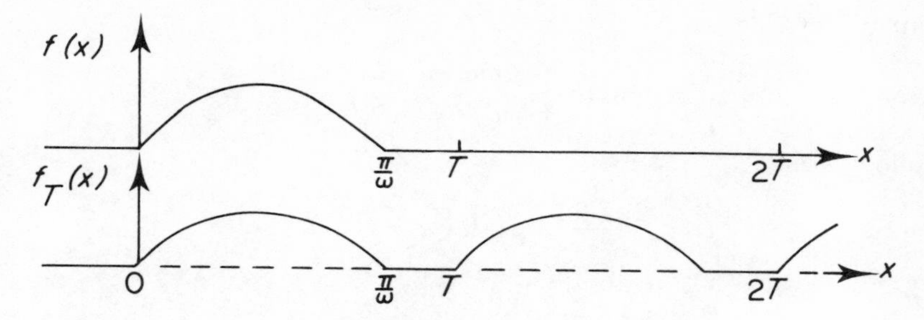

Fig. II.29. A periodic sinusoidal arc

Since $\sin(\omega x) = \mathrm{Im}(\mathrm{e}^{i\omega x})$ and $\mathrm{e}^{i\pi} = -1$, it follows that

$$\mathscr{L}[f(x)] = \mathrm{Im} \int_0^{\pi/\omega} \mathrm{e}^{i\omega x}\, \mathrm{e}^{-sx}\, \mathrm{d}x = \mathrm{Im}\left[\frac{\mathrm{e}^{(\mathrm{i}\omega - s)\ x}}{\mathrm{i}\omega - s}\right]_0^{\pi/\omega}$$

$$= \mathrm{Im}\, \frac{1}{\mathrm{i}\omega - s}\left(\mathrm{e}^{(\mathrm{i}\omega - s)\pi/\omega} - 1\right)$$

$$= \mathrm{Im}\, \frac{-s - \mathrm{i}\omega}{s^2 + \omega^2}\left(-\mathrm{e}^{-s\pi/\omega} - 1\right)$$

$$= \frac{\omega}{s^2 + \omega^2}\left(\mathrm{e}^{-s\pi/\omega} + 1\right).$$

By periodic continuation we obtain, from (II.134),

$$\mathscr{L}[f_T(x)] = \frac{1 + \mathrm{e}^{-s\pi/\omega}}{1 - \mathrm{e}^{-sT}} \cdot \frac{\omega}{s^2 + \omega^2}. \tag{II.137}$$

Example III(a)

Full-wave rectification of an alternating current (Fig. II.30)

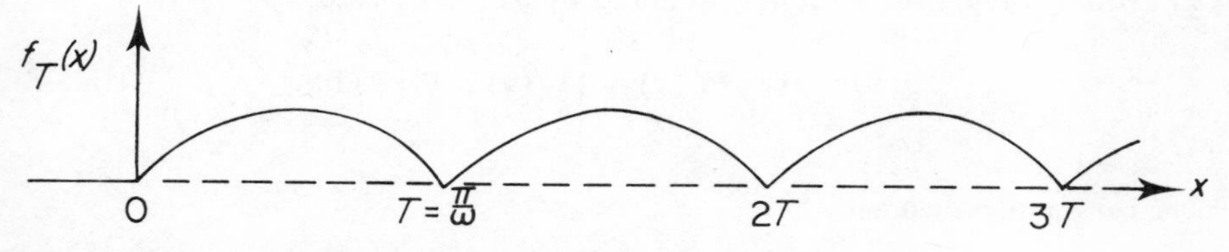

Fig. II.30. A full-wave rectified alternating current

For $T = \pi/\omega$.

$$f_T(x) = \begin{cases} |\sin(\omega x)| & \text{for } x \geq 0 \\ 0 & \text{for } x \leq 0 \end{cases}$$

it follows, since

$$\frac{e^\alpha + 1}{e^\alpha - 1} = \coth \frac{\alpha}{2},$$

that

$$\mathscr{L}[F_T(x)] = \frac{\omega}{s^2 + \omega^2} \coth\left(\frac{s\pi}{2\omega}\right).$$

Example III(b)

Half-wave rectification of an alternating current. Fig. II.31.

For $T = 2\pi/\omega$

$$f_T(x) = \begin{cases} \max(0, |\sin \omega x|) & \text{for } x \geq 0 \\ 0 & \text{for } x \leq 0 \end{cases}$$

$$\mathscr{L}[f_T(x)] = \frac{\omega}{s^2 + \omega^2} \frac{1}{1 - e^{-s\pi/\omega}}.$$

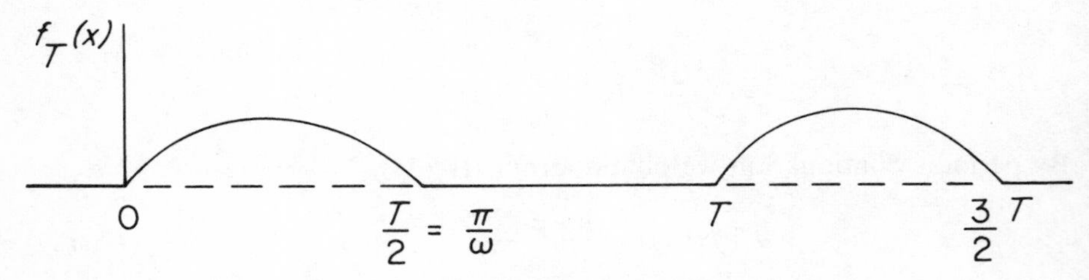

Fig. II.31. A half-wave rectified alternating current

39. The Convolution Theorem and Integral Equations of the First Kind

For the Laplace transformation a product in the image space corresponds to a 'convolution' in the original space. For two admissible functions $f(x)$ and $g(x)$ the *convolution product*, denoted by a $*$, is defined by

$$q(x) = f(x) * g(x) = \int_0^x f(u)g(x-u)\,du. \tag{II.138}$$

The convolution theorem

Let $f(x)$ and $g(x)$ be admissible functions. For their convolution product $q(x)$

defined by (II.138) the Laplace transform is given by

$$\mathscr{L}[q] = \mathscr{L}[f * g] = \mathscr{L}[f] . \mathscr{L}[g]. \tag{II.139}$$

Proof. Since values of f, g, and q appear here only for positive value of the argument x, we can take

$$f(x) = g(x) = q(x) = 0 \quad \text{for } x < 0,$$

and define the convolution product, and the Laplace transforms $\mathscr{L}[f]$ and $\mathscr{L}[g]$ too, by

$$q(x) = f(x) * g(x) = \int_{-\infty}^{\infty} f(u)\, g(x-u)\, \mathrm{d}u$$

$$\mathscr{L}[f] = \int_{-\infty}^{\infty} f(x)\, \mathrm{e}^{-sx}\, \mathrm{d}x, \qquad \mathscr{L}[g] = \int_{-\infty}^{\infty} g(x)\, \mathrm{e}^{-sx}\, \mathrm{d}x.$$

Since for large s the integrals are absolutely convergent, we can change the order of integration in

$$\mathscr{L}[q] = \int_{-\infty}^{\infty} q(x)\, \mathrm{e}^{-sx}\, \mathrm{d}x = \int_{-\infty}^{\infty} \left\{ \int_{-\infty}^{\infty} f(u) g(x-u)\, \mathrm{d}u \right\} \mathrm{e}^{-sx}\, \mathrm{d}x$$

$$= \int_{-\infty}^{\infty} \left\{ \int_{-\infty}^{\infty} g(x-u)\, \mathrm{e}^{-sx}\, \mathrm{d}x \right\} f(u)\, \mathrm{d}u.$$

In the inner integral we put $x = u + v$, and so it becomes

$$\mathrm{e}^{-su} \int_{-\infty}^{\infty} g(v)\, \mathrm{e}^{-sv}\, \mathrm{d}v = \mathrm{e}^{-su}\, \mathscr{L}[g];$$

and so

$$\mathscr{L}[q] = \int_{-\infty}^{\infty} \mathrm{e}^{-su}\, \mathscr{L}[g] f(u)\, \mathrm{d}u = \mathscr{L}[g]\, \mathscr{L}[f].$$

An integral equation of the first kind (distortion correction)

Through an apparatus K an input $i(x)$ is converted into an output $j(z)$, Fig. II.32, or, in another interpretation, a 'true' intensity distribution $i(x)$ is

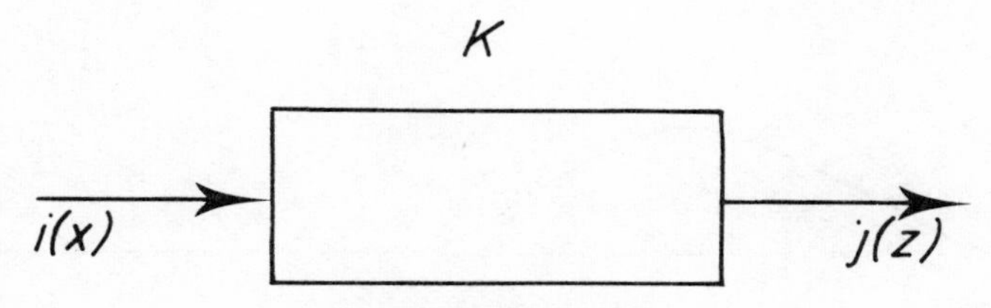

Fig. II.32. A transfer system

distorted by a measuring instrument to an 'observed' intensity distribution $j(z)$. It is required to deduce the true distribution $i(x)$ from the observed distribution $j(z)$. The distortion may arise, for example, in optical observation through a lens system, Fig. II.33. A true intensity 1 at the point $x = x_0$ may produce an intensity $k(|z - x_0|)$ at the point z, where k is a non-negative, known, smooth function depending on the apparatus. For example, in the Fraunhofer diffraction image of a single rectangular slit, we have

$$k(u) = c_1\left(\frac{\sin c_2 u}{u}\right)^2$$

where c_1, c_2 are constants. Altogether, the true intensity distribution $i(x_0)$ produces at the point z the intensity distribution

$$\int k(z - x_0)\, i(x_0)\, \mathrm{d}x_0$$

and we have to determine $i(x)$ for the measured $j(z)$ from the integral equation

$$j(z) = \int_{-\infty}^{\infty} k(z - x)\, i(x)\, \mathrm{d}x. \tag{II.140}$$

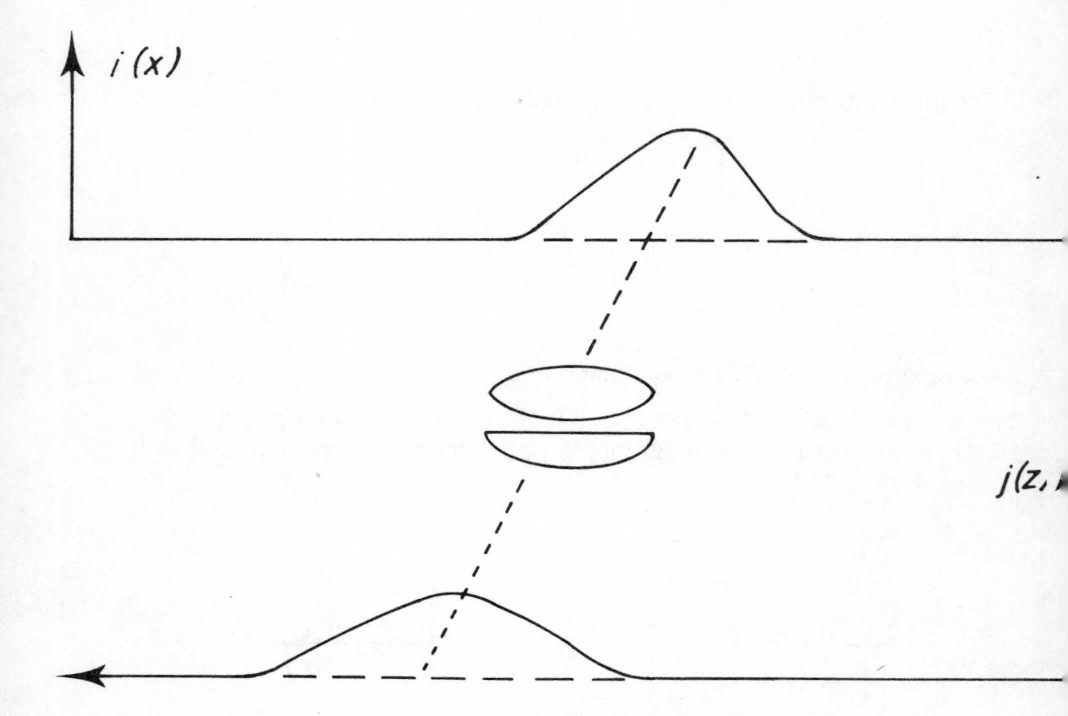

Fig. II.33. Mapping and distortion

Now suppose that both $i(x)$ and $j(z)$ differ from zero only in a finite interval. By shifting the origin of the x-axis and/or the z-axis, we can make $i(x)$ and $j(z) = 0$ for negative x and z; we can then write $\int_0^\infty$ instead of $\int_{-\infty}^\infty$,

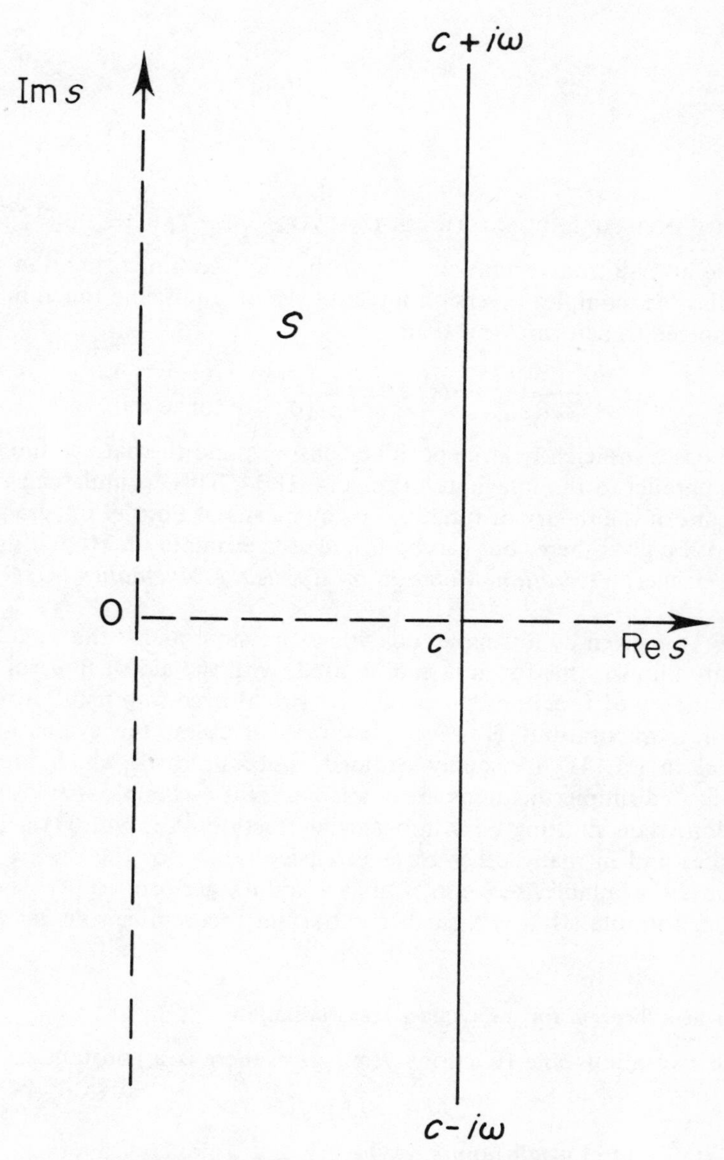

Fig. II.34. The complex inversion intergral

and we have the equation $j = k * i$. Writing the Laplace transforms as

$$\mathscr{L}[i] = I(s), \qquad \mathscr{L}[j] = J(s), \qquad \mathscr{L}[k] = K(s)$$

we have, by (II.139)

$$\mathscr{L}[j] = \mathscr{L}[k] \cdot \mathscr{L}[i], \quad \text{or } \mathscr{L}[i] = \frac{\mathscr{L}[j]}{\mathscr{L}[k]},$$

so

$$i(x) = \mathscr{L}^{-1}\left[\frac{J(s)}{K(s)}\right].$$

40. The inverse Laplace transformation and tables

For the inverse transformation $\mathscr{L}^{-1}$ of the Laplace transformation there is a formula, the 'complex inversion integral'. If an admissible function $f(x)$ has the Laplace transform $F(s)$, then

$$\frac{1}{2\pi i}\int_{c-i\infty}^{c+i\infty} F(s)\,\mathrm{e}^{sx}\,\mathrm{d}s = \begin{cases} f(x) & \text{for } x > 0, \\ 0 & \text{for } x < 0, \end{cases} \tag{II.141}$$

where c is a sufficiently large positive constant, and the path of integration is a line parallel to the imaginary axis, Fig. II.34. This formula can be proved by means of the theory of functions or by means of Fourier integrals; a proof will not be given here, but can be found, for example, in H. S. Carslaw and J. C. Jaeger: *Operation Methods in Applied Mathematics,* Oxford Univ. Press, 1941.

If $F(s)$ is given by a 'simple' analytic expression, and if the singularities of $F(s)$ are known, this formula can be used, with the aid of integral theorems in the theory of functions, to calculate $f(x)$, if necessary using formulae for numerical quadrature. However, in practical cases, the evaluation of the integral in (II.141) is usually avoided, since in cases which are not too complicated simpler methods are available. If, for example, $F(s)$ is a rational function, then splitting $F(s)$ into partial fractions (cf. No. 37) achieves the purpose, and in many other cases extensive tables for the reverse transformation are available, see, e.g., Carslaw and Jaeger, op. cit.

From formula (II.141) a further important theorem can be derived.

Uniqueness theorem for the Laplace transformation

If, for two admissible functions $f(x)$, $g(x)$ there is a constant s_0 such that

$$\mathscr{L}[f] = \mathscr{L}[g] \quad \text{for } s > s_0,$$

then $f(x) = g(x)$ at all points x where f and g are continuous.

This theorem, too, can be proved by functional analysis results without using the complex inversion integral (see the book by I. S. Sokolnikoff and R. M. Redheffer, *Mathematics of Physics and Modern Engineering,* McGraw-Hill, 1966, p. 217).

A table now follows of some selected functions and their Laplace transforms, including some transcendental functions such as Bessell functions, which will serve to illustrate the range of application of the Laplace transformation.

Particular examples of Laplace transformations

$f(x)$	$\mathscr{L}[f(x)] = F(s) = \int_0^\infty e^{-sx} f(x)\,dx$
I. General rules	
$f(cx)$	$\frac{1}{C} F\left(\frac{s}{c}\right), \quad c > 0$
$x^n f(x)$	$(-1)^n \frac{d^n F}{ds^n}, \quad n$ an integer ≥ 0
$e^{cx} f(x)$	$F(s-c), \quad c \geq 0$
$f'(x)$	$sF(s) - f(0)$
$f^{(n)}(x)$	$s^n F(s) - \sum_{\nu=0}^{n-1} s^{n-\nu-s} f^{(\nu)}(0)$
$\int_0^x f(u)\,du$	$\frac{1}{s} F(s)$
$f(x+c)$	$e^{cs}[F(s) - \int_0^c e^{-su} f(u)\,du]. \quad c \geq 0$
$\int_0^x f_1(u) f_2(x-u)\,du$	$F_1(s) \cdot F_2(s)$, where $F_\nu(s) = \mathscr{L}[f_\nu(x)], \quad \nu = 1, 2$
$\begin{cases} f(x-T) & \text{for } x \geq T \\ 0 & \text{for } 0 < x < T \end{cases}$	$e^{-Tc} F(s), \quad c > 0, T \geqslant 0$
$f_T(x) = \sum_{k=0}^{\infty} f(x-kT)$	$\frac{1}{1-e^{-sT}} F(s)$

II. Particular function pairs

		Formulae valid for $\operatorname{Re} s > \sigma$ where $\sigma =$
$\delta(x)$	1	
x^n	$\dfrac{n!}{s^{n+1}}$, $\quad n$ an interger ≥ 0	0
$x^{\nu-1}$	$\dfrac{\Gamma(\nu)}{s^\nu}$ $\quad \operatorname{Re} \nu > 0$	0
$x^n \, e^{cx}$	$\dfrac{n!}{(s-c)^{n+1}}$, $\quad n$ an integer > 0	$\operatorname{Re} c$
$a \, e^{cx}$	$\dfrac{a}{s-c}$	$\operatorname{Re} c$
$e^{kx} \cos(\omega x)$ $e^{kx} \sin(\omega x)$ $e^{kx} \cosh(\omega x)$	$\dfrac{s-k}{(s-k)^2+\omega^2}$ $\dfrac{\omega}{(s-k)^2+\omega^2}$ $\dfrac{s-k}{(s-k)^2-\omega^2}$ $\quad$ } More precise data about the range of validity can be found in the specialist literature	$\operatorname{Re} k$
$e^{kx} \sinh(\omega x)$	$\dfrac{\omega}{(s-k)^2-\omega^2}$	
$e^{kx} \dfrac{\sin(cx)}{x}$	$\tan^{-1}\left(\dfrac{c}{s-k}\right)$	
$\dfrac{e^{kx}}{\sqrt{x}}$	$\dfrac{\sqrt{\pi}}{\sqrt{s-k}}$	$\operatorname{Re} k$
$J_0(cx)$	$\dfrac{1}{\sqrt{s^2+c^2}}$	0
$J_0(2\sqrt{cx})$	$\dfrac{1}{s} e^{-c/s}$, $\quad c > 0$	0
$J_\nu(cx)$	$\dfrac{1}{\sqrt{s^2+c^2}} [\sqrt{s^2+c^2}-s]^\nu$, $\quad \operatorname{Re} \nu > -1$	
$\dfrac{J_\nu(cx)}{x}$	$\dfrac{1}{\nu}\left[\dfrac{c}{s+\sqrt{s^2+c^2}}\right]^\nu$, $\quad \operatorname{Re} \nu > -1$	0
$N_0(cx)$	$-\dfrac{2}{\pi}\dfrac{1}{\sqrt{s^2+c^2}} \ln\left[\dfrac{1}{c}(s+\sqrt{s^2+c^2})\right]$	

III

Boundary-value problems and, in particular, eigenvalue problems

1. Initial-value problems and boundary-value problems

As regards the adjustment of the n free constants in the general solution of an nth-order differential equation to the special physical or technical circumstances of a particular problem, one distinguishes three basic types of problem: 1. initial-value problems; 2. boundary-value problems and in particular, 3. eigenvalue problems.

We speak of an *initial-value problem* when all the conditions relate to the same value of x, or to one value of t for processes taking place in time. The latter is particularly the case in problems from dynamics.

For $x_0 = x$ then, we are given

$$y(x_0) = y_0, \qquad y'(x_0) = y_0', \qquad y''(x_0) = y_0'', \ldots, y^{(n-1)}(x_0) = y_0^{(n-1)} \quad \text{(III.1)}$$

cf. the examples in Chapter II. 14 on the equation for forced oscillations.

Boundary-value problems, in contrast to initial-value problems, usually arise only with differential equations of at least the second order. While initial conditions relate to a single point x_0, we speak of boundary conditions when in the equations given for the determination of the integration constants, the values of the unknown function and its derivatives at *at least two points* appear.

For differential equations of the second order the following nomenclature is customary;

A problem with the conditions	is called
$y(x_0) = y_0, \qquad y(x_1) = y_1$	a boundary-value problem of the first kind
$y'(x_0) = y_0', \qquad y'(x_1) = y_1'$	a boundary-value problem of the second kind

A problem with the conditions	is called
$\left.\begin{array}{l} c_0 y(x_0) + d_0 y'(x_0) = a \\ c_1 y(x_1) + d_1 y'(x_1) = b \end{array}\right\}$	a boundary-value problem of the third kind, or a Sturmian boundary-value problem

But other kinds of boundary conditions can arise; for example, periodicity:

$$y(x_0) = y(x_1), \qquad y'(x_0) = y'(x_1), \ldots, y^{(n-1)}(x_0) = y^{(n-1)}(x_1)$$

or integral conditions:

$$\int_{x_0}^{x_1} y(x) p(x) \, \mathrm{d}x = a,$$

where the constant a and the function $p(x)$ are given, or 'non-linear boundary conditions' such as $y(x_0) \cdot y'(x_0) = 1$.

A non-linear boundary condition also appears in (III.6). A boundary-condition is said to be linear when it is linear in all the values of y and its derivatives which appear in the condition. A boundary-value problem is linear when the differential equation and the boundary conditions are linear.

§1. EXAMPLES OF LINEAR BOUNDARY-VALUE PROBLEMS

2. A BEAM. THE SEVERAL FIELDS OF THE DIFFERENTIAL EQUATION

Boundary-value problems arise in numerous problems in statics. As an example we may mention the calculation of the deflection $y(x)$ of a beam of constant bending stiffness which is encastré at both ends and carries a load uniformly distributed over part of its length, Fig. III.1.

Since the system is (multiply) statically underdetermined, to calculate the bending-moment distribution (needed for calculating the stress) we have to bring in considerations of the elastic deformation. This leads to a fourth-order differential equation

$$y^{\mathrm{IV}} = \frac{p(x)}{EJ}. \tag{III.2}$$

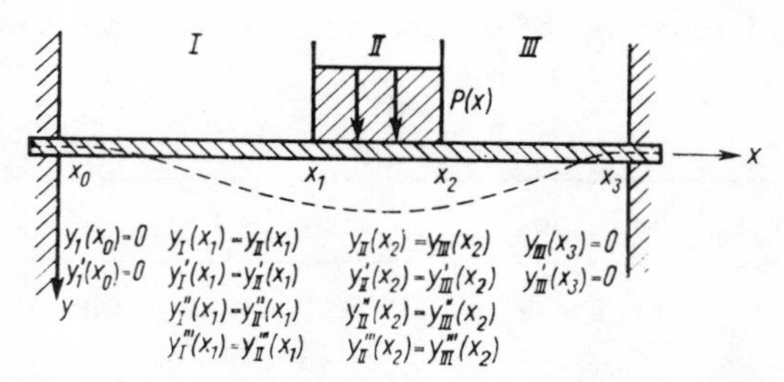

Fig. III.1. A beam encastré at both ends with a load uniformly distributed over part of its length

Because of the discontinuities in the load density $p(x)$ at the points x_1 and x_2 we have to apply this differential equation three times separately for the three fields or regions I, II, III. We thus obtain three general solutions each with four free constants, and therefore 12 free constants in all. The adjustment of the solutions in the particular case takes place by means of the 12 boundary conditions, which have been written out in Fig. III.1 in the appropriate places.

Deflection of a beam loaded at some intermediate point. As a simple example of a boundary-value problem with several sub-intervals we calculate the deflection of a beam, 3 m long, supported at each end, and loaded as shown in Fig. III.2 by a 3 t force. The weight of the beam itself may be neglected. Let J, the moment of inertia, about the neutral axis, of the beam's cross-section, and E, the modulus of elasticity of the material of the beam, be constant, so that $\alpha = EJ$, the bending stiffness is also constant. Because of the discontinuity in the shear force and the sudden change of the bending moment at the point where the applied force acts, we divide the beam into two intervals I and II, apply the differential equation separately to each, solve and adjust the solutions to the whole problem by means of the boundary conditions.

The differential equation for the flexure reads $-\alpha y'' = M$.

The forces A and B at the two ends can be written down immediately:

$$A = 2\ \text{t}, \qquad B = 1\ \text{t}.$$

The bending-moment distribution for the two intervals is then (taking $1m$ as the unit of length, and 1 t as the unit of force)

$$M(x) = \begin{cases} 2x & \text{for } 0 \le x \le 1 \quad \text{(interval I)} \\ 3 - x & \text{for } 1 \le x \le 3 \quad \text{(interval II)} \end{cases}$$

So the differential equations for the two intervals read:

$$y_{\text{I}}'' = 2x \qquad \text{for interval I}$$

$$y_{\text{II}}'' = -(3 - x) \quad \text{for interval II}$$

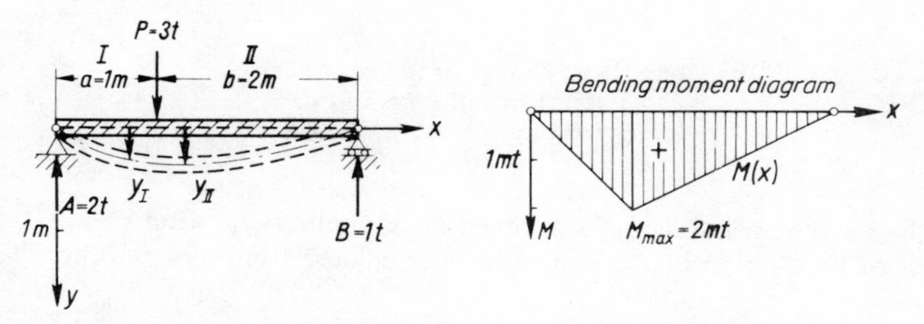

Fig. III.2. Beam with hinged supports and a single load

A double integration yields the general solutions

$$\alpha y_{\mathrm{I}} = -\frac{x^3}{3} + C_1 x + C_2 \qquad \text{for interval I}$$

$$\alpha y_{\mathrm{II}} = -\frac{3x^2}{2} + \frac{x^3}{6} + C_3 x + C_4 \quad \text{for interval II.}$$

The boundary conditions require that:

$$\left.\begin{array}{l} y_{\mathrm{I}}(0) = 0 \\ y_{\mathrm{II}}(3) = 0 \end{array}\right\} \quad \text{no bending at the supports}$$

$$\left.\begin{array}{l} y_{\mathrm{I}}(1) = y_{\mathrm{II}}(1) \\ y_{\mathrm{I}}'(1) = y_{\mathrm{II}}'(1) \end{array}\right\} \quad \text{deflection and slope at the load point continuous}$$

A short calculation shows that the constants have the values

$$C_1 = \frac{5}{3}, \qquad C_2 = 0, \qquad C_3 = \frac{19}{6}, \qquad C_4 = -\frac{1}{2}.$$

Hence the solution reads

$$y_{\mathrm{I}} = \frac{1}{6EJ}(10x - 2x^3)$$

$$y_{\mathrm{II}} = \frac{1}{6EJ}(x-3)(x^2 - 6x + 1).$$

In order to avoid showing the beam constants in the graph, $Y = y \cdot 6EJ$ has been plotted as the ordinates. We obtain the two cubical parabolas $Y_{\mathrm{I}}(x)$ and $Y_{\mathrm{II}}(x)$; $Y_{\mathrm{I}}(x)$ is valid only in the field I, $Y_{\mathrm{II}}(x)$ only in field II. The particular solution is represented in the diagram by the composite heavy line.

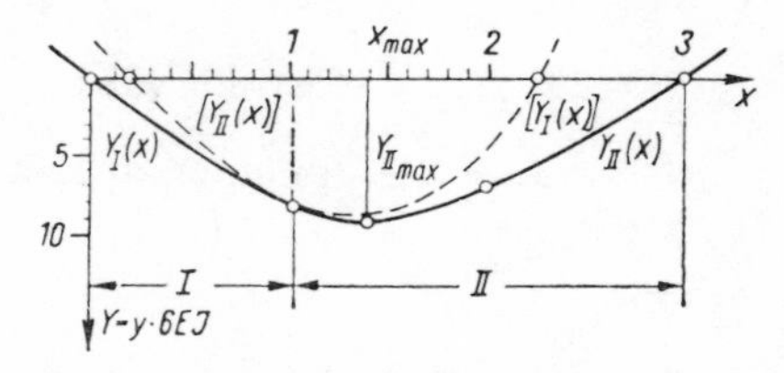

Fig. III.3. Flexure of the beam. The curve consists of parts of two different cubical parabolas in the two fields

The maximum deflection is attained at the point $x_{\max}$—and so not at the point of application of the load, and is calculated from $Y_{\mathrm{II}\max}$ to be

$$y_{\max} = Y_{\mathrm{II}\max} \cdot \frac{1}{6Ej} \approx 8.71 \cdot \frac{1}{6Ej} \approx \frac{1.45}{EJ}.$$

3. THE NUMBER OF SOLUTIONS IN LINEAR BOUNDARY-VALUE PROBLEMS

Typical cases occurring in linear boundary-value problems may be illustrated by means of a simple example. The differential equation

$$y'' + y = 0$$

has the general solution

$$y = C_1 \cos x + C_2 \sin x;$$

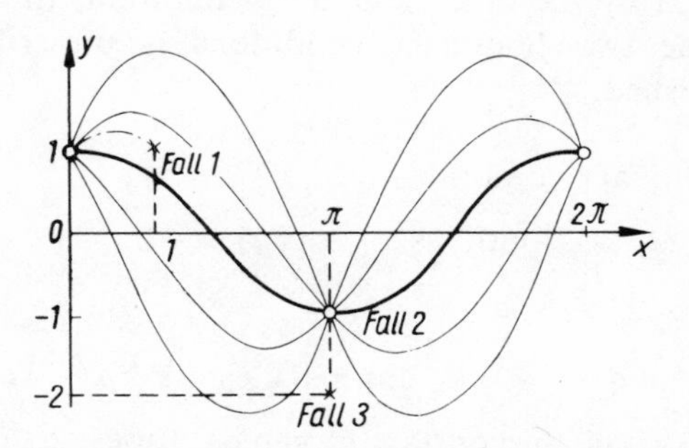

Fig. III.4. Linear boundary-value problems having no solution, one solution, or infinitely many solutions

Case 1. Let the boundary conditions be

$$y(0) = 1, \qquad y(1) = 1.$$

To determine the constants we substitute these in the general solution

$$1 = C_1 \cos 0 + C_2 \sin 0 = C_1 \cdot 1 + C_2 \cdot 0, \quad \text{therefore } C_1 = 1$$

and

$$1 = C_1 \cos 1 + C_2 \sin 1, \quad \text{therefore } C_2 = \frac{1 - \cos 1}{\sin 1} \approx 0.5463.$$

Hence the particular solution (shown by a dot-and-dash line in Fig. III.4) reads

$$y = \cos x = \frac{1 - \cos 1}{\sin 1} \sin x \approx \cos x + 0.5463 \sin x.$$

Case 2. The boundary conditions

$$y(0) = 1, \qquad y(\pi) = -1$$

give

$$1 = C_1 \cos 0 + C_2 \sin 0 = C_1 \cdot 1 + C_2 \cdot 0, \qquad \text{so } C_1 = 1, \text{ and}$$
$$-1 = C_1 \cos \pi + C_2 \sin \pi = 1 \cdot (-1) + C_2 \cdot 0, \qquad \text{i.e., } C_2 \text{ is undetermined.}$$

The constant C_2 can thus take any value, the equation is always satisfied. C_2 cannot be uniquely determined by the boundary conditions. We obtain as the solution $y = \cos x + C_2 \sin x$ not a single curve, but a one-parameter family of curves. Because of the remaining infinite indeterminateness such solutions are mostly unusable in practice, even though compared with the general solution there has been a reduction of the solution manifold.

Figure III.4 shows that, in our example, one of the boundary conditions $y(\pi) = -1$ accidentally coincides with the node of the family of solutions already determined by the other boundary condition; this means that when either one of the two boundary conditions is prescribed, the other is automatically satisfied.

Case 3. The boundary conditions

$$y(0) = 1, \qquad y(\pi) = -2$$

lead to

$$C_1 = 1 \quad \text{and} \quad -2 = C_1 \cos \pi + C_2 \sin \pi = 1(-1) + C_2 \cdot 0.$$

There is no way in which the constant C_2 can be chosen to satisfy this equation, i.e., there is no solution which will satisfy this boundary condition. This circumstance is immediately understandable if we look for the point in Fig. III.4 which is fixed by this boundary condition.

In No. 8 this circumstance will be investigated more closely.

§2. EXAMPLES OF NON-LINEAR BOUNDARY-VALUE PROBLEMS

4. THE DIFFERENTIAL EQUATION OF A CATENARY

As an example with an integral as the boundary condition the determination of a catenary may be mentioned; this is the curve assumed by a homogeneous, ideally flexible cable attached to two fixed points (x_1, y_1) and (x_2, y_2) and hanging under the influence of gravity only; see Fig. III.5.

If γ is the specific weight, F the cross-sectional area of the cable, and s the arc-length along the cable measured from some fixed point, then the weight of an arc-element of length Δs (see Fig. III.6) is

$$G = \gamma F \Delta s.$$

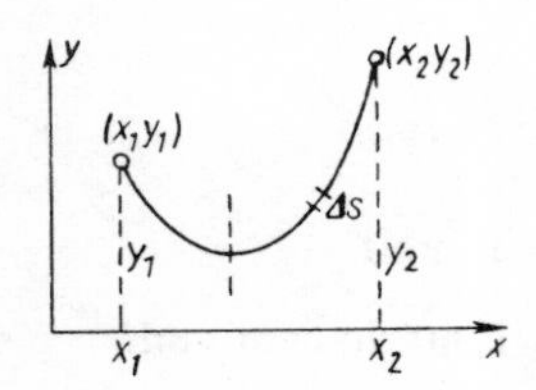

Fig. III.5. Boundary-value problem for a cable suspended from two points

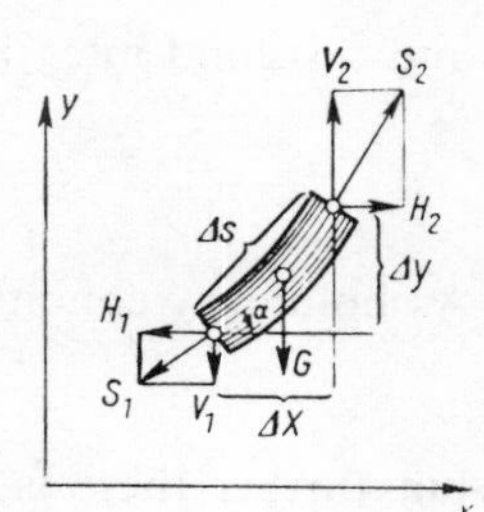

Fig. III.6. The forces acting on an element of the cable

On this element act the forces S_1 and S_2, with horizontal and vertical components H_1, V_1 and H_2, V_2 respectively, and the weight G. The conditions for equilibrium give

$$H_1 = H_2 = \text{const} = H, \qquad V_1 + G - V_2 = 0.$$

Since V is a function of x, we can write

$$V(x) + G - V(x + \Delta x) = 0.$$

From the curtailed Taylor expansion for the second term

$$\gamma F \Delta s + V(x) = V(x + \Delta x) \approx V(x) + V'(x)\Delta x$$

we get, on passage to the limit as $\Delta x \to 0$,

$$V'(x) = \gamma F \frac{\mathrm{d}s}{\mathrm{d}x}.$$

Now

$$V(x) = H \tan \alpha = Hy'(x), \quad \text{and so } V'(x) = Hy''(x).$$

Substitution yields $Hy'' = \gamma F \, \mathrm{d}s/\mathrm{d}x$. Since $\mathrm{d}s/\mathrm{d}x = \sqrt{1 + y'^2}$, we obtain as the differential equation of the catenary

$$y'' = a\sqrt{1 + y'^2}. \tag{III.3}$$

where

$$\frac{\gamma F}{H} = a.$$

Solution of the differential equation and adjustment to the boundary conditions. For the function $z(x) = y'(x)$ the variables separate:

$$a \, \mathrm{d}x = \frac{\mathrm{d}z}{\sqrt{1 + z^2}}.$$

Integration gives

$$z = y' = \sinh a(x + C_1) \tag{III.4}$$

and a second integration yields

$$y = C_2 + \frac{1}{a} \cosh a(x + C_1).$$

As boundary conditions we have, firstly,

$$y(x_1) = y_1, \qquad y(x_2) = y_2, \tag{III.5}$$

but further there is the condition that the arc-length between the points of suspension must be equal to the actual length l of the cable, Fig. III.7. So, as an auxiliary condition, we have the arc-length integral

$$\int_{x_1}^{x_2} \sqrt{1 + y'^2}\, dx = l. \tag{III.6}$$

There are thus *three* boundary conditions, but on the other hand, there is, as well as the two free constants C_1 and C_2, the unknown parameter a in the differential equation (III.3) still to be determined; a is connected with the horizontal tension H.

On substituting for y' from (III.4) and using

$$\cosh^2\varphi - \sinh^2\varphi \equiv 1, \tag{III.7}$$

we find the integral (III.6) simplifies to

$$\int_{x_1}^{x_2} \cosh a(x + C_1)\, dx = \frac{1}{a} [\sinh a(x + C_1)]_{x_1}^{x_2} = l.$$

For an actual case with definite values of the co-ordinates and a definite length l of the cable, we thus have three transcendental equations from which to determine C_1, C_2 and a.

For cables with only a shallow sag we can approximate to the catenary by a parabola, but for large sags we have to work with the hyperbolic functions. The system of three equations can easily be transformed into three new equations from which the unknowns a, C_1, C_2 can be successively determined. From (III.5) and (III.8) we obtain, by using the addition formula for hyperbolic functions,

$$\begin{aligned} y_2 - y_1 &= \frac{1}{a} [\cosh a(x_2 + C_1) - \cosh a(x_1 + C_1)] \\ &= \frac{2}{a} \sinh a \frac{x_2 - x_1}{2} \cdot \sinh a \left(\frac{x_2 + x_1}{2} + C_1\right), \end{aligned}$$

$$\begin{aligned} l &= \frac{1}{a} [\sinh a(x_2 + C_1) - \sinh a(x_1 + C_1)] \\ &= \frac{2}{a} \sinh a \frac{x_2 - x_1}{2} \cdot \cosh a \left(\frac{x_2 + x_1}{2} + C_1\right). \end{aligned}$$

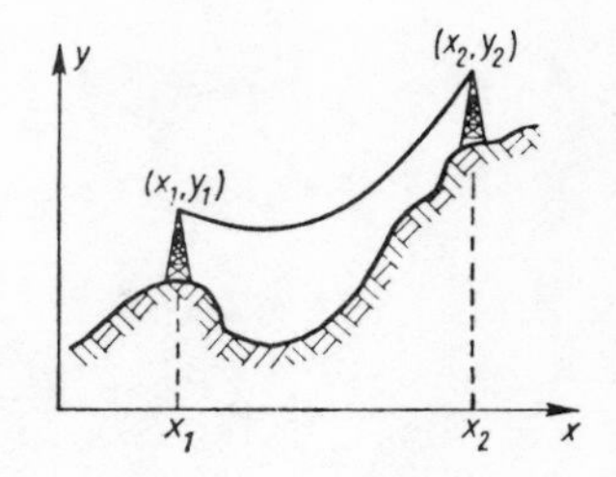

Fig. III.7. Catenary between two points of suspension of different heights

Now squaring each of these and subtracting to get $l^2 - (y_2 - y_1)^2$, and using (III.7), we find that C_1 drops out, and we have a transcendental equation in which only the number a is still unknown:

$$\frac{2}{a} \sinh \left(a \frac{x_2 - x_1}{2} \right) = \sqrt{l^2 - (y_2 - y_1)^2}. \tag{III.9}$$

If both points of suspension are at the same height, the square root on the right is, of course, simply equal to l.

A numerical example

$$x_1 = y_1 = 0, \qquad x_2 = 200\,[\text{m}], \qquad y_2 = 100\,[\text{m}], \qquad l = 250\,[\text{m}].$$

With

$$\frac{x_2 - x_1}{2}\, a = 100a = z$$

(III.9) becomes

$$\frac{\sinh z}{z} = 1.1456.$$

By means of tables of functions (such as, e.g., L. M. Milne-Thomson and L. J. Comrie, *Standard Four-figure Mathematical Tables*, Macmillan, 1948) or a pocket calculator, we find $z = 0.9155$, from which the other data which determine the catenary, a, C_1, C_2 can be found.

Figure III.7, which is drawn to scale, shows the catenary so determined:

$$y = -122.71 + 109.23 \cosh\,[0.009155(x - 53.72)].$$

5. THE DIFFERENTIAL EQUATION $y'' = y^2$

For investigating the phenomena in non-linear boundary-value problems we

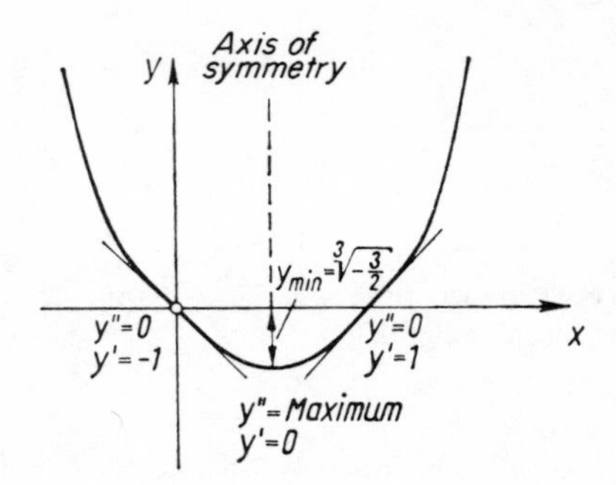

Fig. III.8. Type of solutions of $y'' = y^2$ through the origin

may take the differential equation†

$$y'' = y^2$$

with the boundary conditions

$$y(0) = 0, \qquad y(A) = B,$$

where A and B are not as yet further determined.

An attempt to integrate the differential equation directly leads to

$$\frac{1}{2} y'^2 = \frac{y^3}{3} + C, \qquad \int \frac{\mathrm{d}y}{\sqrt{C_{12} + \frac{2}{3}y^3}} = x + C_2.$$

This is an elliptic integral; the solution cannot be given in closed form in terms of elementary functions.

In this case, however, as in many others, it is possible to obtain a general idea of the solutions by looking directly at the differential equation itself.

To do this, we consider first not the given boundary-value problem, but an initial-value problem for the already fixed point

$$y(0) = 0$$

and first of all with the second initial condition, say,

$$y'(0) = -1.$$

†The function $u(x) = -y(x)$ which satisfies the differential equation $u'' = -u^2$ can be interpreted as the steady-state temperature distribution in a rod (with x as the longitudinal co-ordinate); namely, $u(x)$ is the difference in temperatures of the rod and its environment, and a quantity of heat proportional to u^2 is generated per unit time in the rod, due to some chemical or other process. The ends of the rod are kept at the constant temperatures $u(0) = 0$, $u(A) = -B$. Depending on the values of A and B, there may be 0, 1 or 2 solutions of the boundary-value problem; in the last case only one of the solutions corresponds to a stable temperature distribution.

The particular solution chosen in this way is easily discussed on the basis of the differential equation.

$$y'' = y^2$$

says that the rate of change of slope of the required function is to be equal to the square of this function, so clearly a concave curve similar to a parabola is under consideration. Further the intermediate solution (III.9a) shows that for every positive y two values differing only in sign are possible for the slope of a particular curve. For the curve we have chosen the initial values show that $C = \frac{1}{2}$; therefore

$$y' = \sqrt{2y^3/3 + 1}.$$

The slope is zero (its minimum!) at

$$y = \sqrt[3]{-3/2} = -1.145.$$

For $y < -1.145$ there are no real values of the slope.

The curve runs symmetrically about the ordinate through the minimum, see Fig. III.8. Next, by a linear deformation of the scales of abscissae and ordinates we can obtain a quantitative connection between the various solution curves. In a new co-ordinate system ξ, η obtained from the x, y-system by the affine deformation

$$x = a\xi, \qquad y = b\eta, \qquad \mathrm{d}x = a\,\mathrm{d}\xi, \qquad \mathrm{d}y = b\,\mathrm{d}\eta,$$

with

$$\frac{\mathrm{d}y}{\mathrm{d}x} = \frac{b}{a}\frac{\mathrm{d}\eta}{\mathrm{d}\xi}, \qquad \frac{\mathrm{d}^2y}{\mathrm{d}x^2} = \frac{b}{a}\frac{\mathrm{d}\;\mathrm{d}\eta/\mathrm{d}\xi}{\mathrm{d}\xi}\frac{\mathrm{d}\xi}{\mathrm{d}x} = \frac{b}{a^2}\frac{\mathrm{d}^2\eta}{\mathrm{d}\xi^2}$$

the differential equation reads

$$\frac{\mathrm{d}^2\eta}{\mathrm{d}\xi^2} = a^2 b \eta^2.$$

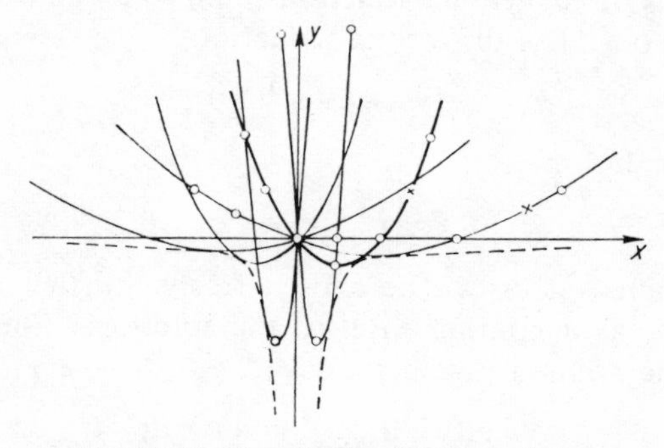

Fig. III.9. Two-fold solubility of the boundary-value problem

If we put

$$a^2 b = 1 \quad \text{or} \quad b = \frac{1}{a^2}, \qquad \text{i.e.,}\ x = a\xi, \qquad y = \frac{\eta}{a^2},$$

we get back to the original differential equation, this time in the ξ,η-coordinates.

If one solution is on hand, then we can get from it another solution by multiplying all the x-values by a and dividing all the y-values by a^2. If we double all the x-values, then we have to divide all the y-values by 4. If $a = -1$, the x-values change sign, the y-values remain unchanged.

The number of solutions of the boundary-value problem. We are now in a position to sketch the family of solution curves which go through the origin.

The family has an envelope, which, shaped like a hyperbola in the third and fourth quadrants, has the x-axis and negative y-axis as asymptotes. The x,y-plane is divided by the envelopes into an upper region and a lower region. The upper region is doubly covered by solutions. In the lower region there are no real solutions going through the origin.

If we now bring the second boundary conditions $y(A) = B$ into consideration, it follows that:

1. If the point $P\ (A, B)$ defined by the second boundary condition lies in the upper region, then there are two different real solutions of the boundary-value problem.
2. If $P\ (A, B)$ lies on the envelope, then there is just one real solution.
3. If $P\ (A, B)$ lies in the lower region, then there is no real solution.

As to whether there are complex solutions or not, no statement can be made without further detailed investigation.

Another example

Another example may be given, in which the solutions can be expressed explicitly in terms of elementary functions, and so a complete discussion can easily be carried out. The differential equation

$$y'' + 6yy'^3 = 0$$

has the general solution

$$x = y^3 + c_1 y + c_2,$$

where the constants c_1, c_2 can be chosen freely. Further, the straight lines $y = \text{const}$ belong, as a limiting case, to the solutions. Solutions are sought which assume the boundary values

$$y(0) = 1, \qquad y(A) = B;$$

only these functions $y = f(x)$ which are single-valued and twice continuously differentiable in the open interval between $x = 0$ and $x = A$ are admissible as

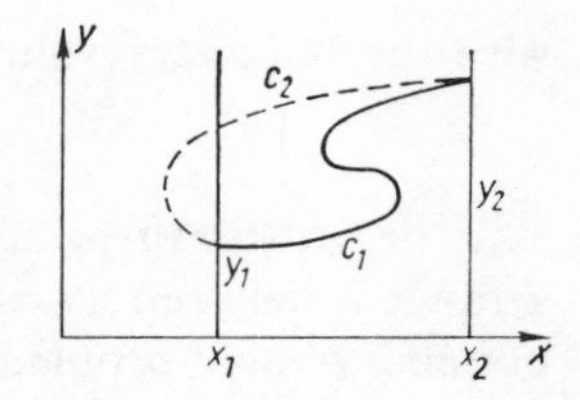

Fig. III.10. On the concept of the solution of a boundary-value problem

solutions. Functions which are graphically represented by curves such as C_1 or C_2 in Fig. III.10 are therefore not to be regarded as solutions of the boundary-value problem with the boundary values

$$y(x_1) = y_1, \qquad y(x_2) = y_2$$

The boundary condition $y(0) = 1$ picks out from the general solution the family of cubical parabolas

$$x = y^3 + c_1 y - 1 - c_1$$

with c_1 as the family-parameter, and as a limiting case the straight line $y = 1$. These curves cover the whole x, y-plane as a single sheet and without gaps; but boundary-value problems with $y(A) = B$, with A and B arbitrary, are by no means always soluble in the sense specified above, but only in cases where the point $x = A$, $y = B$ lies in one of the closed, shaded regions $\mathfrak{B}_1$, $\mathfrak{B}_2$ (without the point $x = 0$, $y = 1$). If the point $x = A$, $y = B$ lies outisde these regions, then solutions of the kind shown in Fig. III.10 arise, and these have been excluded. The regions $\mathfrak{B}_1$, $\mathfrak{B}_2$ are bounded by the solution curves $x = y^3 - 3y + 2$, $x = y^3 - 1$ and the curve of reversal points at which the tangents to the solution curves are perpendicular to the x-axis: $x = -2y^3 + 3y^2 - 1$ (this curve is marked by bold dashes in Fig. III.11).

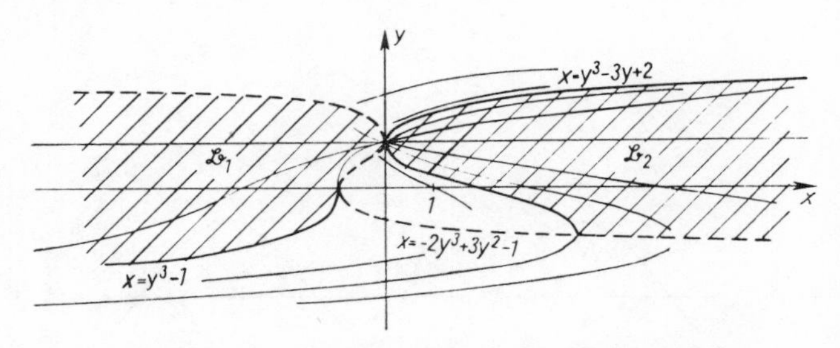

Fig. III.11. The circumstances for solubility in a non-linear boundary-value problem

6. A Countable Infinity of Solutions of a Boundary-Value Problem with the Equation $y'' = -y^3$

How complicated the circumstances regarding coverage can be is shown by the differential equation

$$y'' = -y^2 \tag{III.10}$$

with the boundary values

$$y(0) = 0, \; y(A) = B.$$

This describes, in principle, for small deflections, the vibrations of a mass arranged between two springs as in Fig. I.46 and equation (I.95). The boundary-value problem can be treated in the same way as the first example in No. 5 was, by starting from the initial values $y(0) = 0$, $y'(0) = -1$. The transformation as above yields

$$x = a\xi, \qquad y = \frac{\eta}{a}.$$

We obtain Fig. III.12. The family covers the whole field actually infinitely many times.

Through a prescribed second boundary-value $y(A) = B$ run infinitely many, but *discrete*, solution curves, not a continuum of solutions. Thus a fixed second boundary-value corresponds to a definite infinite sequence of initial slopes at the origin

$$y'(0) = y'_{0,n} \quad (n = 0, 1, 2, \ldots).$$

Conversely, however, arbitrary over-determined boundary-value problems of the form

$$y(0) = 0, \qquad y'(0) = d, \qquad y(a) = b$$

are not normally soluble in spite of the infinite-fold coverage of the field, because to a particular initial slope at the origin there corresponds one and only one solution.

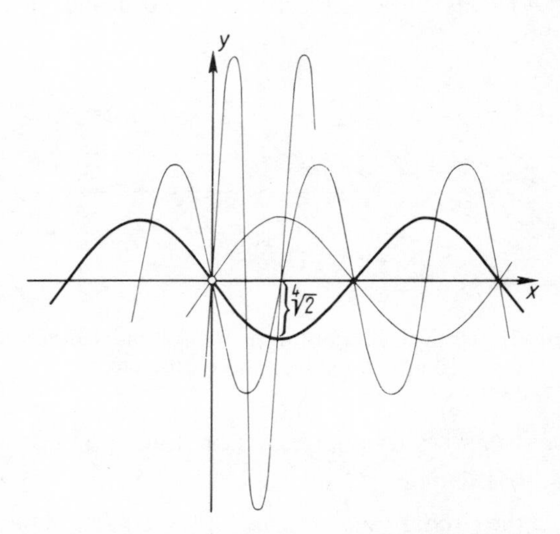

Fig. III.12. A non-linear boundary-value problem with a countable infinity of solutions

What, in fact, the criterion is in order that a differential equation

$$y'' = f(x, y, y') \tag{III.11}$$

of higher than the first degree with boundary-values prescribed at two points shall have, say, n solutions has up to now been investigated only under additional restrictive assumptions about the function f. Concerning the results, the report in E. Kamke, *Differentialgleichungen, Lösungsmethoden und Lösungen*, vol. 1, 6th ed., Leipzig 1959, pp. 277–281 may be consulted. In contrast, we are much more fully acquainted with boundary-value problems for *linear* differential equations; cf. the general theorem in No. 8. In the preceding decades predominantly linear differential equations have been applied in engineering. It was largely customary, by neglecting small quantities or by considering only small regions, to make linear approximations. To give a few instances, one may recall the use of 'Hooke's law', magnetization curves, thermionic valve characteristics, and the treatment of small vibrations of mechanical systems. The technical advances of our time compel us more and more to take the area of non-linear differential equations into our considerations. In solving differential equations of this kind, however, we are often obliged to have recourse to approximation methods.

§3. THE ALTERNATIVES IN LINEAR BOUNDARY-VALUE PROBLEMS WITH ORDINARY DIFFERENTIAL EQUATIONS

7. Semi-homogeneous and fully homogeneous boundary-value problems

Using the same notation (II.12) as for a differential equation, we can write linear boundary-conditions in the form

$$U_\mu[y] = V_\mu, \qquad (\mu = 1, 2, \ldots, n). \tag{III.12}$$

where the U_μ denote given differential expressions; for example, the boundary conditions might read

$$U_1[y]_{x=0} \equiv y(0) = 1,$$

$$U_2[y]_{x=1} \equiv y''(1) + 2y'(1) = 0.$$

In the elastic support of a beam, for example, boundary conditions of the form

$$U_1[y]_{x=0} \equiv y(0) + ay'(0) = k.$$

appear. If the differential equation and the boundary conditions are *inhomogeneous*, then either the differential equation or the boundary conditions can be made homogeneous. After this transformation the problem is said to be *semi-homogeneous*. In general, however, it is not possible to make such a problem a *fully homogeneous* one, i.e., a problem in which both the differential equation and the boundary conditions are homogeneous.

If $\psi(x)$ is a particular solution of the differential equation (II.12), then we put $y = \psi + z$. Then for z we have the homogeneous differential equation

$$L[z] = 0.$$

Conversely, if ψ^* is a function which satisfies the inhomogeneous boundary-condition

$$U_\mu[\psi^*] = V_\mu$$

but which need not satisfy the differential equation, then by putting $y = \psi^* + w$ the boundary-value problem is transformed into a new problem for w, in which the boundary conditions are homogeneous:

$$U_\mu[w] = 0.$$

Example

Let the boundary-value problem presented be

$$y'' + y = 1, \qquad y(0) = 1, \qquad y(1) = 2.$$

The particular solution $y = 1$ of the inhomogeneous differential equation is easily guessed. Putting $y = 1 + z$ we obtain for z the boundary-value problem

$$z'' + z = 0, \qquad z(0) = 0, \qquad z(1) = 1$$

with a homogeneous differential equation.

On the other hand, a particular function which satisfies the original boundary conditions is, for example, $\psi^* = 1 + x$. Putting $y = 1 + x + w$ it then follows that

$$w'' + w = -x,$$

and the boundary conditions become homogeneous: $w(0) = 0$, $w(1) = 0$.

8. The General Alternative

A criterion for a linear boundary-value problem

$$L[y] = 0, \qquad U_\mu = V_\mu \qquad (\mu = 1, \ldots, n)$$

to be solvable will now be derived. Let the differential equation have the general solution

$$y = C_1 y_1 + C_2 y_2 + \cdots + C_n y_n,$$

where $y_1, y_2, \ldots, y_n$ is a fundamental system. The adjustment to the linear boundary conditions is made by substituting

$$U_\mu[y] = C_1U_\mu[y_1] + C_2U_\mu[y_2] + \cdots + C_nU_\mu[y_n] = V_\mu.$$

so that a system of n linear equations for the n unknowns $C_1, \ldots, C_n$ is obtained. The solubility of these equations depends on the behaviour of the determinant of the coefficients

$$\Delta = \begin{vmatrix} U_1[y_1] & U_1[y_2] & \ldots & U_1[y_n] \\ U_2[y_1] & U_2[y_2] & \ldots & U_2[y_n] \\ \vdots & \vdots & & \vdots \\ U_n[y_1] & U_n[y_2] & \ldots & U_n[y_n] \end{vmatrix} \tag{III.13}$$

If $\Delta \neq 0$, the boundary-value problem has a unique solution, whatever the V_μ may be.

If $\Delta = 0$, the boundary-value problem is, in general, not soluble. Solutions exist only for certain V_μ, and if there are solutions at all, then there are infinitely many of them (cf. the example in No. 3, the second case).

In a *fully homogeneous problem*, all the $V_\mu = 0$, and then there are the following possibilities:

1. If $\Delta \neq 0$, there is only the trivial solution.
2. If $\Delta = 0$ the boundary-value problem has 'non-trivial' solutions.

This case of an *eigenvalue problem* is gone into in §5.

Example

In the semi-homogeneous boundary-value problem

$$y'' - y = 0$$

with the boundary conditions

$$y(0) + y'(0) = V_1 = 2, \qquad y'(0) + y(1) = V_2 = 4$$

and the fundamental system

$$y_1 = e^z, \qquad y_2 = e^{-x}$$

we obtain from the homogeneous parts of the boundary conditions, which alone have to be considered, the determinant

$$\Delta = \begin{vmatrix} 2 & 0 \\ 1+e & 1+\dfrac{1}{e} \end{vmatrix} = 2\left(1+\frac{1}{e}\right) \neq 0.$$

Hence the boundary-value problem has a unique solution for all V_μ; and, in particular, for $V_1 = V_2 = 0$ there is only the trivial solution

$$y \equiv 0.$$

§4. SOLVING BOUNDARY-VALUE PROBLEMS BY MEANS OF THE GREEN'S FUNCTION

9. Some very simple examples of Green's functions

1. There are formulae for the solutions of boundary-value problems; one of them will be worked for the case of the particularly simple boundary-value problem

$$-y'' = f(x), \qquad y(0) = y(l) = 0 \tag{III.14}$$

($-y''$ and not y'' has been written in order that later the Green's function shall be ≥ 0). Let $l > 0$ be given, and let $f(x)$ be a function, possibly continuous, defined in the interval $[0, l]$. Solutions of the differential equation can be formed by a double integration of $f(x)$; but we can also set, instead of a double integral, a single integral of the following type. Let

$$y^* = -\int_0^x (x - \xi) f(\xi)\, d\xi \tag{III.15}$$

be a particular solution of the differential equation; we then use the rule for differentiating such an integral

$$\frac{d}{dx}\int_{u(x)}^{v(x)} g(t, x)\, dt = \int_{u(x)}^{v(x)} \frac{\partial g(t, x)}{\partial x}\, dt + g(v(x), x)v'(x) - g(u(x), x)u'(x) \tag{III.16}$$

(see, e.g. L. A. Pipes, *Applied Mathematics for Engineers and Physicists*, McGraw-Hill 1958, p. 320), and so we have

$$y^{*\prime} = -\int_0^x f(\xi)\, d\xi, \qquad y^{*\prime\prime} = -f(x).$$

Accordingly, the general solution of the differential equation reads

$$y = \int_0^x (\xi - x) f(\xi)\, d\xi + c_1 + c_2 x \tag{III.17}$$

with c_1, c_2 as free, arbitrary constants. The boundary condition $y(0) = 0$ requires that $c_1 = 0$, and $y(l) = 0$ gives

$$c_2 = -\frac{1}{l}\int_0^l (\xi - l) f(\xi)\, d\xi. \tag{III.18}$$

Hence

$$y(x) = \int_0^x (\xi - x) f(\xi)\, d\xi - \int_0^l \frac{x}{l}(\xi - l) f(\xi)\, d\xi.$$

If the second integral is written in the form

$$\int_0^l \cdots = \int_0^x \cdots + \int_x^l \cdots,$$

then we have

$$y(x)=\int_0^x \frac{\xi}{l}(l-x)f(\xi)\,\mathrm{d}\xi+\int_x^l \frac{x}{l}\quad(l-\xi)f(\xi)\,\mathrm{d}\xi. \tag{III.19}$$

We can recombine these two integrals into a single one by introducing a so-called 'Green's function', $G(x,\xi)$, which is defined differently for $x \geq \xi$ and for $x \leq \xi$:

$$G(x,\xi)=\begin{cases}\frac{\xi}{l}(l-x) & \text{for } \xi \leq x\\[2ex] \frac{x}{l}(l-\xi) & \text{for } \xi \geqq x.\end{cases} \tag{III.20}$$

We can then write

$$y(x)=\int_0^l G(x,\xi)f(\xi)\,\mathrm{d}\xi \tag{III.21}$$

The Green's function is represented in Fig. III.13 by height contours. It is continuous in the square $0 \leq x, \xi \leq 1$, but the partial derivative $\partial G/\partial x$ makes a jump of magnitude -1, for constant ξ, as x goes through $x=\xi$ in the direction of increasing x.

2. Now let the boundary-value problem (III.14) be altered merely in one of the boundary conditions; the calculation for

$$-y''=f(x), \qquad y(0)=y'(l)=0 \tag{III.22}$$

goes through exactly as before up to formula (III.17) with $c_1=0$; but $y'(l)=0$ now requires that

$$c_2=\int_0^l f(\xi)\,\mathrm{d}\xi=\int_0^x f(\xi)\,\mathrm{d}\xi+\int_x^l f(\xi)\,\mathrm{d}\xi,$$

and the solution reads

$$y(x)=\int_0^x \xi f(\xi)\,\mathrm{d}\xi+\int_x^l xf(\xi)\,\mathrm{d}\xi. \tag{III.23}$$

For this we can again give the unified formula (III.21) if we define the Green's

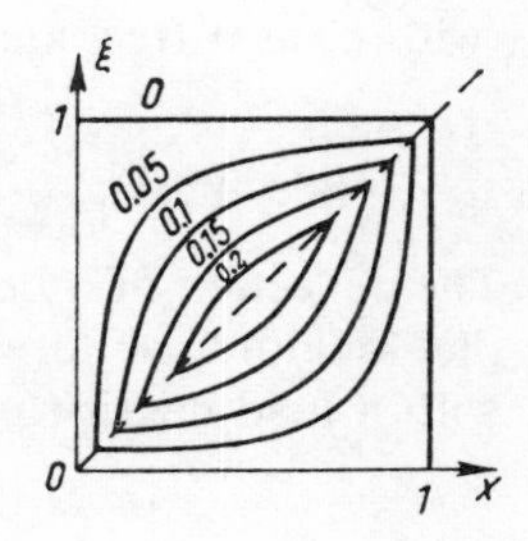

Fig. III.13. Height contours for the Green's function (III.20)

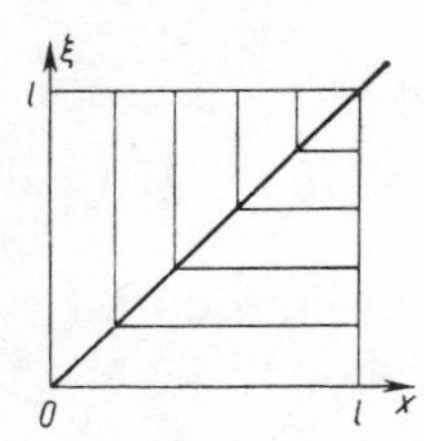

Fig. III.14. Height contours for the Green's function (III.24)

function this time by

$$G(x, \xi) = \begin{cases} \xi & \text{for} \quad \xi \leqq x \\ x & \text{for} \quad \xi \geqq x. \end{cases} \tag{III.24}$$

G is represented in Fig. III.14 by height contours. Here again G is continuous, but $\partial G/\partial x$ makes a jump of magnitude -1, as in the previous example.

3. Finally we work through a case where the boundary-value problem is soluble only for certain functions $f(x)$, and we have to work with a 'Green's function in the extended sense'. Such a case occurs when the boundary conditions are determined by periodicity:

$$y(0) - y(l) = y'(0) - y'(l) = 0. \tag{III.25}$$

The second boundary-condition requires that

$$\int_0^l f(\xi)\, d\xi = 0. \tag{III.26}$$

This boundary-value problem is therefore soluble only if the given function $f(x)$ satisfies this condition (III.26), which will from now on be assumed to be the case. The first boundary condition yields for c_2 the value given by (III.18). The number c_1 remains undetermined; if $y(x)$ is a solution of the boundary-value problem, then so also is $y(x) + c_1$, for *any* c_1. The result then is as follows.

The boundary-value problem with the conditions (III.25) is soluble if and only if $f(x)$ satisfies the equation (III.26), and it then has infinitely many solutions

$$y(x) = c_1 + \int_0^l G(x, \xi) f(\xi)\, d\xi,$$

(with c_1 as a free constant), where $G(x, \xi)$ is given by (III.20).

10. THE GREEN'S FUNCTION AS AN INFLUENCE FUNCTION

The deflection of a beam, supported as in Fig. I.2, (with the x-co-ordinate in the longitudinal direction) which has bending stiffness $\alpha(x) = EJ(x)$ and with a load density $p(x)$, satisfies the fourth-order differential equation

$$[\alpha(x) y''(x)]'' = p(x) = f(x); \tag{III.27}$$

or, if the bending stiffness is constant,

$$\alpha y^{\mathrm{IV}}(x) = f(x).$$

If the beam is simply supported at the points $x = 0$ and $x = l$ (so that there are no moments at these points), then the boundary conditions are

$$y(0) = y(l) = y''(0) = y''(l) = 0. \tag{III.28}$$

If now instead of the loading with density $p(x)$ there is only a single load of magnitude 1 at the point $x = \xi$, then the beam undergoes a flexure which can be calculated without more ado by the method in No. 2, and which will be denoted by $G(x, \xi)$. $G(x, \xi)$ is therefore the 'Green's function' of the boundary-value problem or the 'influence function', and it gives the deflection at the point x when a load 1 acts at the point ξ. We can now plausibly argue as follows. If at the point $x = \xi$ there acts not the load 1 but $f(\xi)\,\Delta\xi$, then the deflection at the point x will be $G(x, \xi)f(\xi)\,\Delta\xi$, and if over the whole length of the beam a load with density $f(\xi)$ is distributed, then the deflection at the point x will be

$$y(x) = \int_0^l G(x, \xi)f(\xi)\,\mathrm{d}\xi \tag{III.29}$$

Thus the solution of the boundary-value problem (III.27). (III.28) appears in the form of this integral, which has the same form as (III.21) except that the Green's function is different. It can easily be calculated from the above definition without trouble, and if $\alpha = EJ$, we find

$$G(x, \xi) = \begin{cases} \dfrac{1}{6EJl}\,\xi(x - l)(x^2 + \xi^2 - 2lx) & \text{for } \xi \le x \\[2ex] \dfrac{1}{6EJl}\,x(\xi - l)(x^2 + \xi^2 - 2l\xi) & \text{for } \xi \ge x \end{cases} \tag{III.30}$$

(for the particular case $\xi = 1$, $l = 3$, we obtain exactly the result of No. 2, except that there the load was chosen to be $P = 3$ t, instead of 1).

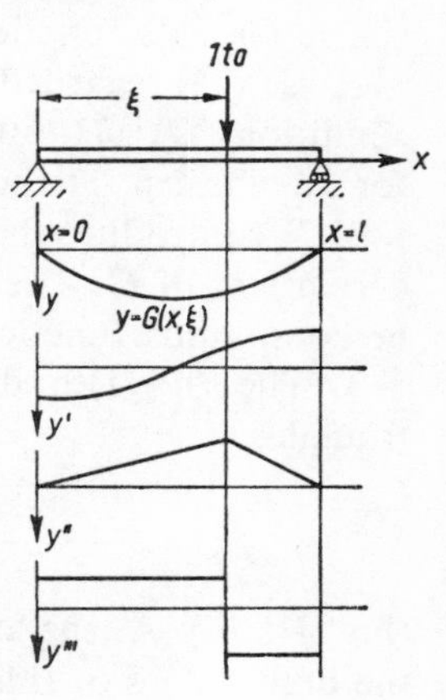

Fig. III.15. The Green's function (for a fixed ξ) as an influence function and its derivatives

If we draw, for a fixed ξ, the Green's function $G(x, \xi) = y$, and the first three derivatives y', y'', y''', then Fig. III.15 is obtained. The second derivative y'' gives, up to a constant factor, the familiar triangular bending-moment diagram for a single load, and the third derivative, which (likewise up to a constant factor) gives the shear force distribution, has at the point ξ a jump of magnitude $1/(Ej)$, which is characteristic of a Green's function.

11. General definition of the Green's function

From now on our discussion is based on the general linear boundary-value problem, in the semi-homogeneous form in which, according to No. 2, it can always be put. For the differential equation (II.12)

$$L[y] \equiv \sum_{\nu=0}^{n} p_\nu(x) y^{(\nu)} = r(x) \tag{III.31}$$

let the boundary conditions be

$$U_\mu[y] = 0, \qquad (\mu = 1, 2, \ldots, n). \tag{III.32}$$

In these, only two boundary points $x = a$ and $x = b$ with $b > a$ are to appear. As always it will also be assumed here that the p_ν and r are continuous in (a, b) and that $p_n(x) \neq 0$ in (a, b). Our object is to construct a Green's function $G(x, \xi)$ which will give the solution of the boundary-value problem in the integral form

$$y(x) = \int_a^b G(x, \xi) r(\xi) \, d\xi, \tag{III.33}$$

(This idea of solving boundary-value problems in integral form by means of a Green's function is much used for partial differential equations too, see, e.g., Chapter V, §2, No. 3).

The Green's function will now be defined by the following requirements.

1. $G(x, \xi)$ regarded as a function of x for fixed ξ shall satisfy the boundary conditions $U_\mu[G] = 0$ and the homogeneous differential equation $L[G] = 0$ for all x except at the point $x = \xi$.

2. We introduce the abbreviations $G^{(\nu)}(x, \xi) = \partial^\nu G(x, \xi)/\partial x^\nu$ for the partial derivatives of G. These derivatives for $\nu = 1, 2, \ldots, n-2$ and G itself are to be continuous functions of x and ξ in the whole square $a \leq x, \xi \leq b$.

3. The $(n-1)$th derivative $G^{(n-1)}(x, \xi)$ shall exist in each of the two triangles

$$D_1(a \leq x < \xi \leq b) \quad \text{and} \quad D_2(a \leq \xi < x \leq b),$$

(Fig. III.16). As the 'critical straight line' g for the point $x = \xi$ is approached, the limit values of this derivative exist as one-sided derivatives (x approaches

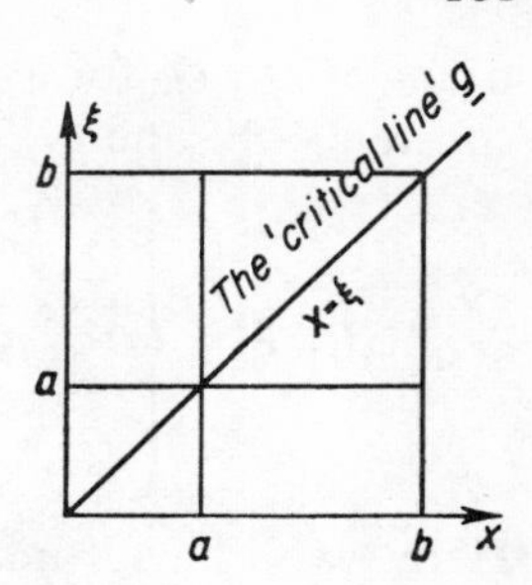

Fig. III.16. On the definition of the Green's function

ξ once from the right, and once from the left):

$$G^{(n-1)}(\xi+0,\xi)=\lim_{\substack{x\to\xi\\x\geq\xi}}G^{(n-1)}(x,\xi),$$
$$G^{(n-1)}(\xi-0,\xi)=\lim_{\substack{x\to\xi\\x\leq\xi}}G^{(n-1)}(x,\xi),\tag{III.34}$$

and these two limit values differ by $1/p_n(\xi)$, i.e., the $(n-1)$th derivative makes a jump at this point:

$$G^{(n-1)}(\xi+0,\xi)-G^{(n-1)}(\xi-0,\xi)=\frac{1}{p_n(\xi)}.\tag{III.35}$$

These limit values are to be independent of the direction in which a point on the critical line is approached, whether in the direction of the x-axis or in the direction of the ξ-axis; i.e.

$$G^{(n-1)}(\xi+0,\xi)=G^{(n-1)}(\xi,\xi-0),\qquad G^{(n-1)}(\xi-0,\xi)=G^{(n-1)}(\xi,\xi+0)\tag{III.36}$$

shall hold (the different directions of approach are shown by arrows in Fig. III.17).

We shall now prove that such a Green's functions which satisfies all these requirements does, under one additional hypothesis, actually exist, and we show how it can be constructed if a fundamental system $y_1(x),\ldots,y_n(x)$ for the homogeneous differential equation is known. Let then

$$L[y_1]=\cdots=L[y_n]=0.$$

Since G is to satisfy the homogeneous differential equation in each of the two

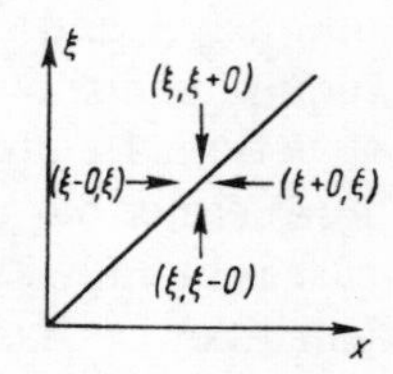

Fig. III.17. On the discontinuity relation for the Green's function

triangles D_1, D_2 for $\xi = \text{const}$, it must have the following form (with different constants $c_j^{(1)}$ and $c_j^{(2)}$ in the two triangles):

$$G(x,\xi) = \begin{cases} \sum_{j=1}^{n} c_j^{(1)} y_i(x) = \sum_{j=1}^{n} (a_j + b_j) y_j(x) \text{ in } D_1, & \text{i.e., for } x < \xi, \\ \sum_{j=1}^{n} c_j^{(2)} y_j(x) = \sum_{j=1}^{n} (a_j - b_j) y_j(x) \text{ in } D_2, & \text{i.e., for } x > \xi. \end{cases} \tag{III.37}$$

Here we have introduced the abbreviations $a_j = \frac{1}{2}(c_j^{(1)} + c_j^{(2)})$ and $b_j = \frac{1}{2}(c_j^{(1)} - c_j^{(2)})$; the a_j and b_j are functions of ξ, still to be determined. We obtain the $b_j(\xi)$ from the required behaviour at the critical line. G is to be continuous, and so we must have $\sum_{j=1}^{n} b_j y_j(\xi) = 0$; in general, the continuity of the ν-th derivative requires that

$$\sum_{j=1}^{n} b_j y_j^{(\nu)}(\xi) = 0 \quad \text{for } \nu = 0, 1, \ldots, n-2. \tag{III.38}$$

The $(n-1)$th derivative, on the other hand, shall, in accordance with (III.35), make a jump

$$\sum_{j=1}^{n} b_j y_j^{(n-1)}(\xi) = -\frac{1}{2p_n(\xi)}. \tag{III.39}$$

(III.38), (III.39) represent n linear equations for the n unknowns b_j; the determinant of the coefficients is again just the Wronskian (II.19) and it is non-zero because the y_j form a fundamental system. The equations therefore have a unique solution and yield the functions $b_j(\xi)$ independent of the boundary conditions.

From the boundary conditions we now have to calculate the $a_j(\xi)$. G is to satisfy the homogeneous boundary conditions, and so we must have

$$U_\mu[G] = \sum_{j=1}^{n} (a_j \pm b_j) U_\mu[y_j] = 0.$$

where the upper sign $+$ is to hold if in the relevant boundary condition the boundary point $x = a$ is used, and the lower sign $-$ if the boundary point $x = b$ is used. Thus we obtain for the $a_j(\xi)$ the linear system of equations

$$\sum_{j=1}^{n} a_j U_\mu[y_j] = \mp \sum_{j=1}^{n} b_j U_\mu[y_j] \qquad (\mu = 1, \ldots, n)$$

It is now a matter of the determinant Δ of this system of equations; Δ is precisely the determinant (III.13); if $\Delta \neq 0$, then the $a_j(\xi)$ can be calculated uniquely, and there exists a Green's function which satisfies all the requirements, even the condition (III.36) which has not been used in the construction. If, on the other hand, $\Delta = 0$, then the system of equations for the $a_j(\xi)$ either has no solution or has infinitely many solutions. This exceptional case is already known to us from No. 8; in this case the boundary-value problem likewise has either no solution or infinitely many solutions; this is the 'eigenvalue case'; the determinant will re-appear, slightly modified, as the eigenvalue-determinant in (III.46).

12. THE SOLUTION FORMULA FOR THE BOUNDARY-VALUE PROBLEM

It now has to be shown that, in the case $\Delta \neq 0$, the function defined in the integral formula (III.33) by means of the Green's function

$$Y(x) = \int_a^b G(x, \xi) r(\xi)\, d\xi$$

actually does solve the boundary-value problem (III.31), (III.32). Repeated differentiation gives

$$Y^{(\nu)}(x) = \int_a^b G^{(\nu)}(x, \xi) r(\xi)\, d\xi \quad \text{for } \nu = 0, 1, \ldots, n-1.$$

But if we form $Y^{(n)}(x)$, we have to remember that $G^{(n-1)}$ has the discontinuity (III.35) at the point $x = \xi$. Accordingly we divide the integral into two integrals, from a to x and from x to b. The differentiation then gives

$$\frac{d}{dx} \int_a^x G^{(n-1)}(x, \xi) r(\xi)\, d\xi = \int_a^x G^{(n)}(x, \xi) r(\xi)\, d\xi + G^{(n-1)}(x, x-0) r(x)$$

and similarly for the integral from x to b, so that altogether

$$Y^{(n)}(x) = \int_a^x G^{(n)}(x, \xi) r(\xi)\, d\xi + G^{(n-1)}(x, x-0) r(x)$$
$$+ \int_x^b G^{(n)}(x, \xi) r(\xi)\, d\xi - G^{(n-1)}(x, x+0) r(x)$$

is obtained. With the jump conditions (III.35) and (III.36) it hence follows that

$$Y^{(n)}(x) = \int_a^b G^{(n)}(x, \xi) r(\xi)\, d\xi + \frac{r(x)}{p_n(x)}.$$

$L[Y]$ can now be formed:

$$L[Y] = \sum_{\nu=0}^{n} p_\nu(x) Y^{(\nu)} = \int_a^b \left\{ \sum_{\nu=0}^{n} p_\nu(x) G^{(\nu)}(x, \xi) \right\} r(\xi)\, d\xi + r(x).$$

The expression within the braces is precisely $L[G]$ and is therefore zero for all $x \neq \xi$; the integral therefore vanishes since, in the integration, the isolated point $x = \xi$ is of no consequence, and there remains only $L[Y] = r(x)$, i.e., Y satisfies the inhomogeneous differential equation (III.31). Since G satisfies the boundary conditions, so does Y; hence Y does satisfy the boundary-value problem.

By means of the solution formula (III.33) linear, and certain types of nonlinear, boundary-value problems can be converted into equivalent problems with 'integral equations'. If, for example, we are presented with the differential equation

$$L[y] = \sum_{\nu=0}^{n} p_\nu y^{(\nu)} = \Phi(x, \lambda, y, y', \ldots, y^{(q)}) \qquad \text{(III.40)}$$

and the homogeneous boundary-conditions

$$U_\mu[y] = 0 \qquad (\mu = 1, \ldots, n) \qquad \text{(III.41)}$$

and if $L[y]$ under these boundary conditions (III.32) has a Green's function $G(x, \xi)$, then y is the solution of the equation

$$y(x) = \int_a^b G(x, \xi)\Phi(\xi, \lambda, y(\xi), y'(\xi), \ldots, y^{(q)}(\xi))\, d\xi \qquad \text{(III.41)}$$

Here λ is brought in merely as a parameter at first. If the derivatives $y', \ldots, y^{(q)}$ appear on the right-hand side, then (III.41) is called an 'integro-differential equation', but if only y and none of its derivatives appear, then (III.41) is called an 'integral equation'.

Of all integral equations the linear equations

$$y(x) = f(x) - \lambda \int_a^b K(x, \xi) y(\xi)\, d\xi, \qquad \text{(III.42)}$$

where λ is a parameter, $f(x)$ is a function, perhaps continuous, defined in (a, b), $K(x, \xi)$ is a given function continuous in the square $a \le x, \xi \le b$ and $y(x)$ is to be found, have been studied in most detail; to present their theory here would take us too far afield. (A detailed presentation can be found in W. Schmeidler: *Integralgleichungen mit Anwendungen in Physik und Technik*, 2nd ed., Leipzig 1955, p. 611, many further references are given there.) In many respects the theory of integral equations is in a more complete and polished state than that of differential equations; one does not have to bother about boundary conditions, and differential equations of different higher orders correspond to the same kind of integral equations.

§5. EXAMPLES OF EIGENVALUE PROBLEMS

13. THE COMPLETELY HOMOGENEOUS BOUNDARY-VALUE PROBLEM

Completely homogeneous boundary-value problems have normally, i.e., if the determinant (III.13) does not vanish right away, only the trivial solution $y \equiv 0$. If, however, in such a problem there is also a parameter λ in the differential equation or in the boundary conditions, then one can ask about those values of this parameter λ for which the boundary-value problem has a non-trivial solution, which is then called an 'eigenfunction'. The corresponding value of λ is called an 'eigenvalue'. In many engineering problems the parameter λ itself appears linearly, namely, as a factor of the unknown function or in some of the differential expressions. If we arrange the homogeneous linear differential equation into expressions with and without the parameter λ we obtain the general form

$$L[y] = \lambda M[y] \tag{III.43}$$

or in greater detail

$$\sum_{\nu=0}^{n} p_\nu(x) y^{(\nu)}(x) = \lambda \sum_{\nu=0}^{n} q_\nu(x) y^{(\nu)}(x) \tag{III.44}$$

with given functions $p_\nu(x)$, $q_\nu(x)$. The boundary conditions read

$$U_\mu[y] = 0 \qquad (\mu = 1, \ldots, n). \tag{III.45}$$

Here too we arrange the expressions into terms with and without λ; in general we can write $U_\mu[y, \lambda] = 0$. In many cases arising in practice λ does not appear explicitly in the boundary conditions; the solutions of the homogeneous equation (III.43) depend on λ, so let $y_1(x, \lambda), \ldots, y_n(x, \lambda)$ be a fundamental system of (III.43); then the general solution of (III.43) reads

$$y = \sum_{j=1}^{n} C_j y_j(x, \lambda).$$

and the C_j are to be determined exactly as in No. 8 so as to satisfy the boundary conditions (III.45). We thus obtain a homogeneous system of n equations for the n unknowns C_j:

$$\sum_{j=1}^{n} C_j U_\mu[y_j(x, \lambda), \lambda] = 0 \qquad (\mu = 1, \ldots, n),$$

This system has a non-trival solution C_j and therefore leads to a non-trivial solution $y(x)$ if and only if the determinant (the 'frequency determinant') vanishes:

$$\Delta(\lambda) = \begin{vmatrix} U_1[y_1(x, \lambda), \lambda] & \ldots & U_1[y_n(x, \lambda), \lambda] \\ \cdots & \cdots & \cdots \\ U_n[y_1(x, \lambda), \lambda] & \ldots & U_n[y_n(x, \lambda), \lambda] \end{vmatrix} = 0. \tag{III.46}$$

This is the equation of conditions for the eigenvalues λ (the 'frequency equation').

More can be said if we apply theorems from the theory of analytic functions. The differential equation depends linearly and therefore holomorphically ('analytically') on λ and so the $y_j(x, \lambda)$, if they have been determined from initial conditions independent of λ, are holomorphic functions of λ. The $U_\mu[y, \lambda]$ also are holomorphic functions of λ (this is true, in particular, if λ does not appear explicitly in the U_μ, and so $\Delta(\lambda)$ is holomorphic in λ, and being an entire function of λ, it is either identically zero (every value of λ is then an eigenvalue), or $\Delta(\lambda)$ has at most a countable infinity of isolated zeros without a point of accumulation at infinity. Therefore the eigenvalues, if they exist at all, can be arranged in order of magnitude of their absolute values and numbered $\lambda_1, \lambda_2, \ldots$ with

$$0 \leq |\lambda_1| \leq |\lambda_2| \leq |\lambda_3| \leq \ldots. \tag{III.47}$$

Such eigenvalue problems arise, for example, in solving stability problems such as the buckling of a column, the lateral deflection of a rod with fixed ends, the oscillations of a cable, etc.

The Euler buckling load.

As an example of a simple eigenvalue problem we have chosen the classical problem of calculating the Euler buckling load of a column. A vertical rod, fixed at its lower end and having the modulus of elasticity E and, say, a constant second moment of area of cross-section J, is loaded by a force P acting along its longitudinal axis. The question is, at what load P will the rod bend sideways from its original position? Let the free length of the rod be l. It is convenient to take the origin to be at the point of action of the force P when the rod is deflected, see Fig. III.18.

The differential equation for the deflection is

$$y'' = -\frac{M(x)}{EJ}. \tag{III.48}$$

From Fig. III.18 it can be seen that the bending moment is $M(x) = Py$, and hence we obtain the homogeneous differential equation

$$y'' = -\frac{Py}{EJ} = -k^2 y \tag{III.49}$$

with the boundary conditions, which also are homogeneous:

$$y = 0 \text{ for } x = 0, \quad \text{and} \quad y' = 0 \text{ for } x = l.$$

The problem is thus fully homogeneous. P is the required parameter. A parameter k^2 also appears (as an abbreviation for P/EJ) and it plays precisely the rôle of an eigenvalue. We are enquiring about that value of P (or of $\lambda = k^2$)

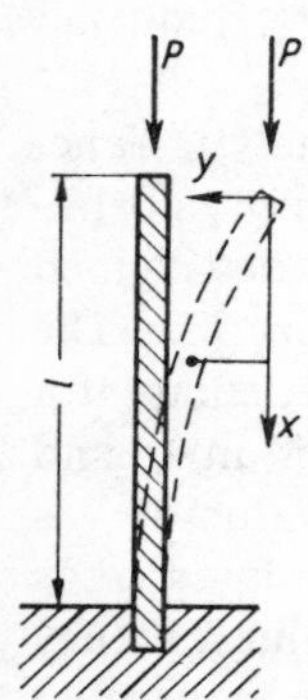

Fig. III.18. Buckling of a column fixed at one end

for which the boundary-value problem has a non-trivial solution, for which, that is, an equilibrium position of the rod deflected from its original rectilinear form is possible.

The general solution of (III.49) reads

$$y = C_1 \cos kx + C_2 \sin kx.$$

From the first boundary condition $y(0) = 0$ it follows that $C_1 = 0$. So there remains as the solution only

$$y = C_2 \sin kx.$$

The second boundary condition gives

$$y' = kC_2 \cos kx; \qquad 0 = kC_2 \cos kl.$$

This equation is satisfied for:

(1) $k = 0$, i.e., the load is zero;

(2) $C_2 = 0$, which gives only the trivial solution, $y \equiv 0$, of the rod remaining straight, which is of no interest here; and

(3) $$kl = \frac{\pi}{2}, \frac{3\pi}{2}, \frac{5\pi}{2}, \ldots, \frac{(2r-1)}{2}\pi \text{ for } r = 1, 2, 3, \ldots. \qquad \text{(III.50)}$$

Hence, with $P = EJk^2$, we obtain a countably infinite sequence of values of the load for which alone, as it seems at first, a lateral deflection appears:

$$P_1 = EJ\left(\frac{\pi}{2l}\right)^2 = \frac{\pi^2 EJ}{4l^2} \text{ (the Euler buckling load)}$$

$$P_2 = EJ\left(\frac{3\pi}{2l}\right)^2 = \frac{9\pi^2 EJ}{4l^2},$$

$$\cdots\cdots\cdots\cdots\cdots\cdots$$

$$P_r = EJ(2r-1)^2 \frac{\pi^2}{4l^2}.$$

Fig. III.19. Buckling shapes of a column fixed at the lower end

In all these cases the magnitude of C_2 has no influence as regards the satisfaction of the equation, but only, as the solution shows, determines the amplitude of the sine-curve form assumed by the rod.

Of all these values of P, however, only the first is of any practical significance. It bears the name of 'the Euler buckling load'. All the other values yield unstable forms of deflection, which can be demonstrated experimentally only by taking special measures. Normally the unstable forms change, on the slightest disturbance, into the stable form. It should be stressed that the trivial solution still remains as another possibility for arbitrary values of the load P.

14. THE CORESPONDING NON-LINEAR BOUNDARY-VALUE PROBLEM

It might seem from our considerations so far as if the elastic deformation of the column occurs only for certain loads. We should have to deduce from this the paradoxical behaviour of the column that, as the load increases, it deforms only for certain quite definite values—those calculated above—into the corresponding shapes, and that for all the intermediate values it snaps back into its original form. This is the mathematical expression of the fact that in the formulation of the problem an approximation has already been made in the form of a linearization of the differential equation for small y. Since

$$\text{curvature} = -\frac{\text{bending moment}}{\text{bending stiffness}},$$

the differential equation should, more accurately, be

$$\frac{y''}{\sqrt{1+y'^2}^3} = -\frac{M(x)}{EJ}. \tag{III.51}$$

This completed differential equation can be solved in closed form with the help of elliptic functions, though we shall not go into this treatment here. Suffice it to say that the maximum deflection y_{max} is then found to be, in principle, represented by a picture as in Fig. III.20. Only curve 1 represent a stable variation, while curves 2 and 3 represent the unstable possibilities which are nevertheless present. If y'^2 is neglected in comparision with 1, then instead of this function which is defined for all values of P and which is three-valued from P_2 to P_3 and, in general, $(r+1)$-valued from P_r to P_{r+1}, we obtain only function values for the discrete P-sequence $P_1, P_2, P_3, \ldots, P_r, \ldots$, where with each P_r alike there is associated an infinite continuum of values of y_{max}. (In

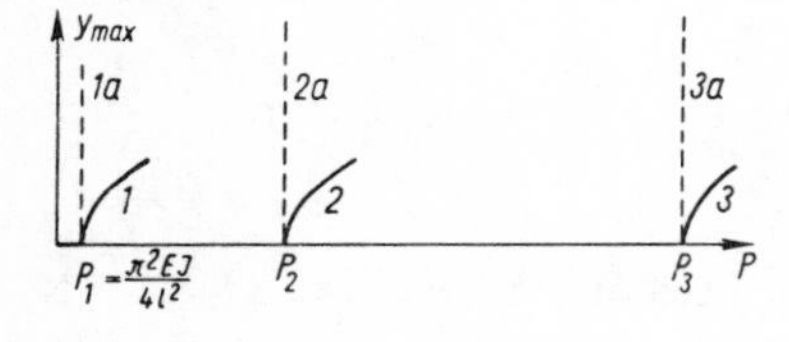

Fig. III.20. The maximum deflection of a column fixed at its lower end as a function of the load

practice, of course, $y_{\max}$ cannot exceed certain limits.) The dotted straight lines 1a, 2a, 3a, in Fig. III.20, represent these conditions.

The problem shows how important it is to investigate in each particular case the significance and limits of admissibility of the linearization of a differential equation. The linearized differential equation produces in the dotted lines 1a, 2a, 3a the tangents to the exact curves 1, 2, 3; it thus provides in a sufficiently narrow strip above the P-axis the correct results, in particular, the eigenvalues are given exactly, i.e., here the buckling loads, as the 'branching points'; but the linearized equation is incapable of giving a correct over-all picture of the phenomena in the buckling problem.

15. Partial differential equations

Suppose the natural frequencies of torsional vibration of a shaft clamped at one end and carrying a flywheel at the other end are to be calculated.

The angle of rotation φ of each cross-section of the shaft relative to the cross-section at the clamped end is, during the vibrations, a function of both the position x and the time t; see Fig. III.21. In the equilibrium state $\varphi(x, t) \equiv 0$. The line of thought leading to the derivation of the equation of motion (III.53) can only be briefly outlined here.

Let the torques M and $M + \Delta M$ act on an element Δx, see Fig. III.22. Let $\Delta\varphi$ be the rotation of the cross-section at $x + \Delta x$ relative to the cross-section at x. Since

$$M = GJ_p \frac{\partial \varphi}{\partial x}$$

where G is the shear modulus, and J_p the polar moment of area of the cross-section, and

$$M + \Delta M = GJ_p \frac{\partial \phi}{\partial x} + GJ_p \frac{\partial^2 \varphi}{\partial x^2} \Delta x + \cdots$$

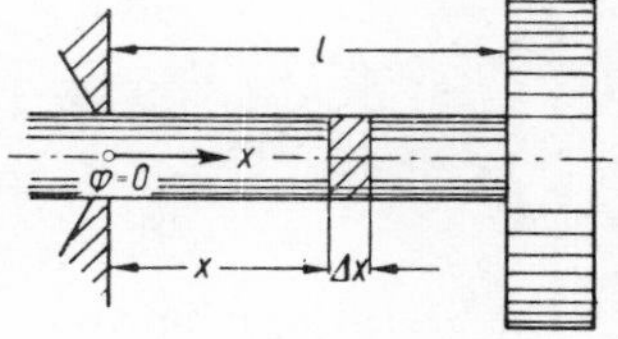

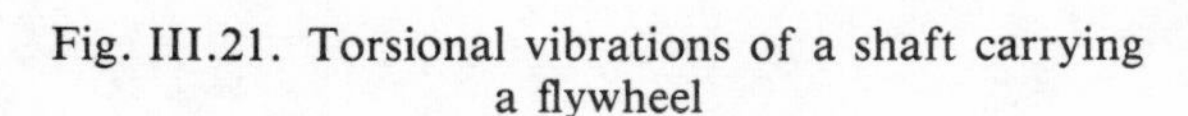
Fig. III.21. Torsional vibrations of a shaft carrying a flywheel

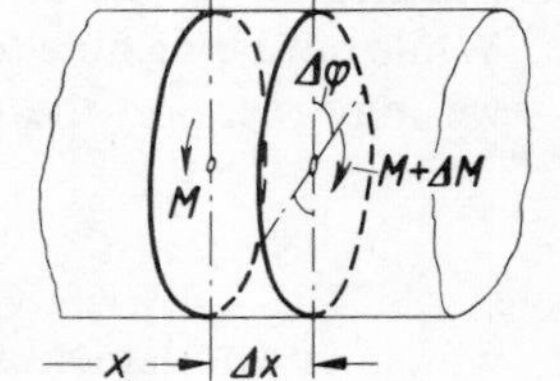

Fig. III.22. Angle of rotation and moments at a selected element of the shaft

(the Taylor expansion), the effective torque for the rotation of the shaft element is the difference

$$\Delta M \approx GJ_p \frac{\partial^2 \varphi}{\partial x^2} \Delta x;$$

this produces the angular acceleration of the shaft element, which has the moment of inertia $\Delta \Theta = \mu J_p \Delta x$, where μ is the density of the material. From the equation of motion

$$\text{torque} = \text{moment of inertia} \times \text{angular acceleration} \tag{III.52}$$

it follows that

$$GJ_p \frac{\partial^2 \varphi}{\partial x^2} \Delta x = \mu J_p \Delta x \frac{\partial^2 \varphi}{\partial t^2},$$

or, writing $G = \mu c^2$, we have the 'wave equation'

$$\frac{\partial^2 \varphi}{\partial t^2} = c^2 \frac{\partial^2 \varphi}{\partial x^2}, \tag{III.53}$$

in which c denotes the velocity of wave propagation (cf. Chapter 5, No. 2).

Boundary conditions. These are that

$$\varphi = 0 \text{ for } x = 0 \quad \text{and} \quad M_l = GJ_p \left(\frac{\partial \phi}{\partial x}\right)_{x=l} = -\Theta \left(\frac{\partial^2 \varphi}{\partial t^2}\right)_{x=l} \quad \text{for } x = l,$$

where Θ is the total moment of inertia of the end piece, here the moment of inertia of the flywheel. The minus sign means that the torque acts in the direction of decreasing φ. If we introduce the further abbreviation $GJ_p/\Theta = A$, then we have the wave equation and the boundary conditions in the following form (where dots denote derivatives with respect to time t, and primes denote derivatives with respect to the position co-ordinate x)

$$\ddot{\varphi} = c^2 \varphi'',$$

$$\varphi(0, t) = 0,$$

$$\ddot{\varphi}(l, t) = -A\varphi'(l, t).$$

The Schrödinger equation

Eigenvalue problems have acquired great significance in theoretical physics. While not being able to explain here the meaning of the individual magnitudes, we may mention the time-dependent Schrödinger equation

$$-\frac{\hbar^2}{2m} \Delta \varphi + V(x)\varphi = \mathrm{i}\hbar \frac{\partial \psi}{\partial t}$$

which is fundamental in quantum mechanics; here $\psi(x, t)$ is a quantity dependent on the position $x=(x_1, x_2, x_3)$ and the time t, $V(x)$ is a given potential dependent on the position, and $\hbar$ and m are constants. Using the formula

$$\psi(x,t)=\varphi(x)e^{-i(E/\hbar)t}$$

for an exponential dependence on the time, we obtain the time-independent Schrödinger equation

$$-\frac{\hbar^2}{2m}\Delta\varphi+(V(x)-E)\varphi=0 \quad \text{for } x\in B \tag{III.54}$$

where E is the 'energy eigenvalue', and B is the region considered in the x_1, x_2, x_2-space. If b is the whole three-dimensional space R^3, then we can demand as the boundary condition for φ that φ shall tend to zero so rapidly that φ shall be quadratically integrable (in the Lebesgue sense) over R^3. If B is a finite region, e.g., a finite cylinder Z, then as the 'first boundary-value problem' we can require that $\varphi=0$ on the boundary ∂B, but other boundary conditions can also be considered.

If we put $V(x)=0$ in (III.54), we obtain the so-called Helmholtz equation

$$\Delta\varphi+\lambda\varphi=0 \quad \text{for } x\in B$$

with the eigenvalue $\lambda=2mE/\hbar^2$; here too the boundary condition $\varphi=0$ on ∂B is often demanded. For a finite region B this is also the mathematical formulation for acoustic vibrations in a region B with fixed walls. (At apertures we usually put $\partial\varphi/\partial n=0$ in the simplest case, n being the outward normal.)

16. THE BERNOULLI SUBSTITUTION FOR NATURAL VIBRATIONS

The general solution of the wave equation (III.53) is

$$\varphi=w_1(x+ct)+w_2(x-ct),$$

where w_1 and w_2 can be arbitrary functions (cf. Chapter V, equation (V.6)). Here, however, the problem is to seek out from the manifold of all solutions those which represent natural vibrations, i.e., we are seeking not an arbitrary vibratory process, but those cases in which all particles of the shaft vibrate with the same frequency. To this corresponds the 'Bernoulli product-substitution', by means of which we try to separate the partial derivatives:

$$\varphi(x,t)=y(x)f(t),$$
$$\ddot{\varphi}(x,t)=y(x)\ddot{f}(t),$$
$$\varphi''(x,t)=y''(x)f(t).$$

Equation (III.53) now becomes

$$y(x)\ddot{f}(t)=c^z y''(x)f(t)$$

and the variables can be separated:

$$c^2 \frac{y''(x)}{y(x)} = \frac{\ddot{f}(t)}{f(t)}. \tag{III.55}$$

This equation between a function of x alone and a function of t alone must hold for all combinations of the arguments x and t. If, for example, we put $t = \text{const}$, then the relation must hold for all x, and reversely. But this can happen only if both sides are equal to one and the same *constant* k. The partial differential equation thus splits up into two *ordinary* differential equations coupled by the parameter k:

$$\ddot{f}(t) = kf(t) \tag{III.56}$$

and

$$y''(x) = \frac{k}{c^2}\, y(x). \tag{III.57}$$

Since oscillatory processes are to be expected in any case for the first equation, a comparison with the differential equation for harmonic motion (II.27) shows that the constant k must turn out to be negative, a fact which the later calculation also confirms. It is convenient, therefore, to put $k/c^2 = -a^2$:

$$\ddot{f}(t) = -a^2c^2f(t),$$

$$y''(x) = -a^2y(x).$$

The boundary conditions have still to be rewritten: we must have

$$\varphi(0, t) = 0 = y(0)f(t)$$

for all t, but $f(t)$ must not be identically zero, since otherwise $\varphi = 0$; so we can divide by $f(t)$ and obtain $y(0) = 0$.

Similarly the second condition

$$\ddot{\varphi}(l, t) = -A\varphi'(l, t), \quad \text{or} \quad y(l)\ddot{f}(t) = -Ay'(l)f(t)$$

becomes, using (III.56),

$$-a^2c^2y(l) = -Ay'(l).$$

Hence we again have an eigenvalue problem for $y(x)$. We seek those values of the parameter $\lambda = a^2$ for which the boundary-value problem

$$y'' + \lambda y = 0$$

$$y(0) = 0$$

$$\lambda y(l) - \frac{A}{c^2}\, y'(l) = 0$$

has a solution $y(x)$ not identically zero. This is a case where the eigenvalues appear in the boundary conditions.

The calculation for the eigenvalue problem. The general solution of the equation for $y(x)$ is

$$y = C_1 \cos ax + C_2 \sin ax.$$

The first boundary condition $y(0) = 0$ gives $C_1 = 0$ and so the solution is $y = C_2 \sin ax$.

From the second boundary condition we find

$$-a^2c^2C_2 \sin al = -AC_2 a \cos al.$$

If C_2 were $= 0$, we would obtain the trivial solution $y(x) \equiv 0$. But if $C_2 \neq 0$, then we have the transcendental equation

$$z \tan z = \frac{Al}{c^2} \tag{III.58}$$

for $z = al$. This equation has an infinite sequence of discrete zeros z_1, z_2, $z_3, \ldots$ which can be determined approximately by plotting the graphs of $z \tan z$ and A/c^2 and finding their points of intersection, as in Fig. III.23. For each of these values of z there is a definite value of a and of k and from (III.56) a natural frequency ω of the shaft (by (II.27) $-k = \omega^2$). To each z_j there corresponds an eigenfunction y_j uniquely determined, up to a constant factor. Figure III.24 shows the first three eigenfunctions.

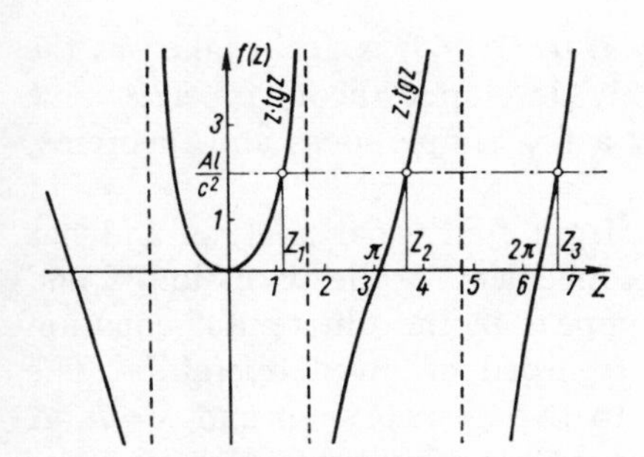

Fig. III.23. Graphical solution of the transcendental equation

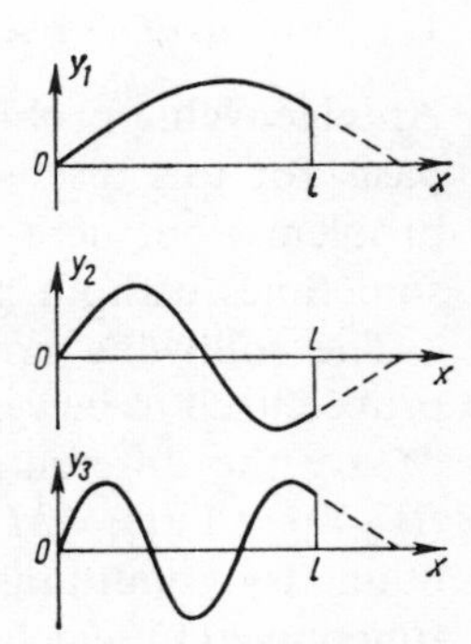

Fig. III.24. The first three eigenfunctions

General vibration of the shaft. The vibrations occurring in general in the shaft are superimpositions of the possible neutral vibrations

$$\sin\left(\frac{z_j}{l}x\right) \cdot [\cos \omega_j(t - t_0)] \quad \text{with } \omega_j = \frac{c}{l} z_j$$

and the t_0 as a free constant. The particular combination depends on the initial state of the shaft, i.e., on the deflection $g(x) = \varphi(x, 0)$ and on the rate of deflection $h(x) = \dot{\varphi}(x, 0)$ at the time $t = 0$. If the initial state can be expanded in terms of the eigenfunctions, i.e., if there are constants g_j and h_j such that

$$g(x)=\sum_{j=1}^{\infty} g_j y_j(x), \qquad h(x)=\sum_{j=1}^{\infty} h_j y_j(x), \tag{III.59}$$

then the solution can be written as a series

$$\varphi(x,t)=\sum_{j=1}^{\infty}\left\{g_j y_j(x)\cos\omega_j t + h_j\frac{y_j(x)}{\omega_j}\sin\omega_j t\right\}. \tag{III.60}$$

An important question arises from this: can given functions, which may be continuous, say, or perhaps continuously differentiable, be expanded in terms of the eigenfunctions of an eigenvalue problem?

If the initial state of the shaft does not coincide with the initial state of a single natural vibration, so that there are therefore several natural vibrations acting together, the resultant vibration is not, in general, periodic, because the ratios of different eigenvalues are, in general, irrational. Thus the function $\sin x+\sin\sqrt{2}x$, for example, is not periodic, in contrast to $\sin x+\sin 2x$. In such cases, $\varphi(x,t)$ with x constant is called an *almost-periodic function* of t.

§6. EIGENVALUE PROBLEMS AND ORTHONORMAL SYSTEMS

17. Self-adjoint and positive-definite eigenvalue problems

An eigenvalue problem in the form (III.43) to (III.45) is again taken as the basis for this discussion. There is a highly developed theory of eigenvalue problems, but here we can mention only a few simple ideas and theorems, sometimes without proof.

The following notation, which comes from functional analysis and has proved itself to be very useful, is the most convenient for the discussion. L and M are the differential operators which appear in the differential equation (III.43) $L[y]=\lambda M[y]$ and are to be regarded as given operators. The boundary conditions (III.45) may relate to two points $x=a$ and $x=b$. A function $u(x)$ which has in (a,b) continuous derivatives of as high an order as those which appear in the differential equation, and which satisfies the boundary conditions, but which is not identically zero, will be called a 'comparison function'. Let $u(x)$ and $v(x)$ be two comparison functions; we introduce two 'inner products' by the definition

$$\langle u,v\rangle=\int_a^b uL[v]\,dx, \qquad (u,v)=\int_a^b uM[v]\,dx. \tag{III.61}$$

Throughout this section the brackets (u,v), $\langle u,v\rangle$ always stand for these integrals. This notation has greatly contributed towards simplification and clarity.

The eigenvalue problem (III.43) (III.45) is said to be 'self-adjoint' if, for two arbitrary comparison functions $u(x)$, $v(x)$, both the inner products are symmetric or 'commutative':

$$\langle u, v\rangle = \langle v, u\rangle, \qquad (u, v) = (v, u). \tag{III.62}$$

The eigenvalue problem (III.43) (III.45) is said to be 'positive-definite' if

$$\langle u, u\rangle > 0, \qquad (u, u) > 0 \tag{III.63}$$

for every comparison function $u(x)$.

Whether in a particular eigenvalue problem the relations (III.62), (III.63) hold or not can generally be ascertained by means of partial integrations, as in the following example.

Example

For the torsional vibrations of a rotating wheel of variable thickness $h(x)$ one obtains the following eigenvalue problem (a detailed treatment is given in C. B. Biezeno and R. Grammel, *Technische Dynamik*, 2nd ed., Vol. II, Berlin-Göttingen-Heidelberg 1953; English trans: *Engineering Dynamics*, Blackie & Son Ltd. 1956).

$$\begin{aligned} -\frac{\mathrm{d}}{\mathrm{d}x}[x^3h(x)y'(x)] &= \lambda x^3h(x)y(x), \\ cy(x_0) - y'(x_0) = 0, &\qquad y(l) = 0. \end{aligned} \tag{III.64}$$

λ is the required eigenvalue, which is proportional to the square of the circular frequency of the natural vibrations, and c is a given positive constant. x_0 and l are to be taken from the dimensions of the wheel. Using the abbreviations

$$\begin{aligned} x^3h(x) = p(x), &\qquad y(x_0) = y_0, \\ y'(x_0) = y_0', &\qquad y(l) = y_l, \end{aligned}$$

we can bring the eigenvalue problem into line with the general notation by

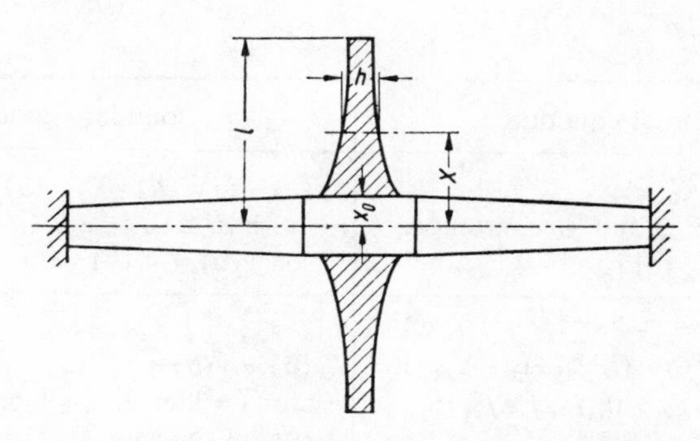

Fig. III.25. Torsional vibrations of a wheel of variable thickness

putting

$$L[y] = -(py')', \qquad M[y] = py,$$
$$U_1[y] = cy_0 - y_0', \qquad U_2[y] = y_l. \tag{III.65}$$

The comparison functions here are all the functions other than $u \equiv 0$ which are twice continuously differentiable in the interval (x_0, l) and which satisfy the boundary conditions. Next, by simple substitution, we determine whether (III.62) and (III.63) hold, transforming the integrals by partial integration if need be. From the definition (III.61)

$$\langle u, v\rangle = -\int_{x_0}^{l} u(pv')'\,\mathrm{d}x = -[u(pv')]_{x_0}^{l} + \int_{x_0}^{l} u'pv'\,\mathrm{d}x.$$

Since u and v satisfy the boundary conditions, $u(l) = 0$ and $v'(x_0) = cv(x_0)$; therefore

$$\langle u, v\rangle = cp(x_0)u(x_0)v(x_0) + \int_{x_0}^{l} p(x)u'(x)v'(x)\,\mathrm{d}x.$$

The right-hand side is now symmetric in u and v, and so $\langle u, v\rangle = \langle v, u\rangle$. Further, it now follows that

$$\langle u, u\rangle = cp(x_0v)\cdot[u(x_0)]^2 + \int_{x_0}^{l} p(x)[u'(x)]^2\,\mathrm{d}x.$$

So $\langle u, u\rangle$ is certainly ≥ 0. $\langle u, u\rangle$ can $= 0$ only if $u(x_0) = 0$ and $u'(x) \equiv 0$; but then $u =$ const., and since $u(x_0) = 0$ we would have $u \equiv 0$, a possibility excluded by the definition of a comparison function. Hence $\langle u, u\rangle > 0$ for every comparison function $u(x)$. The argument for the expression $M[y]$ is simpler. Here we have

$$(u, v) = \int_{x_0}^{l} upv\,\mathrm{d}x,\ (u, u) = \int_{x_0}^{l} pu^2\,\mathrm{d}x,$$

and we can see at once the symmetry $(u, v) = v, u)$ and the positive-definiteness

Differential equation	Boundary conditions
$-(f_0(x)y')' = \lambda g_0(x)y$ $f_0 > 0$, $g_0 > 0$; f_0' and g_0 continuous	$y(a) - c_1y'(a) = 0$, $y(b) + c_2y'(b) = 0$ with $c_1 \geq 0$, $c_2 \geq 0$ or $y(a) = y'(b) = 0$
$(f_0(x)y'')'' - (f_1(x)y') + f_2(x)y = \lambda g_0(x)y$ $f_0 > 0$, $f_j \geq 0$, $g_0 > 0$, f_0'', f_1', f_2, g_0 continuous	$y(a) = y(b) = y^{(i)}(a) = y^{(j)}(b) = 0$ for $i = 1$ or 2; $j = 1$ or 2 or (in the case $f_1(b) = 0$): $y(a) = y'(a) = y''(b) = y'''(b) = 0$.

$(u, v) > 0$ for all comparison functions. The eigenvalue problem (III.64) is therefore self-adjoint and positive-definite.

For applications it is convenient to know from the outset that they are self-adjoint and positive-definite. That is the case for the eigenvalue problems in the table on page 196; these include many stability problems and vibration problems for beams.

18. ORTHOGONALITY OF THE EIGENFUNCTIONS

From now on the eigenvalue problem (III.43) (III.45) will be assumed to be self-adjoint and positive-definite. Then the number 0 cannot be an eigenvalue. For, if there were an eigenfunction, not identically zero, corresponding to $\lambda = 0$, with $L[y] = 0$, we would have

$$\langle y, y \rangle = \int_a^b yL[y] \, \mathrm{d}x = 0,$$

but we know that $\langle y, y \rangle > 0$ holds for every comparison function, and so also for y. Also, it cannot be the case that every number is an eigenvalue, and so the eigenvalues, if they exist at all, can be arranged in order of magnitude of their absolute values, as in (III.47). Now let

$$y_1, y_2, \ldots$$

denote the eigenfunctions corresponding to the eigenvalues $\lambda_1, \lambda_2, \ldots$ (and not, as hitherto, a fundamental system). As a further simplification we shall write Lu and Mu instead of $L[u]$ and $M[u]$; then we have

$$Ly_j = \lambda_j My_j. \tag{III.66}$$

If we multiply (III.66) by y_k and integrate over the inveral (a, b), then it follows, using the notation (III.61), that

$$\langle y_k, y_j \rangle = \lambda_j (y_k, y_j). \tag{III.67}$$

Interchanging the subscripts,

$$\langle y_j, y_k \rangle = \lambda_k (y_j, y_k);$$

since the inner products are symmetric, we obtain, on subtracting these two equations,

$$(\lambda_j - \lambda_k)(y_j, y_k) = 0. \tag{III.68}$$

If, now, $\lambda_j \neq \lambda_k$, it follows that

$$(y_j, y_k) = 0. \qquad \text{(III.69)}$$

By (III.67) we also have

$$\langle y_j, y_k \rangle = 0 \qquad \text{(III.70)}$$

Two functions $u(x)$, $v(x)$ are said to be 'orthogonal' to one another if

$$\int_a^b u(x)v(x)\,\mathrm{d}x = 0 \qquad \text{(III.71)}$$

and to be 'orthogonal in the generalized sense' if

$$(u, v) = \int_a^b uMv\,\mathrm{d}x = 0. \qquad \text{(III.72)}$$

The following theorem therefore holds.

Theorem

If the eigenvalue problem (III.43) (III.45) *is self-adjoint, then two eigenfunctions corresponding to two different eigenvalues are orthogonal to one another in the generalized sense.*

$$(y_j, y_k) = \langle y_j, y_k \rangle = 0 \quad \text{holds for } \lambda_j \neq \lambda_k. \qquad \text{(III.73)}$$

If we put $y_k = y_j$ in (III.67), we obtain

$$\langle y_j, y_j \rangle = \lambda_j (y_j, y_j). \qquad \text{(III.74)}$$

The quotient

$$Ru = \frac{\langle u, u \rangle}{(u, u)} \qquad \text{(III.75)}$$

formed from a comparison function is called the 'Rayleigh quotient'. Because of the positive-definiteness (III.63) the denominator cannot vanish. By (III.63) again the numerator and denominator in R are positive, and so $R > 0$ for all comparison functions. By (III.74),

$$\lambda_j = Ry_j, \qquad \text{(III.76)}$$

and so we have the following

Theorem

If the eigenvalue problem (III.43) (III.45) *is positive-definite, then all the existing eigenvalues are real and, indeed, positive.*

19. ORTHONORMAL SYSTEMS

In the ordinary point-space of three or n dimensions we can introduce rectangular co-ordinates; we can take mutually perpendicular ('orthogonal') unit vectors $\mathfrak{e}_j$ as a 'basis', so that every arbitrary vector $\mathfrak{v}$ can be 'expanded' in terms of these unit vectors, i.e., can be expressed as a linear combination of them

$$\mathfrak{v} = \sum_{j=1}^{n} v_j \mathfrak{e}_j \quad \text{with } v_j = (\mathfrak{v}, \mathfrak{e}_j). \tag{III.77}$$

In the 'space of functions', for which we ought to indicate clearly which functions $v(x)$ we would like to admit, the eigenfunctions $y_j(x)$ may, in quite general cases, be suitable as a basis for the functions $v(x)$ to be expressible in the form

$$v(x) = \sum_{j=1}^{\infty} v_j y_j(x) \quad \text{with } v_j = (v, y_j). \tag{III.78}$$

We would then have the function $v(x)$ 'expanded' in terms of the eigenfunctions, and the constants v_j would correspond to the co-ordinates; the v_j are then called the coefficients of the expansion, or the 'Fourier coefficients'. One very important special case is the expansion of a function $f(x)$ which is periodic with period 2π,

$$f(x + 2\pi) = f(x), \tag{III.79}$$

in terms of the trigonometrical functions $\cos mx$ and $\sin mx$, which are the eigenfunctions of

$$y'' + \lambda y = 0, \qquad y(0) - y(2\pi) = 0, \qquad y'(0) - y'(2\pi) = 0. \tag{III.80}$$

In the former case of a point-space, the basis vectors have length l and are mutually perpendicular, i.e.,

$$(\mathfrak{e}_j, \mathfrak{e}_k) = \delta_{jk} = \begin{cases} 1 \text{ for } j = k \\ 0 \text{ for } j \neq k. \end{cases} \tag{III.81}$$

If for an arbitrary vector $\mathfrak{v}$ we put $\mathfrak{v} = \sum_{k=1}^{n} v_k \mathfrak{e}_k$ with constants v_j, and form the inner product of both sides with the unit vector $\mathfrak{e}_j$, then we have

$$(\mathfrak{v}, \mathfrak{e}_j) = \sum_{k=1}^{n} v_k (\mathfrak{e}_k, \mathfrak{e}_j) = \sum_{k=1}^{n} v_k \, \delta_{kj} = v_j,$$

i.e., the fact that the expansion coefficients v_j have the simple form already given in (III.77) depends on the equation (III.81). We say that the vectors e_j form an 'orthonormal system' (ONS, for short).

We now ask whether there are orthonormal systems in the function space. Let $z_1(x)$, $z_2(x)$, ... be a finite or infinite set of given functions defined in an interval (a, b) and such that, with the 'metric' introduced earlier, the inner product (w_1, w_2) can be formed, where w_1, w_2 are linear combinations of the form $w = \sum_{j=1}^{p} a_j z_j$ with a finite number of the z_j. It is convenient also to introduce the norm $\|z\|$ of an element $z = z(x)$ by

$$\|z\| = +\sqrt{(z, z)}. \tag{III.82}$$

Concerning the metric it is, for the present, assumed only that $\|z\| \geq 0$ and that $\|z\| = 0$ only if $z(x) \equiv 0$. Then the 'orthonormalization process' introduced by Erhard Schmidt enables us to introduce a new system of functions Z_j which is an ONS, and is such that each Z_j can expressed as a linear combination of a finite number of the $z_k(x)$. It is assumed that the $z_k(x)$ are not identically zero and are linearly independent of one another. In the process auxiliary functions $\varphi_j(x)$ are introduced. We put:

$$Z_1 = \frac{z_1}{\|z_1\|}, \qquad \text{and so} \qquad \|Z_1\| = 1$$

$$\varphi_2 = z_2 - Z_1(z_2, Z_1), \qquad \text{and so} \qquad (\varphi_2, Z_1) = 0$$

$$Z_2 = \frac{\phi_2}{\|\varphi_2\|}, \qquad \text{and so} \qquad \|Z_2\| = 1,$$

and continue in this way. Suppose it is already known that $Z_1, Z_2, \ldots, Z_{m-1}$ are orthonormal; we then put

$$\varphi_m = z_m - \sum_{j=1}^{m-1} Z_j(z_m, Z_j), \qquad Z_m = \frac{\varphi_m}{\|\varphi_m\|} \qquad (m = 1, 2, \ldots). \tag{III.83}$$

Now φ_m cannot be identically zero, for if it were, z_m would be a linear combination of $z_1, \ldots, z_{m-1}$; but it was assumed that the z_j are linearly independent. We then have

$$(\varphi_m, Z_k) = (z_m, Z_k) - \sum_{j=1}^{m-1} \underbrace{(Z_j, Z_k)}_{\delta_{jk}}(z_m, Z_j) = 0 \text{ for } k = 1, \ldots, m-1.$$

Thus the z_j do indeed form an ONS.

We now return to the consideration of eigenvalue problems. It can happen that one of the eigenvalues λ is a multiple root, and that several linearly independent eigenfunctions z_1, z_2, ... correspond to it. Since linear combinations of eigenfunctions which correspond to the *same* eigenvalue λ are again eigenfunctions corresponding to λ (providing that they are not identically zero), we can by using the orthonormalization process just described replace the $z_1, z_2, \ldots$ by eigenfunctions $Z_1, Z_2, \ldots$ with $(Z_j, Z_k) = \delta_{jk}$ in the sense that every linear combination of the z_j can also be written as a linear combination of the Z_j. Since, further, eigenfunctions y, z which correspond to *different* eigenvalues always satisfy $(y, z) = 0$ by (III.73), it follows that for every eigenvalue problem of the kind considered in No. 18 an ONS of eigenfunctions $y_j(x)$ can always be selected with

$$(y_j, y_k) = \delta_{jk} = \begin{cases} 0 \text{ for } j \neq k \\ 1 \text{ for } j = k. \end{cases} \tag{III.84}$$

In the eigenvalue problem (III.80) for the trigonometrical functions, the eigenfunction $y_1 = \text{const.}$ corresponds to the eigenvalue $\lambda_1 = 0$, and all the other eigenvalues $\lambda = m^2$ (for $m = 1, 2, \ldots$) are double roots; here the inner product is

$$(y_j, y_k) = \int_a^b y_j(x) y_k(x) \, \mathrm{d}x \tag{III.85}$$

with $a = 0$, $b = 2\pi$. The eigenfunctions

$$\frac{1}{\sqrt{2\pi}}, \quad \frac{1}{\sqrt{\pi}} \cos mx, \quad \frac{1}{\sqrt{\pi}} \sin mx \qquad (m = 1, 2, \ldots) \tag{III.86}$$

form an ONS.

Orthogonal systems with polynomials

A number of important orthogonal function systems which crop up in applications can be obtained in the following way. A positive 'weight function' $p(x)$ is given in an interval (a, b); a may $= -\infty$ or $b = +\infty$ or both. The powers $1, x, x^2, \ldots$ are to be replaced by a system of polynomials $Q_0(x), Q_1(x), \ldots$ by using the Erhard Schmidt orthonormalization process with the metric

$$(u, v) = \int_a^b p(x) u(x) v(x) \, \mathrm{d}x \tag{III.87}$$

In the simplest case of a weight function $p(x) = 1$ and the interval $(-1, 1)$, we obtain as the $Q_n(x)$ the Legendre polynomials $P_n(x)$, which will be discussed in more detail in Chapter IV, §1. Here we restrict ourselves to presenting in the adjacent table some further frequently used cases. In each case the first few polynomials are given, the differential equation which the polynomials satisfy, other representations of them, and the 'generating function', the significance of which is explained, in the case of the Legendre polynomials, in Chapter IV, 2.

Interval (a, b)	Weight function $p(x)$	Name	Notation and representation	Differential equation
$(-1, 1)$	$\frac{1}{\sqrt{1-x^2}}$	Chebyshev polynomials of the first kind	$T_n(x)$ $= \cos(n \operatorname{arc} \cos x)$ $= G_n\left(0, \frac{1}{2}, \frac{1-x}{2}\right)$	$(x^2-1)y'' + xy'$ $- n^2 y = 0$
$(-1, 1)$	1	Legendre polynomials (Spherical harmonics of the first kind)	$P_n(x)$ $= \frac{1}{2^n n!} \frac{d^n(x^2-1)^n}{dx^n}$ $= G_n\left(1, 1, \frac{1-x}{2}\right)$	$(x^2-1)y'' + 2xy'$ $- n(n+1)y = 0$
$(-1, 1)$	$\sqrt{1-x^2}$	Chebyshev polynomials of the second kind	$S_n(x)$ $= \frac{\sin[(n+1) \operatorname{arc} \cos x]}{\sqrt{1-x^2}}$ $= (n+1)G_n\left(2, \frac{3}{2}, \frac{1-x}{2}\right)$	$(x^2-1)y'' + 3xy'$ $- n(n+2)y = 0$
$(0, 1)$	$(1-x)^{p-q}$ $\cdot x^{q-1}$ $p > q-1$ $q > 0$	Jacobi or hyper-geometric polynomials	$G_n(p, q, x)$ $= \frac{x^{1-q}(1-x)^{q-p} z^{(n)}}{q(q+1) \ldots (q+n-1)}$ $z = x^{q+n-1}(1-x)^{p+n-q}$	$x(1-x)y''$ $+ [q-(p+1)x]y'$ $+ (p+n)ny = 0$
$(0, \infty)$	e^{-x}	Laguerre polynomials	$L_n(x) = e^x \frac{d^n(x^n e^{-x})}{dx^n}$	$xy'' + (1-x)y'$ $+ ny = 0$
$(-\infty, \infty)$	e^{-x^2}	Hermite polynomials	$H_n(x)$ $= (-1)^n e^{x^2} \frac{d^n(e^{-x^2})}{dx^n}$	$y'' - 2xy'$ $+ 2ny = 0$

Generating function	Normalizing integral	The first polynomials
$\dfrac{1-rx}{1-2rx+r^2} = \sum_{n=0}^{\infty} T_n(x) r^n$	$\int_{-1}^{1} \dfrac{T_n^2(x)\,\mathrm{d}x}{\sqrt{1-x^2}} = \begin{cases} \pi & \text{for } n=0 \\ \dfrac{\pi}{2} & \text{for } n \geqq 1 \end{cases}$	$T_0 = 1,\ T_1 = x$ $T_2 = 2x^2 - 1$ $T_3 = 4x^2 - 3x$ $T_4 = 8x^4 - 8x^2 + 1$
$\dfrac{1}{\sqrt{1-2rx+r^2}} = \sum_{n=0}^{\infty} P_n(x) r^n$	$\int_{1}^{1} P_n^2(x)\,\mathrm{d}x = \dfrac{2}{2n+1}$	$P_0 = 1,\ P_1 = x$ $P_2 = \frac{1}{2}(3x^2 - 1)$ $P_3 = \frac{1}{2}(5x^3 - 3x)$ $P_4 = \frac{1}{8}(35x^4 - 30x^2 + 3)$
$\dfrac{1}{1-2rx+r^2} = \sum_{n=0}^{\infty} S_n(x) r^n$	$\int_{a}^{1} \sqrt{1-x^2}\, S_n^2(x)\,\mathrm{d}x = \dfrac{\pi}{2}$	$S_0 = 1,\ S_1 = 2x$ $S_2 = 4x^2 - 1$ $S_3 = 8x^3 - 4x$ $S_4 = 16x^4 - 12x^2 + 1$
$\left(\dfrac{r-1+W}{x}\right)^{q-1} \cdot \left(\dfrac{r+1-W}{1-x}\right)^{p-q} \cdot \dfrac{1}{(2r)^{p-1} W} = \sum_{n=0}^{\infty} \binom{q+n-1}{n} G_n r^n$ so $W^2 = (1-r)^2 + 4rx$	$\int_0^1 (1-x)^{p-q} x^{q-1} G_n^2\,\mathrm{d}x = \dfrac{n!\Gamma^2(q)}{(p+2n)} \cdot \dfrac{\Gamma(p-q+n+1)}{\Gamma(p+n)\Gamma(q+n)}$	$G_0 = 1,\ G_1 = 1 - \dfrac{p+1}{q} x$ $G_2 = 1 - 2\dfrac{p+2}{q} x + \dfrac{(p+2)(p+3)}{q(q+1)} x^2$ $G_3 = 1 - 3\dfrac{p+3}{q} x + 3\dfrac{(p+3)(p+4)}{q(q+1)} x^2 - \dfrac{(p+3)(p+4)(p+5)}{q(q+1)(q+2)} x^3$
$\dfrac{\mathrm{e}^{-xr/(1-r)}}{1-r} = \sum_{n=0}^{\infty} \dfrac{L_n(x)}{n!} r^n$	$\int_0^{\infty} \mathrm{e}^{-x} L_n^2(x)\,\mathrm{d}x = (n!)^2$	$L_0 = 1,\ L_1 = -x + 1$ $L_2 = x^2 - 4x + 2$ $L_3 = -x^3 + 9x^2 - 18z + 6$ $L_4 = x^4 - 16x^3 + 72x^2 - 96x + 24$
$\mathrm{e}^{-r^2+2rx} = \sum_{n=0}^{\infty} \dfrac{H_n(x)}{n!} r^n$	$\int_{-\infty}^{\infty} \mathrm{e}^{-x^2} H_n^2(x)\,\mathrm{d}x = 2^n n! \sqrt{\pi}$	$H_0 = 1,\ H_1 = 2x$ $H_2 = 4x^2 - 2$ $H_3 = 8x^3 - 12x$ $H_4 = 16x^4 - 48x^2 + 12$

Now let $y_1, y_2, \ldots$ be an arbitrary ONS. It is required to approximate a given function $v(x)$, say a continuous function, as 'well as possible' by means of a linear combination of the first p functions y_j of the system with coefficients a_j which are to be chosen suitably:

$$\Phi = \sum_{j=1}^{p} a_j y_j \tag{III.88}$$

We must first of all define what we mean by 'as well as possible'. Sometimes an approximation in the *Chebyshev sense* is desirable, i.e., an approximation such that the maximum deviation of $| v - \Phi |$ in the interval (a, b) shall be as small as possible. Here, however, we demand that a sort of 'mean deviation' shall be as small as possible, namely that

$$\| v - \Phi \| = \min \tag{III.89}$$

or, equivalently,

$$(v - \Phi, v - \Phi) = \min. \tag{III.90}$$

If the metric (III.85) is used, this is equivalent to the method of least mean square error:

$$\int_a^b (v - \Phi)^2 \, dx = \min. \tag{III.91}$$

It is now assumed that the metric is such that there is symmetry, i.e., $(f, g) = (g, f)$, as was required by (III.62) for self-adjoint eigenvalue problems. For brevity we write

$$(v, y_i) = v_i. \tag{III.92}$$

Then, using the orthogonality (III.84), we have that

$$(\Phi, \Phi) = \left(\sum_{j=1}^{p} a_j y_j, \sum_{k=1}^{p} a_k y_k \right) = \sum_{j=1}^{p} a_j^2$$

and

$$(v - \Phi, v - \Phi) = (v, v) - 2 \sum_{j=1}^{p} a_j v_j + \sum_{j=1}^{p} a_j^2 = (v, v) - \sum_{j=1}^{p} v_j^2 + \delta \tag{III.93}$$

where

$$\delta = \sum_{j=1}^{p} (a_j - v_j)^2;$$

Hence $\| v - \Phi \|^2$ will be least if we choose $a_j = v_j$, since then δ will vanish.

The expressions

$$w(x) = \sum_j v_j y_j(x) \tag{III.94}$$

are called the 'Fourier coefficients of $v(x)$ relative to the ONS of the y_j'.

Theorem

Under the stated assumptions, the function $v(x)$ is best approximated in the sense of (III.89) *by a linear combination Φ of the first p functions $y_j(x)$ of an ONS if Φ is taken to be the pth partial sum of the Fourier series for $v(x)$ relative to the $y_j(x)$.*

21. On the Expansion Theorem

If $v(x)$ can be expanded in terms of the functions y_j in a uniformly convergent series

$$v = \sum_j b_j y_j \tag{III.95}$$

and if the metric is defined by an integral of the form (III.61) or (III.85), both sides of (III.95) can be multiplied by a fixed y_k and the order of integration and summation on the right-hand side can be interchanged; then, by (III.84) only the term b_k is left on the right:

$$v_k = (v, y_k) = b_k, \tag{III.96}$$

i.e., the series (III.95) is then the Fourier series of $v(x)$ relative to the y_j.

This result contains as a special case the formulae for the Fourier coefficients in ordinary harmonic analysis (the expansion of a 2π-periodic function $v(x)$ in terms of $\sin jx$ and $\cos jx$, $j = 0, 1, 2, \ldots$). By (III.94) the Fourier series $w(x)$ for the ONS (III.86) is

$$w(x) = v_0 \frac{1}{\sqrt{2\pi}} + \sum_{j=1}^{\infty} \left[v_j' \frac{1}{\sqrt{\pi}} \cos jx + v_j'' \frac{1}{\sqrt{\pi}} \sin jx \right]$$

with

$$v_0 = \frac{1}{\sqrt{2\pi}} \int_0^{2\pi} v(x)\,\mathrm{d}x, \qquad \begin{Bmatrix} v_j' \\ v_j'' \end{Bmatrix} = \frac{1}{\sqrt{\pi}} \int_0^{2\pi} v(x) \begin{Bmatrix} \cos jx \\ \sin jx \end{Bmatrix} \mathrm{d}x.$$

In the theory of trigonometrical series it is proved that for extensive classes of functions, e.g., if $v(x)$ is continuous and of bounded variation, then $w(x)$ coincides with $v(x)$ (see, e.g., Mangoldt–Knopp: *Einführung in die höhere Mathematik*, vol. III, 10th ed., Stuttgart 1958, p. 536); but in the case of an arbitrary eigenvalue problem, as in No. 17, the following questions arise.

1. For a given function $v(x)$ which we assume, say, to be continuous and piecewise continuously differentiable, we can calculate the Fourier coefficients $v_j = (v, y_j)$ and construct the series (III.94) $\sum_j v_j y_j$ formally. Does this series converge in the interval $[a, b]$, does it converge uniformly and absolutely, may one differentiate the series term by term one or more times?

2. If the series (III.94) converges and therefore defines a function $w(x)$, does $w(x)$ coincide with $v(x)$, i.e., is $v(x)$ really expanded in terms of the $y_j(x)$?

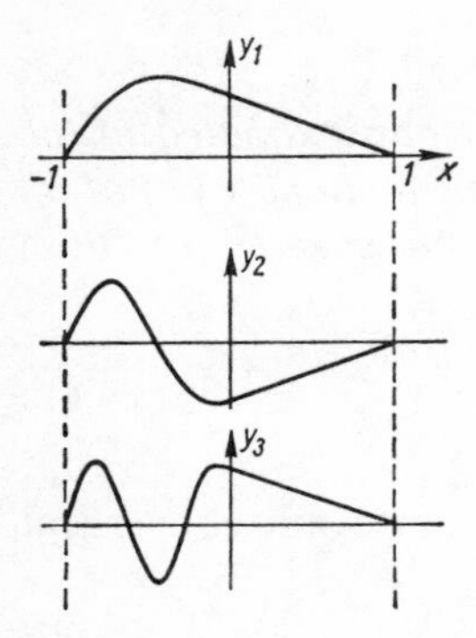

Fig. III.26. On the expansion theorem

That it is in no way possible to answer these questions with a simple general affirmative may be shown by a simple example. The eigenvalue problem

$$y'' + \lambda p(x)y = 0; \qquad y(-1) = y(1) = 0$$

with

$$p(x) = \begin{cases} 1 \text{ for } & -1 \leq x < 0 \\ 0 \text{ for } & 0 \leq x \leq 1 \end{cases}$$

has a countably infinite sequence of eigenvalues λ_j, with easily calculable eigenfunctions, the first three of which are shown in Fig. III.26. All the eigenfunctions run linearly in the interval $0 \leqq x \leqq 1$, and therefore every sum if it converges, can only yield a linear function as the sum. If we start from a function $v(x)$ which is not a linear function in (0, 1), its Fourier series, even if it converges, can certainly not represent $v(x)$.

The questions 1 and 2 lie very deep, and to answer them would take us beyond the scope of this small volume. An insight into certain facts can, however, easily be gained, for example, it follows from (III.93) that, for $a_j = v_j$, i.e., for $\delta = 0$, 'Bessel's inequality'

$$(v, v) \geq \sum_j v_j^2 \tag{III.97}$$

holds, and hence, in particular, the sum of the right-hand side converges; this therefore implies that the Fourier coefficients must decrease sufficiently rapidly as j increases. But if we go beyond the special case of a second-order differential equation, which can be interpreted well geometrically, then even the existence of infinitely many eigenvalues, or even a single eigenvalue, can be proved only by laborious arguments. We shall therefore cite here only one result from the theory, which, although it is by no means the most general of its kind, nevertheless suffices for many applications.

An expansion theorem

In a self-adjoint positive-definite eigenvalue problem with the 'self-adjoint differential equation'

$$\sum_{\nu=0}^{m}(-1)^{\nu}\frac{d^{\nu}}{dx^{\nu}}[p_{\nu}(x)y^{(\nu)}]=\lambda q(x)y \tag{III.98}$$

and 2m mutually linearly independent boundary-conditions at the points $x=a$ *and* $x=b$, *let the* $p_\nu(x)$ *be continuously differentiable at least* ν *times, let* $q(x)$ *be continuous, and* $p_m(x)>0$, $q(x)>0$, $m>0$. *This eigenvalue problem has infinitely many positive eigenvalues.*

Let $v(x)$ *be an arbitrary comparison function; then* $v(x)$ *can be expanded in terms of the eigenfunctions* $y_j(x)$ *in a series* (III.78), *for which the series of absolute values is uniformly convergent in* (a, b), *and which can be differentiated term by term* $(m-1)$ *times.*

(For a detailed proof, see, e.g., L. Collatz: *Eigenwertaufgaben mit technischen Anwendungen*, Leipzig 1963, pp. 131–145.)

22. THE THEOREM ON INCLUSION OF AN EIGENVALUE BETWEEN TWO QUOTIENTS

This theorem, very useful at times for the numerical calculation, and especially for a survey of the position, of the eigenvalues, will be proved here only for quite a special case, but it also holds under much more general hypotheses (see, e.g., L. Collatz: *Eigenwertaufgaben*: J. Albrecht: Verallgemeinerung eines Einschleszungssatzes von L. Collatz, *Z. angew. Math. Mech.* **48** (1968), T43–T46; F. Goerisch: Weiterentwicklung von Verfahren zur Berechnung von Eigenwertsschranken, Diss. Clausthal 1978 and Internat. *Ser. Num. Math.* **39** (1978), 86–100.)

Let the eigenvalue problem be the one presented in No. 17:

$$-(f_0(x)y')'=\lambda g_0(x)y, \quad \text{in an interval } I=(a,b)$$

$$y(a)-c_1y'(a)=y(b)+c_2y'(b)=0 \quad \text{or} \quad y(a)=y'(b)=0.$$

In I let $f_0>0$, $g_0>0$, let f_0' and g_0 be continuous, and let $c_1\geq 0$, $c_2\geq 0$. It may be assumed that there is a positive eigenfunction $y(x)$ in the open interval I (with the corresponding eigenvalue λ), and let $v(x)$ be a positive comparison function in I (the existence of such an eigenfunction $y(x)$ can actually be proved). From No. 17 the eigenvalue problem is self-adjoint, i.e., $(y, w)=(w, y)$, or

$$\int_a^b [y(f_0w')'-w(f_0y')']\,dx=0.$$

We now introduce the quotient

$$\Phi(x)=\frac{-(f_0(x)w')'}{g_0(x)w}$$

and let w be chosen so that $\Phi(x)$ lies between finite limits $\Phi_{\min}$ and $\Phi_{\max}$:

$$\Phi_{\min} \le \Phi(x) \le \Phi_{\max}.$$

If in the integral above we put

$$(f_0 w')' = -\Phi g_0 w \quad \text{and} \quad -(f_0 y)' = \lambda g_0 y$$

then it follows that

$$\int_a^b g_0(x) w(x) y(x) [-\Phi(x) + \lambda] \, \mathrm{d}x = 0.$$

The factor $g_0 wy$ is continuous and positive in I; the continous function $\lambda - \Phi(x)$ cannot therefore be positive throughout, nor negative throughout (a, b), but there must be a point $x = \xi$ in I at which $\lambda = \Phi(\xi)$, i.e.,

$$\Phi_{\min} \le \lambda \le \Phi_{\max}. \tag{III.98a}$$

A numerical example. Buckling load of a bar of variable bending stiffness

When applying the inclusion theorem we can often improve the bounds numerically very considerably by introducing parameters a_1, a_2, ... and choosing these so that the maximum of the quotient Φ becomes as small as possible. By a new calculation, possibly with other values of the parameters a_ν, we try to make the minimum of the quotient Φ as large as possible (numerically we have a non-linear optimization problems on hand). The method will be illustrated by a simple example (with one additional parameter a_1).

Suppose the eigenvalue problem is

$$-y'' = \lambda(3 + \cos x)y, \qquad y(\pm\pi) = 0.$$

Physically, we can interpret this as a buckling problem for a bar of variable bending stiffness, written in dimensionless form as

$$EJ = \frac{\text{const}}{3 + \cos x}$$

(see Fig. III.27). For the quotients

$$\Phi[w] = \frac{-w''}{(3 + \cos x)w}$$

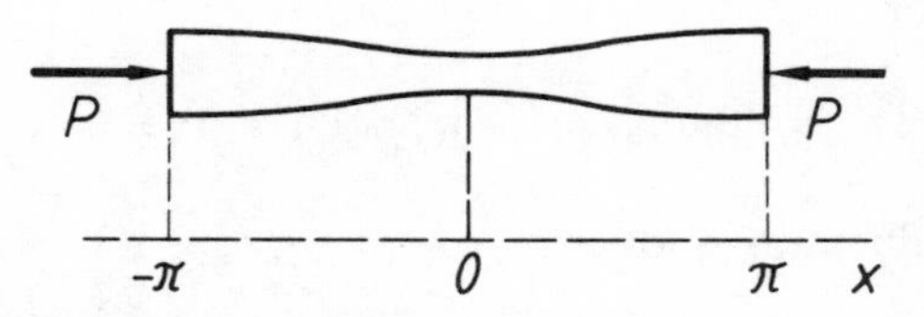

Fig. III.27. Buckling of a bar of variable bending stiffness

we obtain, with the simplest substitution $w = \cos(x/2)$ the expression

$$\Phi\left[\cos\left(\frac{x}{2}\right)\right] = \frac{1}{4(3 + \cos x)}$$

and hence

$$\Phi_{\min} = \frac{1}{16} \le \lambda \le \Phi_{\max} = \frac{1}{8}.$$

We can now improve the bounds considerably by including a term $\cos(3x/2)$ and a parameter a_1, i.e., by working with the function

$$w(x) = a_1 \cos\left(\frac{x}{2}\right) + \cos\left(\frac{3x}{2}\right).$$

Writing $s = [\cos(x/2)]^2$ for brevity (then we have $0 \le s \le 1$) we obtain

$$\Phi = \frac{1}{8}\hat{\Phi} \quad \text{with } \hat{\Phi} = \frac{a_1 - 27 + 36s}{(a_1 - 3 + 4s)(1 + s)}.$$

If we plot $\hat{\Phi}$ for several values of the parameter a_1 as in Fig. III.28, we have to select a value of a_1 for which the curve for $\hat{\Phi}(s)$ is as nearly constant as possible. This occurs for the value

$$a_1 = 29 + \sqrt{868} \approx 58.462.$$

We obtain

$$0.56727 \le \hat{\Phi} \le 0.57385$$

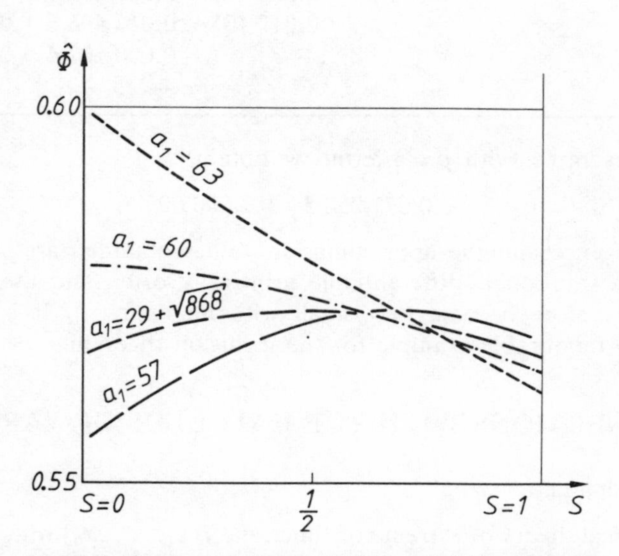

Fig. III.28. The quotient is to be as nearly constant as possible

and for λ the bounds

$$0.0709 \leq \lambda \leq 0.07174.$$

We have thus bracketed one eigenvalue for which a positive eigenfunction exists in the interior of the interval. This corresponds to the least eigenvalue, and hence to the required buckling load.

By taking in further terms the accuracy can be considerably improved. With the formula (I thank Mr. Uwe Grothkopf for the calculations on a computer)

$$y \approx w = \sum_{\nu=1}^{p} b_\nu \cos\left[\left(\nu - \frac{1}{2}\right)x\right]$$

the following table is obtained. The error estimate here is

$$|\lambda - \hat{\lambda}| \leq h$$

with

$$\hat{\lambda} - h = \Phi_{\min}, \qquad \hat{\lambda} + h = \Phi_{\max}$$

further b_1 is put equal to 1, and in each case the value of the last used coefficient a_ν is given.
Calculation with p terms:

	$p=1$	$p=2$	$p=3$	$p=4$
Approximation value $\hat{\lambda}$ for λ	0.093 75	0.071 34	0.071 250 5	0.071 250 472
h	0.031 25	0.000 43	0.000 002 6	0.000 000 008
b_2	—	0.017 105	0.017 496 6	0.017 497 33
b_3	—	—	0.000 102 4	0.000 103 27
b_4	—	—	—	0.000 000 30

Thus, for example, with $p = 3$ terms we obtain

$$|\lambda - 0.071\,250\,5| \leq 0.000\,002\,6.$$

The convergence and the approximation values obtained are so good here because the solution is differentiable arbitrarily often and the Fourier coefficients therefore become very small with increasing ν. It is therefore a particularly favourable example for the inclusion theorem.

§7. CONNECTIONS WITH THE CALCULUS OF VARIATIONS

23. Some simple examples

In the classical theory of extrema of functions $f(x_1, \ldots, x_n)$ which are defined in a domain B of the n-dimensional point-space R^n the problem is to find the points P in B at which f assumes a maximum, or a minimum, or, in general, a stationary value. In the calculus of variations the question about extrema is generalized to the case of a functional I which depends not only on a finite number of variables x_ν but also on a function $\varphi(x)$, and this functional I is to be a maximum or minimum or is to take a stationary value. The functional may also depend on several functions and their derivatives; and there may be special subsidiary conditions or constraints. Here we discuss only the simplest case, and for more complicated cases the extensive literature must be referred to (see, e.g., L. A. Pars, *Calculus of Variations*, Wiley, New York, 1963); I. M. Gelfand and S. V. Fomin, *Calculus of Variations*, Prentice-Hall, Englewood Cliffs, 1963 (trans. R. A. Silverman); R. Weinstock, *Calculus of Variations*, with applications to physics and engineering, McGraw-Hill, New York, 1952).

As an introduction we select two very well-known examples.

1. A surface of revolution

Fig. III.29 shows as the curve C the graph in the x,y-plane of a function $\varphi(x)$, which satisfies the boundary conditions

$$\varphi(x_1) = y_1, \qquad \varphi(x_2) = y_2 \tag{III.99}$$

with given positive values y_1, y_2. Let $\varphi(x)$ be > 0 in the interval $J = [x_1, x_2]$. The curve C is now rotated about the x-axis to obtain a surface of revolution R as in Fig. III.30. It is a known result from the calculus that R has the area

$$M = 2\pi I$$

with

$$I = I[\varphi] = \int_{x_1}^{x_2} \varphi(x)\sqrt{1 + [\varphi'(x)]^2}\,\mathrm{d}x. \tag{III.100}$$

Functions $\varphi(x)$ which satisfy the boundary conditions (III.99) and belong to the class $C^1[J]$, i.e., which are continuous and have continuous first derivatives in J may be called 'admissible' functions.

The question now is: what is the smallest value of $I[\varphi]$ when $\varphi(x)$ runs through the set of admissible functions; which admissible function $u(x)$ gives

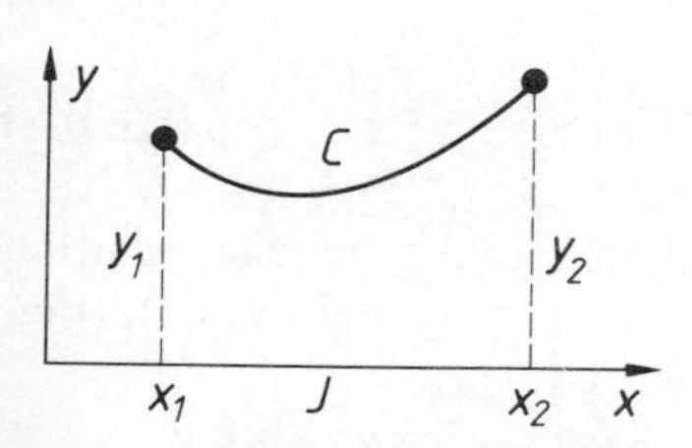

Fig. III.29. The meridian curve of a surface of revolution

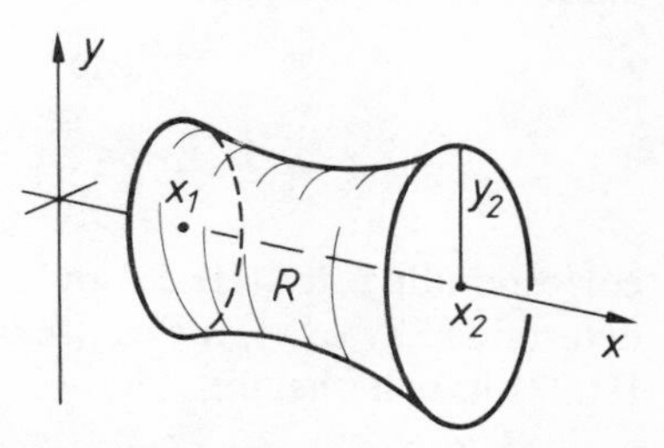

Fig. III.30. Problem of a surface of revolution of minimum area

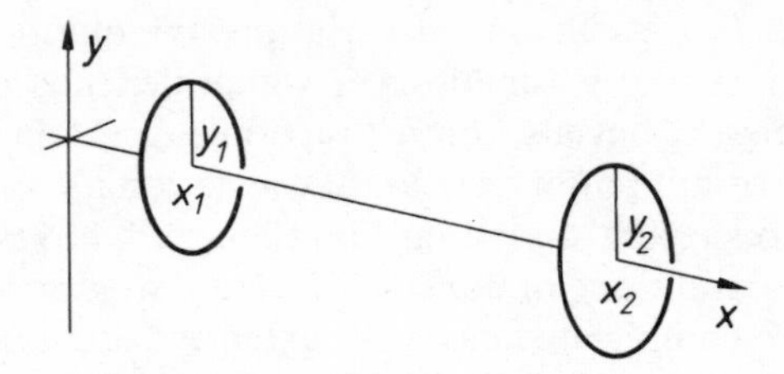

Fig. III.31. Variational problem without an admissible solution

I the smallest possible value; and finally how can u be determined or, at least, calculated approximately?

The variational problem in this case, as can be 'seen' directly and as we can also prove, does not by any means always have a solution; for example, if y_1, y_2 are "small" and the distance $x_2 - x_1$ is 'large', the values of I have for all admissible values the lower limit

$$\text{Inf } I[\varphi] = y_1^2 + y_2^2$$

$(= (1/(2\pi)) \cdot$ (the surface area of the two discs in Fig. III.31)); but there is no admissible function $\varphi(x)$ for which the limit value is assumed.

The treatment of this variational problem is continued in No. 24.

2. The catenary.

In III, No. 4 the catenary was determined with the aid of the equilibrium conditions by means of a boundary-value problem. Here the problem will be presented as a variational problem. The same notation as in No. 4 is used. Suppose a homogeneous, ideally flexible, thin cable hanging under the influence of gravity is attached to the points (x_1, y_1), (x_2, y_2) in the x,y-plane, Fig. III.5, and we ask about the function $y(x)$ describing the equilibrium figure; as in No. 4 let γ be the specific weight, F the constant area of cross-section, l the length, and $G = \gamma Fl$ the weight of the cable.

Now let $\varphi(x)$ be an 'admissible', i.e., an arbitrary continuously differentiable, function defined in $[x_1, x_2]$ which satisfies the conditions

$$\varphi(x_1) = y_1, \qquad \varphi(x_2) = y_2 \tag{III.99}$$

$$\int_{x_1}^{x_2} \sqrt{1 + \varphi'^2}\, \mathrm{d}x = l \tag{III.101}$$

corresponding to (III.5) and (III.6). If the cable takes up the position determined by $\varphi(x)$, the centre of gravity S with the co-ordinates (x_s, φ_s), (Fig. III.32) has the height

$$\varphi_s = \frac{1}{G} \int_{x_1}^{x_2} \varphi(x)\gamma F \sqrt{1 + (\varphi'(x))^2}\, \mathrm{d}x.$$

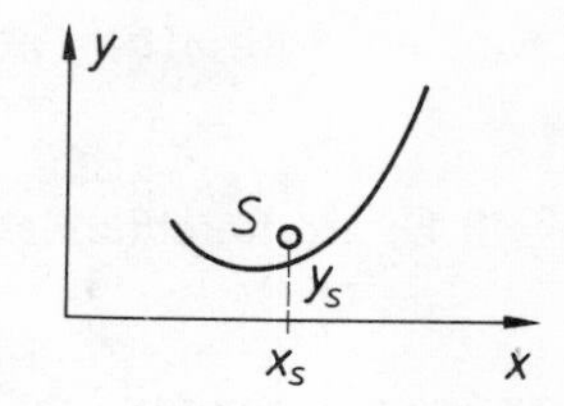

Fig. III.32. Centre of gravity of an element

Now the equilibrium position is characterized by a minimum of the potential energy, in this case by the lowest position of the centre of gravity, i.e., by the smallest possible value of φ_s or by the minimum of the expression (III.100). Thus we now have a 'variational' problem with the subsidiary condition (III.101). The problem is therefore more difficult than in Example 1.

24. THE EULER EQUATION OF THE CALCULUS OF VARIATIONS IN THE SIMPLEST CASE

Example 1 belongs to the following general class of problems. An integral

$$I[\varphi] = \int_a^b F(x, \varphi, \varphi')\,dx \tag{III.107}$$

is considered, where $F(x, u, v)$ is a function given in $a \le x \le b$, $-\infty < u, v < +\infty$ which is twice continuously differentiable with respect to all its arguments. (This assumption can be somewhat moderated.) Further, let two values y_a, y_b be prescribed. A function $\varphi(x)$ defined in $[a, b]$ which is twice continuously differentiable is said to be 'admissible' if it satisfies the boundary conditions

$$\varphi(a) = y_a, \qquad \varphi(b) = y_b \tag{III.103}$$

An admissible function $y(x)$ is then sought which makes the integral (III.102) as small as possible, i.e., for which the relation

$$I[y] \le I[\varphi] \tag{III.104}$$

holds for all admissible φ. (III.103) describes the conditions of the 'boundary-value problem of the first kind', but other boundary conditions as in (III.12) can be prescribed. In Example 1 of No. 23 we have

$$x_1 = a, \qquad x_2 = b, \qquad F(x, u, v) = u\sqrt{1 + v^2}.$$

Here we shall establish only one necessary condition for the variational problem (III.102) to (III.104) to have any solution. To do this, let us assume that there is a solution $y(x)$ of the variational problem. Then

$$\varphi(x) = y(x) + \varepsilon\eta(x)$$

is also an admissible function, if $\eta(x)$ is a twice continuously differentiable function with $\eta(a) = \eta(b) = 0$ and if ε is an arbitrary real constant. Then

$$I[y + \varepsilon\eta] = \Phi(\varepsilon)$$

will be a differentiable function of ε, and since

$$I[y + \varepsilon\eta] \geq I[y]$$

it assumes the minimum value $I[y]$ for $\varepsilon = 0$. The derivative with respect to ε must therefore vanish at the point $\varepsilon = 0$:

$$\Phi'(0) = \left(\frac{\mathrm{d}}{\mathrm{d}\varepsilon} I[y + \varepsilon\eta]\right)_{\varepsilon = 0} = 0.$$

By the rules of the differential calculus

$$\frac{\mathrm{d}I[y + \varepsilon\eta]}{\mathrm{d}\varepsilon} = \int_a^b \left[\eta \frac{\partial F}{\partial u}(x, y + \varepsilon\eta, y' + \varepsilon\eta') + \eta' \frac{\partial F}{\partial v}(x, y + \varepsilon\eta, y' + \varepsilon\eta')\right] \mathrm{d}x;$$

$$\Phi'(0) = \left(\frac{\mathrm{d}}{\mathrm{d}\varepsilon} I[y + \varepsilon\eta]\right)_{\varepsilon = 0} = \int_a^b \left[\eta \frac{\partial F}{\partial u}(x, y, y') + \eta' \frac{\partial F}{\partial v}(x, y, y')\right] \mathrm{d}x.$$

The integral over the second term of the sum is transformed by integrating by parts:

$$\int_a^b \eta' \frac{\partial F}{\partial v}(x, y, y')\,\mathrm{d}x = \eta \frac{\partial F}{\partial v}\bigg|_a^b - \int_a^b \eta \frac{\mathrm{d}}{\mathrm{d}x}\frac{\partial F}{\partial v}(x, y, y')\,\mathrm{d}x.$$

Since $\eta(a) = \eta(b) = 0$, the integrated part vanishes, and it follows that

$$\Phi'(0) = \int_a^b \eta(x) \cdot E[y]\,\mathrm{d}x = 0, \qquad \text{(III.105)}$$

where

$$E[\varphi] = \frac{\partial F}{\partial \phi}(x, \varphi, \varphi') - \frac{\mathrm{d}}{\mathrm{d}x}\frac{\partial F}{\partial \phi'}(x, \varphi, \varphi'),$$

we have written $\partial F/\partial\phi$ and $\partial F/\partial\phi'$ instead of $\partial F/\partial u$ and $\partial F/\partial v$, a change which probably will not be misunderstood.

Now the equation (III.105) must hold for every arbitrary continuously differentiable function $\eta(x)$ such that $\eta(a) = \eta(b) = 0$. This is possible only if the factor $H = E[y]$ multiplying η vanishes identically in (a, b). For, if H were not equal to zero at a point $a < \hat{x} < b$, say, if H were > 0, then because of the continuity of H we would also have $H > 0$ in a small neighbourhood of $\hat{x}$, say for $|x - \hat{x}| \leq c$ with a sufficiently small c. But then (III.105) would be infringed for the function

$$\hat{\eta}(x) = \begin{cases} \eta(x) + (x - \hat{x} - c)^3(x - \hat{x} + c)^3 & \text{in } U \\ \eta(x) & \text{otherwise} \end{cases}$$

($\hat{\eta}$ satisfies the conditions imposed for η). Therefore it must necessarily be that

$$E[y] = \frac{\partial F}{\partial \varphi}(x, y, y') - \frac{\mathrm{d}}{\mathrm{d}x}\frac{\partial F}{\partial \varphi'}(x, y, y') = 0. \qquad \text{(III.106)}$$

This equation is called the 'Euler equation' for the variational problem defined by (III.102), (III.103).

The connections with the calculus of variations form the basis for various numerical methods; cf. *V*, §5, No. 15.

Example: the surface of revolution

In the Example 1 of No. 23 with $F=\varphi\sqrt{1+\varphi'^2}$ we have

$$\frac{\partial F}{\partial \varphi}=\sqrt{1+\varphi'^2}, \qquad \frac{\partial F}{\partial \varphi'}=\frac{\varphi\varphi'}{\sqrt{1+\varphi'^2}},$$

and a solution of the variational problem must satisfy the Euler equation (III.106):

$$\sqrt{1+\varphi'^2}-\frac{\mathrm{d}}{\mathrm{d}x}\frac{\varphi\varphi'}{\sqrt{1+\varphi'^2}}=0. \qquad \text{(III.107)}$$

By differentiating out, this equation can be brought into the form

$$\varphi\cdot\varphi''=1+\varphi'^2,$$

and since x does not appear explicitly, the order of the equation can be reduced by 1 by the method II §1, No. 3, and the solution of the differential equation can be obtained. But generally, for an integral $I[\varphi]$ in (III.102) in which the integrand F does not depend explicitly on x, an integral of the Euler equation can be given. For, if $\partial F/\partial x=0$, it follows, of course, that

$$\frac{\mathrm{d}F}{\mathrm{d}x}=\frac{\partial F}{\partial \varphi}\varphi'+\frac{\partial F}{\partial \varphi'}\varphi''.$$

Now by (III.106)

$$\frac{\partial F}{\partial \varphi}=\frac{\mathrm{d}}{\mathrm{d}x}\left(\frac{\partial F}{\partial \varphi'}\right);$$

therefore

$$\frac{\mathrm{d}F}{\mathrm{d}x}=\varphi'\frac{\mathrm{d}}{\mathrm{d}x}\left(\frac{\partial F}{\partial \varphi'}\right)+\varphi''\frac{\partial F}{\partial \varphi'}=\frac{\mathrm{d}}{\mathrm{d}x}\left[\varphi'\frac{\partial F}{\partial \varphi'}\right],$$

and by integrating we get

$$F-\varphi'\frac{\partial F}{\partial \varphi'}=\text{const.} \qquad \text{(III.108)}$$

In the present case this gives

$$\varphi\sqrt{1+\varphi'^2}-\varphi'\cdot\frac{\varphi\varphi'}{\sqrt{1+\varphi'^2}}=\text{const}=c$$

or

$$\varphi=c\sqrt{1+\varphi'^2}.$$

Here, the variables can be separated as in I, §1, No. 4 and we obtain, as in III, §2, No. 4, the catenaries

$$\varphi(x) = c \cdot \cosh \frac{x + c_1}{c}$$

with c, c_1 as free constants. [These catenaries are contained in III, §2, No. 4 for $C_2 = 0$, $a = 1/c$.] The constants c, c_1 have to be adjusted to the boundary conditions (III.99), and it has to be checked whether there is a solution $\varphi(x)$ which is not negative in J. If $x_2 - x_1$ is too great in comparison with y_1 and y_2, the requirement that $\varphi(x) \geq 0$ in J cannot be satisifed, and the variational problem has no admissible solution.

25. Free boundary-value problems and the calculus of variations

'Free boundary-value problems', which have become very topical in recent times and which appear in many very different applications, crop up particularly with partial differential equations (cf. Chapter V, §4); they have in many cases been successfully handled by virtue of their connections with the calculus of variations. This connection may be illustrated here in a model case of a boundary-value problem with a simple ordinary differential equation. We shall need a little more from the calculus of variations than has been proved in No. 24.

In the x, y-plane let a cable be suspended from the point $x = 0$, $y = 1$ and let part of it lie on a table the surface of which is given by $y = 0$ (see Fig. III.33). Let the cable touch the table with a horizontal tangent at the point $x = c$. In the interval $J = (0, c)$ let the shape of the cable, which is defined by $y(x)$, satisfy the differential equation (III.3) of a catenary. The 'free boundary' $x = c$ is not known in advance, but for it we obtain one more boundary condition; namely, for the second-order differential equation we have the three boundary conditions:

$$y(0) = 1,$$
$$y(c) = 0,$$
$$y'(c) = 0.$$

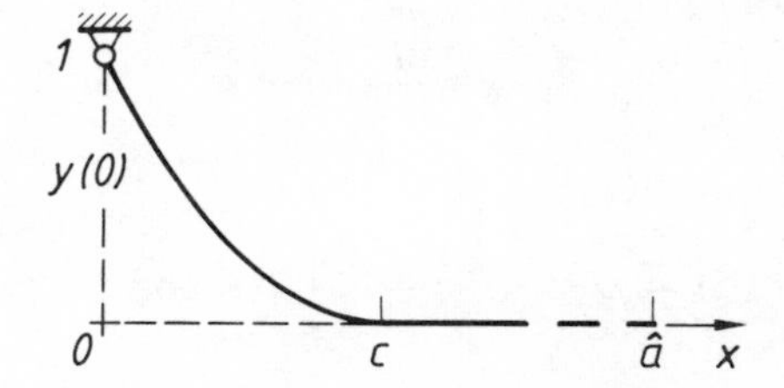

Fig. III.33. A free boundary-value problem with a cable

The function $y(x)$ satisfies the differential equation (III.3) in the interval J, and for $x \geq c$ it satisfies the equation $y(x) \equiv 0$. Free boundary-value problems of this kind are also difficult to handle numerically, because it is not known in advance in what domain the differential equation is valid. They can, however, in many cases be reduced to a variational problem in a fixed domain. To do this, here as in many similar cases, we must at least know an upper bound a for the free boundary: $\hat{a} \geq c$; we can choose a very large and need only to be certain that $\hat{a} \geq c$.

Instead of the catenary we shall work here as a simpler model case with a parabola with a linear differential equation, although we could also work with the catenary. We have the problem

$$\text{(D)} \qquad y'' = 1, \qquad y(0) = 1, \qquad y(c) = y'(c) = 0. \qquad \text{(III.109)}$$

This problem is now compared with the variational problem (V)

$$\text{(V)} \qquad I[z] = \int_0^a [z'(x))^2 + 2z(x)] \, dx = \min \quad \text{for } z(x) \in M, \qquad \text{(III.110)}$$

where M is the set of admissible functions $z(x)$, a function being admissible if it satisfies the conditions

$$\left.\begin{array}{l} z(0) = 1, z(x) \geq 0 \text{ in } J_\nu = [0, \hat{a}], \quad \text{there is a } b \in J_\nu \text{ with } z(b) = 0 \\ z(x) \in C^1[J_\nu], \text{ i.e., } z(x) \text{ is continuously differentiable in } J_\nu \\ z(x) \text{ is twice continuously differentiable in } J_\nu \text{ except possibly} \\ \qquad \text{for one point } x = c \text{ with } c \geq b. \end{array}\right\} \qquad \text{(III.111)}$$

(It can be shown that it suffices for z to be once continuously differentiable.) It is precisely the singularity of the second derivative at the position of the free boundary that is typical of free boundary-value problems. Under these conditions the following lemma holds.

Lemma

If $w(x)$ is a solution of the variational problem (V), *then $w(x)$ is also a solution of the free boundary-value problem* (D).

Proof. Since $w(x)$ itself is an admissible function, there is at least one zero x_0 of $w(x)$ in the interval J. There is then also a smallest positive zero s of $w(x)$; at $x = s$ the function $w(x)$ has a differentiable minimum, and so

$$w(s) = w'(s) = 0, \qquad w(x) > 0 \quad \text{for } 0 \leq x < s.$$

By hypothesis

$$I[w] \leq I[z] \quad \text{for all admissible } z(x)$$

from which it follows that $w(x) \equiv 0$ for $s \leq x \leq a$; for if $w(x)$ were > 0 for some $x = x_1$ with $s < x_1 \leq a$, then by changing $w(x)$ to zero in $s \leq x \leq a$ we could reduce the value of the integral I.

Now the problem (V) differs from the 'classical' variational problem considered in No. 24 by the subsidiary condition $z(x) \geq 0$ in J_ν and by the existence of *a* zero b. However, the whole of the proof given in No. 24 can be carried through with the additional subsidiary condition, and we find that $w(x)$ must satisfy the Euler equation. The integrand in (III.110) has now to be chosen so that the corresponding Euler equation reads $w'' - 1 = 0$. The function $w(x)$ is therefore a solution of the free boundary problem.

Accordingly, the tried and tested methods of the calculus of variations, such as Ritz's method, are available for the numerical treatment of free boundary-value problems. But this No. 25 was intended only to sketch out the connections and possibilities.

§8. BRANCHING PROBLEMS

26. BRANCHING PROBLEMS WITH ORDINARY DIFFERENTIAL EQUATIONS OF THE SECOND ORDER

The eigenvalue problems considered in No. 13 are linear problems, which in applications are frequently only to be regarded as first approximations. This has already been illustrated in No. 14 by the example of the buckling load of a column with a rectilinear axis:

If we substitute for the bending moment $M(x) = Py$ (as in Fig. III.14) into (II.51) and in the transition from (III.48) to (III.49), then (III.51) becomes

$$\frac{y''}{\sqrt{1+y'^2}^3} = -\lambda y(x) \quad \text{with } \lambda = \frac{P}{EI} \tag{III.112}$$

while the boundary conditions

$$y(0) = y'(1) = 0 \tag{III.113}$$

remain unchanged. If we write

$$\left.\begin{aligned} F &= \frac{y''}{\sqrt{1+y'^2}^3} + \lambda y = 0 \\ S_1 &= y(0) = 0 \\ S_2 &= y'(1) = 0 \end{aligned}\right\} \tag{III.114}$$

then the problem is of the type

$$\begin{cases} F(x, y, y', y'', \lambda) = 0 \\ S_\mu(x, y, y', \lambda) = 0 \quad (\mu = 1, 2). \end{cases} \tag{III.115}$$

Here F and S_μ are given functions of their arguments, and the boundary conditions $S_\mu = 0$ relate to, say, the two points $x = a$ and $x = b$.

27. NON-LINEAR EIGENVALUE PROBLEMS AND BRANCHING PROBLEMS

In the general case one arrives at the following distinction of cases:

Case I. The differential equation and the boundary conditions are linear and homogeneous, and therefore are of the form

$$F = \sum_{\nu=0}^{2} p_\nu(x, \lambda) y^{(\nu)} = 0 \tag{III.116}$$

$$S_\mu = c_{1,\mu}(\lambda) y(a) + c_{2,\mu}(\lambda) y'(a) + c_{3,\mu}(\lambda) y(b) + c_{4,\mu}(\lambda) y'(b) = 0 \quad (\mu = 1, 2).$$

(If at least one of the given functions p_ν, $c_{\nu,\mu}$ depends non-linearly on λ, then a 'non-linear eigenvalue problem' is occurring; see Case II.)

Again a number $\lambda = \lambda_0$ is called an 'eigenvalue' if there is a non-identically zero function $y_0(x)$ such that the pair $(\lambda_0, y_0(x))$ satisfies the conditions (III.115). As with linear eigenvalue problems, it is here again true that:

1. If $y(x)$ is an eigenfunction corresponding to the eigenvalue λ, then so is $cy(x)$, where c is any real constant.
2. If $y_j(x)$ for $j = 1, \ldots, r$ are eigenfunctions corresponding to the same eigenvalue λ, *then*

$$\sum_{j=1}^{r} c_j y_j(x) \tag{III.117}$$

also is an eigenfunction corresponding to the eigenvalue λ. The eigenfunctions corresponding to the same λ form a 'linear subspace'.

Case II. The differential equation or the boundary conditions or both are not linear. Every pair $(\lambda, y(x))$ which satisfies the equations (III.115) is called a solution of the problem. We can gain a general idea of the manifold of solutions in the following way. First we associate with every function $y(x)$ some scalar number, e.g., a norm or $y_{\max}$, depending on the problem in hand, and further, we associate with every solution $(\lambda, y(x))$ a point in a λ-y_{mac}-- co-ordinate system. These points then lie in general on certain curves (the 'branches' of a hypergraph). In the example in No. 26, $y(x) = 0$ with an arbitrary λ is always a solution of the equations (III.114). Consequently $y = 0$ with λ arbitrary is a branch of the hypergraph. From this branch at certain values of λ_ν, the eigenvalues of the linearized problem, other branches branch off. The set of all these branches is the 'branching diagram'. The 'branching points' are the 'corners' of the hypergraph. In Chapter III, No. 14 such a branching diagram is drawn in Fig. III.20.

28. Example. The branching diagram for a Urysohn integral equation

The calculation of branching problems for non-linear differential equations is often only possible numerically. Here, therefore, an example with a non-linear integral equation will be given, although this equation does not come from some application, but because we can find the solutions in closed form, and so we can be certain that we have found the complete branching diagram.

The non-linear integral equation

$$Ty(x) = \int_{-\pi}^{\pi} \{py(t) + q(\cos x)(y(t))^2\}\, dt = \hat{\lambda} y(x) = \lambda\pi y(x) \qquad \text{(III.118)}$$

is given for the determination of the unknown function $y(x)$, required in an interval $I = [a, b]$. We have put $\hat{\lambda} = \lambda\pi$ in order to remove the factor π from the calculation. The numbers p and q are given, real, non-zero constants. For $\lambda = 0$ there is the solution $y \equiv 0$.

If $\lambda \neq 0$ any possible solution must be of the form

$$y(x) = a + b \cos x.$$

An elementary calculation shows that

$$\int_{-\pi}^{\pi} y(t)\, dt = 2\pi a, \int_{-\pi}^{\pi} (y(t))^2\, dt = \pi(2a^2 + b^2)$$

and on substituting these values in the integral equation we find

$$2ap + q(2a^2 + b^2)\cos x = \lambda(a + b \cos x).$$

Comparison of coefficients gives

$$(\lambda - 2p)a = q(2a^2 + b^2) - \lambda b = 0.$$

Therefore either

Case I. $a = 0$ and then $(qb - \lambda)b = 0$, which gives the two straight lines

$$G_1 : a = b = 0, \qquad \lambda \text{ arbitrary}$$

and

$$G_2 : a = 0, \qquad \lambda = qb, \text{ or}$$

Case II. $a \neq 0$, ellipse E: $\lambda = 2p$, $q(2a^2 + b^2) - 2pb = 0$. Figure III.34 shows the solution branches G_1, G_2, E of the branching diagram (i.e., of the hypergraph) drawn in the λ-a-b-co-ordinate system; three branching points P_1, P_2, P_3 occur (the corners of the hypergraph:

	λ	a	b
P_1	0	0	0
P_2	$2p$	0	0
P_3	$2p$	0	$2q$

Many other examples of branching diagrams as hypergraphs for differential and integral equations can be found in the paper by L. A. Collatz, Verz-

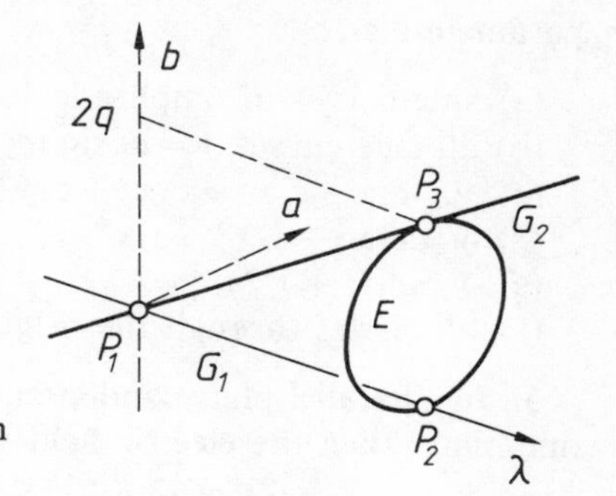

Fig. III.34. Branching diagram for a Urysohn integral equation

weigungsdiagramme und Hypergraphen, *Intern. Ser. Num. Math.* vol. 38 (1977), 9–41.

§9. MISCELLANEOUS PROBLEMS AND SOLUTIONS ON CHAPTERS II AND III

29. PROBLEMS

1. Set up the general solutions of the following differential equations:

(a) $y'' + 4y' + 5y = x + e^x \cos x$;
(b) $y''' + 4y'' + 5y' = e^{2x} \cos x$;
(c) $y''(1 + y') = 1$;

(d) $x^2 y''' - 11y' - \dfrac{5}{x} y = 1$;

(e) $y^{IV} + 4y''' + 8y'' + 8y' + 4y = e^x$;
(f) $y'' = e^{ay}$;
(g) $y'' + ayy'^3 = 0$.

2. By making the power-series substitution $y = \sum_{\nu=0}^{\infty} a_\nu x^\nu$, find the solution, uniquely determined by the initial value $y(0) = 1$, of the differential equation

$$y'' + y'\left(x + \frac{2}{x}\right) + 2y = 0$$

(cf. Example 2 on Chapter I.) The fact that here only one initial condition needs to be given is due to the singularity of the differential equation at $x = 0$.

3. The same for $y'' = y(x^2 + 2)$, with the initial conditions

$$y(0) = 1,\ y'(0) = 0.$$

4. The following multi-parameter families of functions are given; the parameters are $c_1, c_2, \ldots, c_k$. Set up a differential equation for each family by differentiating the equation of the family k times and eliminating the

parameters $c_1, c_2, \ldots, c_k$:

(a) sine curves of amplitude 1, $y = \sin(c_1 x + c_2)$;
(b) all sine curves $y = c_1 \sin(c_2 x + c_3)$;
(c) $y = c_1 x + c_2 x^2 + c_3 x^3 + c_4 x^4$;
(d) $y = c_1 x^2 + c_2 x^3 + c_3 x^4$;
(e) $y = c_1(1 + x^{c_2})$;
(f) all circles through the origin: $x^2 + c_1 x + y^2 + c_2 y = 0$.

5. In a parallel-plate condenser let $\varphi(x)$ be the potential at a distance x from one plate; then the electric field strength has the value

$$|\mathfrak{E}(x)| = |-\operatorname{grad}\varphi| = |\varphi'(x)|.$$

If one plate forms the anode, and the other the hot cathode, in a diode, the electrons emitted from the cathode produce a space charge with the density $\rho(x) = c/\sqrt{\varphi(x)}$ (see B. L. Bleaney and B. Bleany, *Magnetism and Electricity*, Oxford, 1957, p. 332), where c is a constant to be determined. The potential equation for the space-charge field $\Delta\varphi = -\rho/\varepsilon$ (with ε as the dielectric constant) here leads to the equation

$$\varphi'' = \frac{d^2\varphi}{dx^2} = -\frac{c}{\varepsilon\sqrt{\varphi}}.$$

For this second-order differential equation and the unknown parameter c appearing in it we have here three boundary conditions: for $x = 0$, let $\varphi = 0$ and $\varphi' = 0$ (field strength = 0), and for $x = d$ (the distance between the plates), let $\varphi = U$, the applied voltage. Calculate the potential c and find expressions for the potential and space-charge density at distance x.

6. To a series electrical oscillatory circuit which has the capacitance

$$C = 10^{-6} \text{ farad} = 10^{-6} \text{ s}/\Omega$$

and the self-inductance $L = 1$ Henry $= 1\,\Omega$s and negligible resistance a periodically varying EMF $U(t)$ is applied (Fig. III.35), which varies linearly between ± 50 volt with the frequency 100 Hertz:

$$U(t) = \begin{cases} 20\,000t - 50 & \text{for } 0 \leq t \leq \dfrac{T}{2} \text{ with } T = 0.01 \text{ s} \\ -20\,000t + 150 & \text{for } \dfrac{T}{2} \leq t \leq T \\ U(t - 0.01) & \text{for all } t \end{cases}$$

Determine the variation of voltage with time in the steady state.

7. A tightly fitting piston falls in a cylinder from a height $h = 1$ m (Fig. III.36). The friction can be neglected. Initially (at time $t = 0$) the pressure on each side of the piston is $p_0 = 1$ bar, and the piston is at rest $\dot{x} = 0$; further let $G = mg = 1$ kN be the weight of the piston, $F = 100$ cm^2 be the area of the

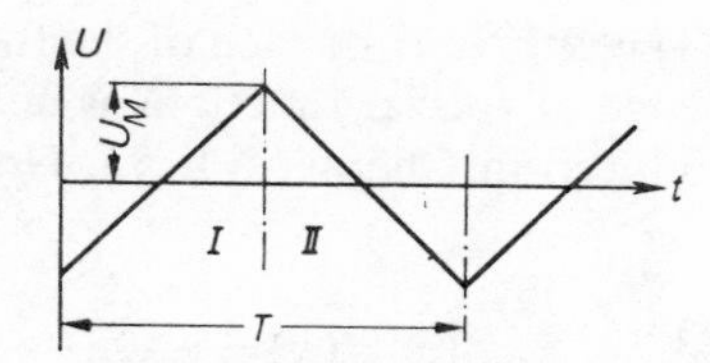

Fig. III.35. Periodically varying EMF applied to an electrical oscillatory circuit

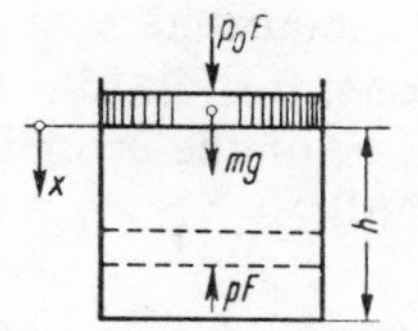

Fig. III.36. Piston falling in a cylinder

piston. $x(t)$ the distance travelled by the piston, and $p(t)$ the pressure in the cylinder at time t. At what height is the velocity of the piston again zero if the process is (a) isothermal, (b) adiabatic?

8. A mass m vibrates between two springs; at a deflection a two further springs come into action (see Fig. III.37); all the springs have the same spring constant c; damping is neglected. Find the period T as a function of the total deflection x_0.

9. A beam, encastré at one end and freely pinned at the other, is loaded as shown in Fig. III.38. Determine the deflection and moment distribution.

10. Solve by means of the Laplace transformation the initial-value problems for the differential equation $y'' - 9y = 0$ with the initial conditions (a) $y(0) = 0$, $y'(0) = 1$; (b) $y(0) = 1$, $y'(0) = 0$.

11. Solve by means of the Laplace transformation:

(a) $y'' - 9y = 2 \sin x, \qquad y(0) = 1, \ y'(0) = 2;$

(b) $y'' - 9y = e^{2x} \cos x, \qquad y(0) = 1, \ y'(0) = 2.$

12. Solve by means of the Laplace transformation the initial-value problem

$$\left.\begin{aligned} \ddot{j}_1 + 3\dot{j}_1 + 2\dot{j}_2 = 0 \\ \ddot{j}_2 + 3\dot{j}_2 - 2\dot{j}_1 = 0 \end{aligned}\right\} \qquad \begin{aligned} j_1(0) = j_2(0) = 0 \\ \dot{j}_1(0) = 2, \ \dot{j}_2(0) = 1 \end{aligned}$$

for the coupled electrical circuits shown in Fig. II.22.

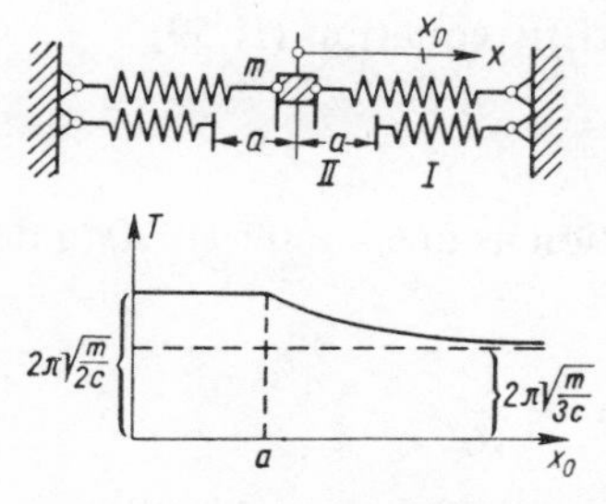

Fig. III.37. An oscillatory system which is non-linear for large deflections

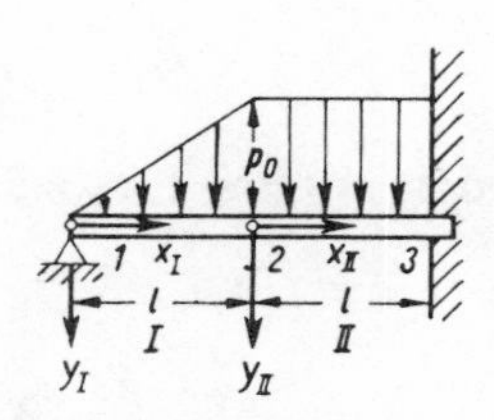

Fig. III.38. A statistically indeterminately supported beam with variable load density

13. Determine the Laplace transform $F(s) = \mathscr{L}[f(x)]$ of each of the following functions; for (c), (d), (e) some knowledge of special functions is needed, in particular, of the error integral Φ introduced in Chapter VI, §2, No. 10, Example 6(d):

(a) $f(x) = \frac{1}{\sqrt{x}} e^{-c/x}$, (b) $f(x) = \ln x$, (c) $f(x) = \frac{1}{\sqrt{1+cx}}$

(d) $f(x) = \frac{\sin(c\sqrt{x}}{x}$ (e) $f(x) = \Phi(c\sqrt{x})$.

14. Show that the function $f(x) = e^{x+x^2}$ has no Laplace transform.

30. Solutions

1. (a) The homogeneous differential equation has by (II.35) the characteristic equation $k^2 + 4k + 5 = 0$ with the zeros $k = -2 \pm i$; therefore by Chapter II, 11, its general solution is $y = e^{-2x}(c_1 \cos x + c_2 \sin x)$. A particular solution of the inhomogeneous equation is found by the rule-of-thumb method of Chapter II, 17 by the substitution

$$y = a_1 + a_2 x + a_3 e^x \cos x + a_4 e^x \sin x$$

(or in place of the last two terms it is convenient to work with complex numbers and use $A_3 e^{(1+i)x}$); the general solution for the inhomogeneous equation then reads

(a) $y = e^{-2x}(c_1 \cos x + c_2 \sin x) + \frac{e^x}{13}\left(\cos x + \frac{2}{3} \sin x\right) + \frac{x}{5} - \frac{4}{25}$;

(b) $y = e^{-2x}(c_1 \cos x + c_2 \sin x) + c_2 + \frac{1}{200} e^{2x}(3 \cos x + 4 \sin x)$;

(c) $y = c_1 - x \pm \frac{1}{3}(2x + c_2)^{3/2}$;

(d) An Euler equation with the characteristic equation (II.59)

$$r^3 - 3r^2 - 9r - 5 = 0$$

and the roots $r_{1,2} = -1$, $r_3 = 5$. General solution as in Chapter II, 20 (a double root!)

$$y = \frac{c_1}{x} + \frac{c_2}{x} \ln x + c_3 x^5 - \frac{1}{16} x.$$

(e) The characteristic equation $k^4 + 4k^3 + 8k^2 + 8k + 4 = 0$ or

$$(k^2 + 2k + 2)^2 = 0$$

here has a pair of complex double roots; general solution

$$y = \frac{1}{25} e^{x} + e^{-x}(c_1 \cos x + c_2 \sin x + c_3 x \cos x + c_4 x \sin x).$$

(f) y'' here is a function of y, therefore multiplication of the differential equation by y'; solution formula (I.68)

$$ay = c_2 x - 2 \ln(1 - c_1 e^{c_2 x}) + \ln \frac{2c_1 c_2^{\,2}}{a} \quad \text{and} \quad ay = -2 \ln\left(c_1 + \sqrt{\frac{a}{2}}\, x\right).$$

(g) x does not appear explicitly, so introduce y' as a function of y as in (II.14); one obtains

$$\frac{a}{3} y^3 + c_1 y - 2x + c_2 = 0 \quad \text{and} \quad y = c_3.$$

2. From the recursion formula $(\nu + 3) a_{\nu+2} = -a_\nu$ (for $\nu = 0, 1, 2, \ldots$) and $a_1 = 0$ it follows that

$$a_{2n} = \frac{(-2)^n n}{(2n+1)!} = \frac{(-1)^n}{1 \cdot 3 \cdot 5 \ldots (2n+1)};$$

$a_3 = a_5 = \ldots = 0$ and hence

$$y = 1 - \frac{x^2}{3} + \frac{x^4}{15} - \frac{x^6}{105} + \frac{x^8}{945} - \frac{x^{10}}{10\,395} + - \cdots.$$

The series converges absolutely for all x.

$$y = \frac{1}{x} \exp\left(-\frac{x^2}{2}\right) \int_0^x \exp\left(\frac{t^2}{2}\right) dt.$$

3. From the recursion formula $(\nu + 1)(\nu + 2) a_{\nu+2} = a_{\nu-2} + 2a_\nu$ (for $\nu = 2, 4, 6, \ldots$ and $a_0 = 1$, $a_1 = a_3 = a_5 \ldots = 0$) it follows at one by induction that

$$0 < \mu!\, a_{2\mu} \leq 1, \quad \text{for } \mu = 0, 1, \ldots,$$

and so the power series converges absolutely for all x.

$$y = 1 + \frac{x^2}{1} + \frac{x^4}{1 \cdot 4} + \frac{x^6}{1 \cdot 4 \cdot 5} + \frac{x^8}{1 \cdot 4 \cdot 5 \cdot 8}$$

$$+ \frac{x^{10}}{1 \cdot 4 \cdot 5 \cdot 8 \cdot 9} + \frac{x^{12}}{1 \cdot 4 \cdot 5 \cdot 8 \cdot 9 \cdot 12} + \cdots.$$

4. (a) $y''(y^2 - 1) = yy'^2$;
(b) $yy''' = y'y''$;
(c) $x^4 y^{\mathrm{IV}} - 4x^3 y''' + 12x^2 y'' - 24xy' + 24y = 0$

or

$$y - \frac{x}{1!} y' + \frac{x^2}{2!} y'' - \frac{x^3}{3!} y''' + \frac{x^4}{4!} y^{\mathrm{IV}} = 0$$

or also

$$x^5\left(\frac{y}{x}\right)^{\text{IV}} = 0.$$

(d)
$$x^3y''' - 6x^2y'' + 18xy' - 24y = 0$$

or

$$\frac{y}{x} - \frac{x}{1!}\left(\frac{y}{x}\right)' + \frac{x^2}{2!}\left(\frac{y}{x}\right)'' - \frac{x^3}{3!}\left(\frac{y}{x}\right)''' = 0$$

or also

$$x^5\left(\frac{y}{x^2}\right)''' = 0;$$

(e) $(xy'' + y')y = y'^2[x + x^{-(xy''/y')}]$;
(f) $y''(x^2 + y^2) + 2(1 + y'^2)(y - xy') = 0$.

5. Multiplication by φ' and integration gives

$$\varphi'^2 = -4\frac{c}{\varepsilon}\sqrt{\varphi} + c_1;$$

the boundary conditions for $x = 0$ require that $c_1 = 0$; then

$$\varphi = \left(\frac{3}{2}\sqrt{-\frac{c}{\varepsilon}}\,x\right)^{4/3}; \qquad c = -\varepsilon\left(\frac{2}{3}\frac{U^{3/4}}{d}\right)^2; \qquad \rho = -\frac{4}{9}\varepsilon U(\mathrm{d}^4x^2)^{-1/3}.$$

So

$$\varphi(x) = U\left(\frac{x}{d}\right)^{4/3}.$$

6. In the two time intervals

I
$$\left(0 \le t \le \frac{T}{2}\right)$$

and

II
$$\left(\frac{T}{2} \le t \le T\right)$$

the following equations hold for the voltages U_{I} and U_{II}:

$$10^{-6}\ddot{U}_{\text{I}} + U_{\text{I}} = 20\,000\,t - 50; \qquad 10^{-6}\ddot{U}_{\text{II}} + U_{\text{II}} = -20\,000\,t + 150$$

with the solutions

$$U_{\text{I}} = 20\,000\,t - 50 + c_1\cos\omega t + c_2\sin\omega t,$$
$$U_{\text{II}} = -20\,000\,t + 150 + c_3\cos\omega t + c_4\sin\omega t \qquad (\omega = 1000/\text{s}).$$

For determining the four constants c_k there are the four transition conditions

$$U_I(0) = U_{II}(T); \qquad \dot{U}_I(0) = \dot{U}_{II}(T);$$

$$U_I\left(\frac{T}{2}\right) = U_{II}\left(\frac{T}{2}\right); \qquad \dot{U}_I\left(\frac{T}{2}\right) = \dot{U}_{II}\left(\frac{T}{2}\right).$$

One obtains $c_1 = -14.94$; $c_2 = -20$; $c_3 = 23.42$; $c_4 = 8.65$. $U(t)$ has the period T.

7. Equation of motion $m\ddot{x} = mg + p_0F - pF$;
(a) isothermal equation: $pF(h - x) = p_0Fh$, and so

$$\ddot{x} = g + \frac{p_0F}{m}\left(1 - \frac{h}{h - x}\right).$$

Multiplication by $\dot{x}$ and integration gives

$$\frac{\dot{x}^2}{2} = f(x) = gx + \frac{p_0F}{m}\left(x + h \ln \frac{h - x}{h}\right) + c_1.$$

$c_1 = 0$ because $\dot{x} = 0$ for $x = 0$; ξ has to be calculated numerically from $f(\xi) = 0$. In the numerical example one finds from $2\xi + \ln(1 - \xi) = 0$ the root $\xi \approx 0.80$ m.

(b) The corresponding calculation for the equation

$$\ddot{x} = g + \frac{p_0F}{m}\left(1 - \frac{h^\varkappa}{(h - x)^\varkappa}\right)$$

(with the value $\varkappa = 1.4$) gives $\xi \approx 0.65$ m.

8. For $x_0 \leq a$, $T = 2\pi\sqrt{m/2c}$. Now let $x_0 > a$. Then the equations of motion are:

$$\text{I: } m\ddot{x} = -2cx - c(x - a) \quad \text{for } a \leq x \leq x_0;$$

$$\text{II. } m\ddot{x} = -2cx \quad \text{for } 0 \leq x < a.$$

Integration with determination of the constants from the boundary conditions (the intermediate time τ is not yet known)

$$x_I(0) = x_0; \quad \dot{x}_I(0) = 0; \quad x_I(\tau) = x_{II}(\tau) = a; \quad \dot{x}_I(\tau) = \dot{x}_{II}(\tau); \quad x_{II}\left(\frac{T}{4}\right) = 0$$

gives

$$T = 2\pi\sqrt{\frac{m}{3c}} - 4\sqrt{\frac{m}{3c}} \text{ arc sin } \frac{2a}{3x_0 - a} + 4\sqrt{\frac{m}{2c}} \text{ arc sin } \frac{\sqrt{2}a}{\sqrt{(x_0 - a)^2 + 2x_0^2}},$$

where arc sin u denotes the principal value lying between $-\pi/2$ and $+\pi/2$ of the infinitely many-valued function arc sin u. Figure III.37 shows T as a function of x_0.

9. The differential equations for the intervals I and II are (co-ordinates x, y_I, y_{II} as in Fig. III.38)

$$y_I^{(IV)} = \frac{p_0}{EJ}\frac{x}{l}; \qquad y_{II}^{(IV)} = \frac{p_0}{EJ},$$

both for $0 < x < l$. The eight constants appearing in the integration have to be determined from the eight boundary and transition conditions

$$y_{\mathrm{I}}(0) = y_{\mathrm{I}}''(0) = y_{\mathrm{II}}(l) = y_{\mathrm{II}}'(l) = 0;$$
$$y_{\mathrm{U}}^{(\nu)}(l) = y_{\mathrm{II}}^{(\nu)}(0) \text{ for } \nu = 0, 1, 2, 3.$$

One finds

$$y_{\mathrm{I}} = \frac{p_0}{1920EJl}(16x^5 - 119x^3l^2 + 228xl^4),$$

$$y_{\mathrm{II}} = \frac{p_0}{1920Ej}(80x^4 + 41x^3l - 197x^2l^2 - 49xl^3 + 125l^4).$$

The expressions for the moments are:

$$M_{\mathrm{I}} = -EJy_{\mathrm{I}}'' = \frac{p_0}{960l}(-160x^3 + 357xl^2),$$

$$M_{\mathrm{II}} = -EJy_{\mathrm{II}}'' = \frac{p_0}{960}(-480x^2 - 123xl + 197l^2).$$

10. The method of Chapter II, No. 37 for the Laplace transform $Y(s) = \mathscr{L}[y(x)]$ gives immediately, using (II.130), in case (a)

$$s^2Y - sy(0) - y'(0) - 9Y = s^2Y - 1 - 9Y = 0,$$

or

$$Y = \frac{1}{s^2 - 9} = \frac{1}{6}\left[\frac{1}{s-3} - \frac{1}{s+3}\right]$$

The inverse transformation, using (II.122), gives $y(x) = \frac{1}{6}(e^{3x} - e^{-3x})$.

(b) From $s^2Y - s - 9Y = 0$, or

$$Y = \frac{s}{s^2 - 9} = \frac{1}{2}\left[\frac{1}{s-3} + \frac{1}{s+3}\right]$$

it follows that $y(x) = \frac{1}{2}(e^{3x} + e^{-3x})$.

11. (a) For the Laplace transform $Y(s) = \mathscr{L}[y6x)]$ we obtain, using (II,130) and (II.123),

$$s^2Y - s\cdot 1 - 2 - 9Y = \frac{2}{s^2 + 1}$$

or

$$Y = \left[\frac{2}{s^2 + 1} + s + 2\right]\frac{1}{s^2 - 9} = \frac{a + bs}{N} + \frac{c + ds}{M},$$

where the constants a, b, c, d are still to be determined, and the denominators are $N = s^2 + 1$, $M = s^2 - 9$: so we have $2 + (s+2)N = (a + bs)M + (c + ds)N$.

The constants a, b, c, d are easily determined by making suitable choices of s. $N=0$ for $s=\pm i$, and we obtain

$$\begin{cases} 2=-10(a+bi) & \text{dor } s=i \\ 2=-10(a-bi) & \text{for } s=-i \end{cases} \quad \text{and hence immediately } b=0,\ a=-\tfrac{1}{5}.$$

Further, $M=0$ for $s=\pm 3$, and we have

$$\begin{cases} 2+50+20=52=10(c+3d) & \text{for } s=3 \\ 2-10=-8=10(c-3d) & \text{for } s=-3 \end{cases} \quad \text{and so } d=1,\ c=11/5.$$

Hence altogether

$$Y=-\frac{1}{5}\frac{1}{s^2+1}+\frac{1}{5}\frac{11+5s}{s^2-9}$$

and

$$y(x)=-\frac{1}{5}\sin x+\frac{1}{15}\left[11\sinh(3x)+15\cosh(3x)\right].$$

(b) Here we have

$$s^2Y-s\cdot 1-2-9Y=\frac{s-2}{(s-2)^2+1}$$

or

$$Y=\left[\frac{s-2}{s^2-4s+5}+s+2\right]\frac{1}{s^2-9}=\frac{as+b}{N}+\frac{cs+d}{M},$$

where $N=s^2-4s+5$ and $M=s^2-9$, and the constants a, b, c, d are determined by the same method as in a).

$$y(x)=\frac{1}{156}(143\,e^{3x}+31\,e^{-3x})+\frac{1}{26}e^{2x}(-3\cos x+2\sin x).$$

12. For the Laplace transforms $J_p=\mathscr{L}[j_p]$, $p=1,2$, we obtain, as in Chapter II, No. 37, Example III,

$$\left.\begin{aligned} s^2J_1-2+3J_1+2sJ_2=0 \\ s^2J_2-1+3J_2-2sJ_1=0 \end{aligned}\right\} \quad \text{Elimination of } J_2 \text{ gives}$$

$$J_1=\frac{2s^2-2s+6}{s^4+10s^2+9}=\frac{as+b}{s^2+1}+\frac{cs+\mathrm{d}}{s^2+9},$$

where the constants a, b, c, d can easily be determined by special choices of s from $2s^2-2s+6=(as+b)(s^2+9)+(cs+d)(s^2+1)$.

$$\left.\begin{aligned} 4-2i=s(ai+b) \quad & \text{for } s=i \\ 4+2i=s(-ai+b) \quad & \text{for } s=-i \end{aligned}\right\} \quad \text{or } a=-1/4,\ b=1/2.$$

$$\left.\begin{aligned} 9b+d=6 \quad & \text{for } s=0 \\ a+c=0 \quad & \text{for } s=\infty \end{aligned}\right\} \quad \text{or } c=1/4,\ d=3/2.$$

Hence

$$J_1 = -\frac{1}{4}\frac{s-2}{s^2+1} + \frac{1}{4}\frac{s+6}{s^2+9} \quad \text{and} \quad j_1(t) = \frac{1}{4}(\cos t - 2\sin t) + \frac{1}{4}\left(\cos(3t) + 2\sin(3t)\right).$$

$d_2(t)$ can now be calculated from J_2 or also directly from $j_1 + 3j_1 - 2j_2 = 0$.

$$j_2(t) = \tfrac{1}{4}\sin t + \tfrac{1}{2}\cos t + \tfrac{1}{4}\sin(3t) - \tfrac{1}{2}\cos(3t).$$

13. (a) $F(s = \sqrt{\frac{\pi}{s}}\,e^{-2\sqrt{cs}}$. (b) $F(s) = -\frac{1}{s}((\ln s) + \gamma)$ where γ is the Euler constant $\gamma = \lim_{n\to a}(\Sigma_{k=1}^{n} - \ln n) = 0.5772\ldots$.

(c) $F(s) = \sqrt{\frac{\pi}{cs}}\,e^{s/c}\left[1 - \Phi\left(\sqrt{\frac{s}{c}}\right)\right]$. (d) $F(s) = \pi\Phi\left(\frac{c}{2\sqrt{s}}\right)$

(e) $F(s) = \frac{c}{s} \cdot \frac{1}{\sqrt{c^2+s^2}}$.

14. $\int_0^\infty e^{-sx}\,e^{x+x^2}\,dx$ does not exist for any (real or complex) s.

Further examples for practice and their solutions are in Chapter VI, §2, No. 9.

IV

Particular differential equations

§1. SPHERICAL HARMONICS

1. SOLUTIONS OF LAPLACE'S EQUATION

Various classes of particular functions appear when one asks about particular solutions of Laplace's equation, which plays a fundamental rôle in indeed many applications. In three-dimensional co-ordinate space x, y, z it reads

$$\Delta u = \frac{\partial^2 u}{\partial x^2} + \frac{\partial^2 u}{\partial y^2} + \frac{\partial^2 u}{\partial z^2} = 0. \tag{IV.1}$$

We shall first enquire about solutions $u(x, y, z)$ which are of a particularly simple form, namely, homogeneous polynomials in x, y, z of degree n. However, we shall write z' instead of z, in order to keep the letter z for another purpose. In spherical co-ordinates r, θ, φ, Fig. IV.1, with

$$\begin{aligned} x &= r \sin\theta \cos\varphi, \\ y &= r \sin\theta \sin\varphi, \\ z' &= r \cos\theta \end{aligned} \tag{IV.2}$$

these solutions have the form

$$u = r^n \cdot Y_n(\theta, \varphi), \tag{IV.3}$$

where the $Y_n(\theta, \varphi)$ are called 'surface harmonics'. Figure IV.2 gives some simple examples of such surface harmonics; here the curves $Y_n(\theta, \varphi) = 0$ have

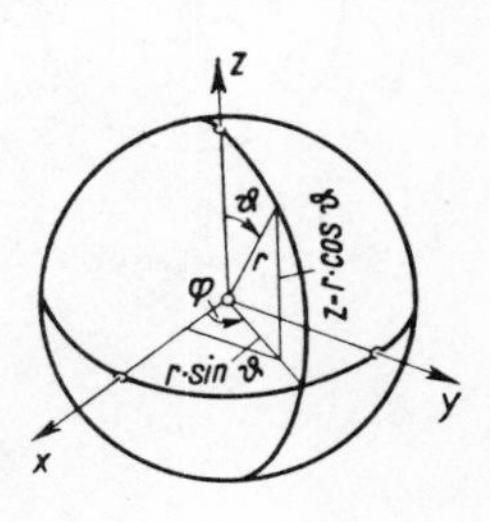

Fig. IV.1. Spherical co-ordinates

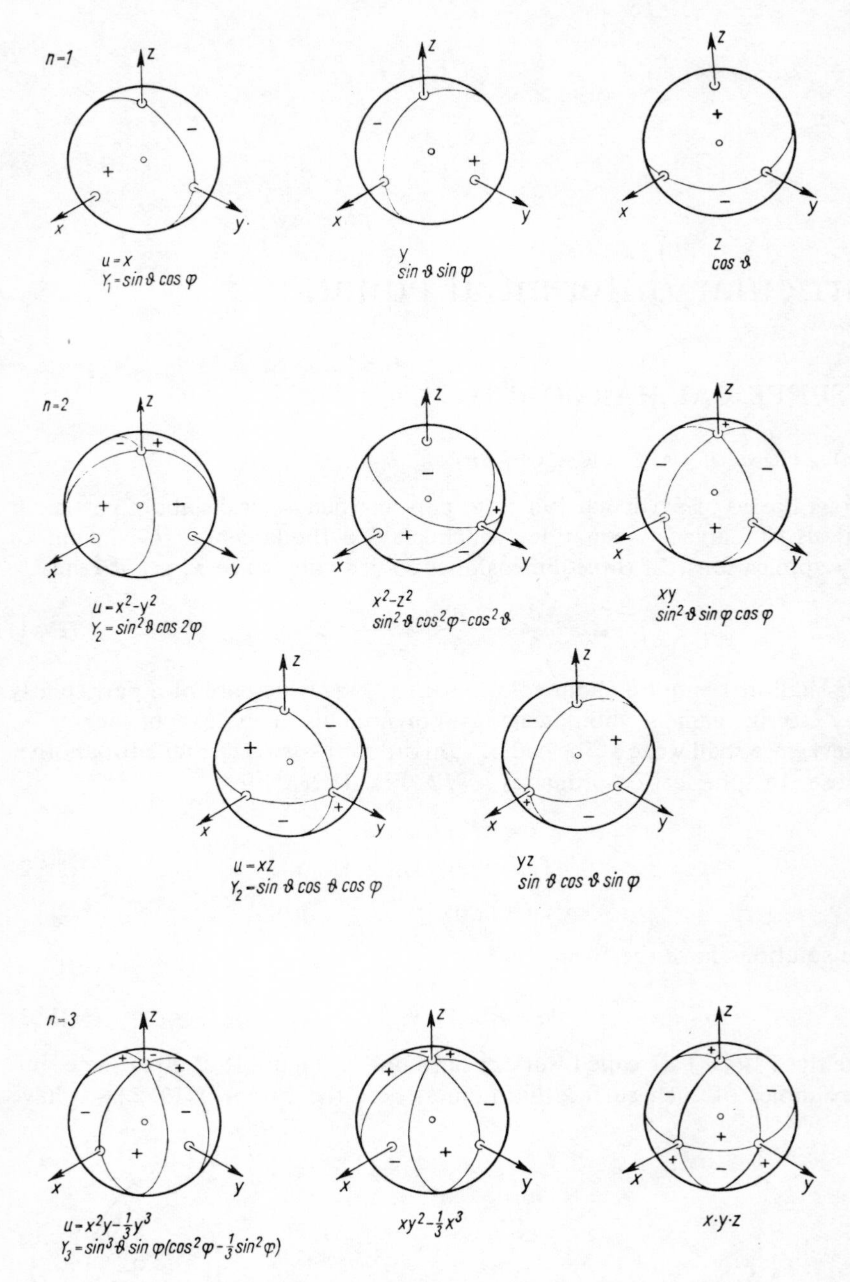

Fig. IV.2. Surface harmonics

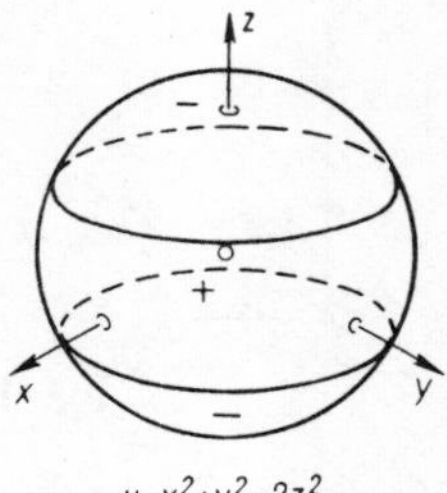

Fig. IV.3 A xonbal harmonic

been drawn on the spherical surface $r = 1$, and the regions in which Y_n is positive or negative are marked + or −. By linear combinations new divisions can be obtained, for example, from $x^2 - z'^2$ and $y^2 - z'^2$ the polynomal

$$x^2 + y^2 - 2z'^2 = r^2(1 - 3\cos^2\theta). \tag{IV.4}$$

The spherical harmonic $1 - 3\cos^2\theta$ is independent of φ and is therefore called a 'zonal harmonic', cf. Fig. IV.3. In order to obtain these, Laplace's equation is first put into spherical co-ordinates. This conversion can be effected in various ways, by means of the Dirichlet integral, or vector calculus, or even directly, albeit rather laboriously, by converting the partial derivatives. The result is

$$r\frac{\partial^2(ur)}{\partial r^2} + \frac{1}{\sin\theta}\frac{\partial}{\partial\theta}\left(\sin\theta\frac{\partial u}{\partial\theta}\right) + \frac{1}{\sin^2\theta}\frac{\partial^2 u}{\partial\varphi^2} = 0. \tag{IV.5}$$

The substitution (IV.3) gives, leaving out the factor r^n

$$(n+1)\,nY_n + \frac{1}{\sin\theta}\frac{\partial}{\partial\theta}\left(\sin\theta\frac{\partial Y_n}{\partial\theta}\right) + \frac{1}{\sin^2\theta}\frac{\partial^2 Y_n}{\partial\varphi^2} = 0. \tag{IV.6}$$

If we are seeking zonal harmonics, the last term on the left drops out. If we now introduce the new variable z by $z = \cos\theta$, $\mathrm{d}z = -\sin\theta\,\mathrm{d}\theta$ and put $Y_n(\theta) = P_n(\cos\theta)$, then we have for $P_n(z)$ the differential equation

$$\frac{\mathrm{d}}{\mathrm{d}z}\left[(1 - z^2)\frac{\mathrm{d}P_n}{\mathrm{d}z}\right] + (n+1)\,nP_n = 0. \tag{IV.7}$$

By (IV.4), for example, for $n = 2$ the polynomial $1 - 3z^2$ is a solution of this equation. The solutions of this differential equation which satisfy the further condition $P_n(1) = 1$ are called *Legendre polynomials* or *spherical harmonics of the first kind*.

2. The generating function

The Legendre polynomials will now again be introduced in an entirely different way. It follows directly from (IV.5) that $u = 1/r$ is a solution of Laplace's equation. Therefore, since Laplace's equation is not changed by a translation of the co-ordinate system, $1/\rho$ is also a solution of (IV.1), where ρ is the

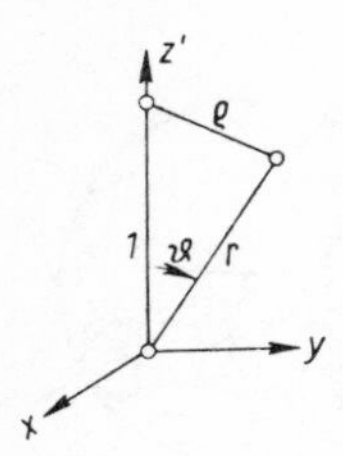

Fig. IV.4. For the generating function of the Legendre polynomials

distance from the point $x = y = 0$, $z' = 1$, see Fig. IV.4. Then $\rho^2 = 1 + r^2 - 2r\cos\theta$ and if we again put $z = \cos\theta$ (z is not to be confused with the space co-ordinate z') we have

$$F(r, z) = \frac{1}{\rho} = \frac{1}{+\sqrt{1 - 2rz + r^2}}. \tag{IV.8}$$

The function $F(r, z)$ defined in this way can be written as a power series in $(2rz - r^2)$ by means of the binomial series; the series converges for $|2rz - r^2| < 1$:

$$F(r, z) = 1 + \frac{1}{2}(2rz - r^2) + \frac{1 \cdot 3}{2 \cdot 4}(2rz - r^2)^2 + \cdots$$
$$+ \frac{1 \cdot 3 \cdots (2k-1)}{2 \cdot 4 \ldots 2k}(2rz - r^2)^k + \cdots. \tag{IV.9}$$

If, in addition,

$$|2rz| + |r^2| < 1, \tag{IV.10}$$

then the powers of $(2rz - r^2)$ can be multiplied out and the resulting terms can be arranged in any order, because the majorant series arising from the expansion of $1/\sqrt{1 - |2rz| - |r|^2}$ is absolutely covergent. In particular, the terms can be arranged in order of powers of r and we then obtain; as the coefficients, certain polynomials in z which may be denoted by $P_n(z)$ and called Legendre polynomials (the equivalence with No. 1 will soon be shown):

$$F(r, z) = \frac{1}{+\sqrt{1 - 2rz + r^2}} = \sum_{n=0}^{\infty} P_n(z)\, r^n. \tag{IV.11}$$

The first few of these polynomials are (z and not x is written for the independent variable because the $P_n(z)$ are often considered in the complex domain):

$$P_0(z) = 1, \qquad P_1(z) = z, \qquad P_2(z) = \tfrac{1}{2}(3z^2 - 1), \qquad P_3(z) = \tfrac{1}{2}(5z^3 - 3z). \tag{IV.12}$$

Many properties of the Legendre polynomials can easily be derived by means of the generating function; Thus:

1. Particular values. In (IV.11) put $z = 1$ (the condition (IV.10) is still always satisfied for sufficiently small r). It follows that

$$F(r, 1) = \frac{1}{+\sqrt{1 - 2r + r^2}} = \frac{1}{1 = r} = \sum_{n=0}^{\infty} r_n = \sum_{n=0}^{\infty} P_n(1)\, r^n.$$

Therefore

$$P_n(1) = 1 \tag{IV.13}$$

In exactly the same way it follows for $z = -1$ that $P_n(-1) = (-1)^n$. Again, for $z = 0$ we have

$$F(r, 0) = \frac{1}{+\sqrt{1 + r^2}} = \sum_{k=0}^{\infty} \binom{-\frac{1}{2}}{k} r^{2k} = \sum_{n=0}^{\infty} P_n(0)\, r^n;$$

$$P_n(0) = \begin{cases} 0 & \text{for odd } n \\ \binom{-\frac{1}{2}}{\frac{n}{2}} = (-1)^{n/2}\, 2^{-n} \binom{n}{\frac{n}{2}} & \text{for even } n \end{cases}$$

cf. the graphical representation of the first few polynomials $P_0, \ldots .\, P_5$ in Fig. IV.5.

2. *The highest coefficient* a_n *of* $P_n(z)$. This occurs in the expansion (IV.9) precisely with the term $z^n r^n$, and such a power appears only in

$$\frac{1 \cdot 3 \ldots (2n - 1)}{2 \cdot 4 \ldots 2n} (2rz - r^2)^n$$

the highest coefficient in the polynomial $P_n(z)$ is therefore equal to

$$a_n = \frac{1 \cdot 3 \ldots (2n - 1)}{2 \cdot 4 \ldots 2n} \cdot 2^n = \frac{(2n)!}{2^n n!} = \frac{1}{2^n} \binom{2^n}{n}. \tag{IV.14}$$

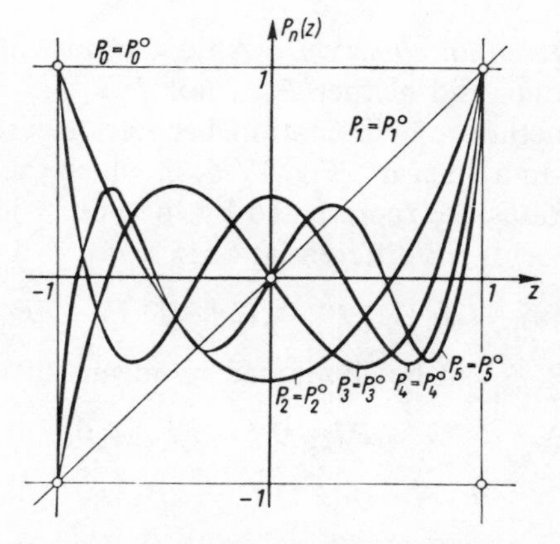

Fig. IV.5. Legendre polynomials

3. Recursion formula Differentiating (IV.11) with respect to r gives

$$\frac{z-r}{\sqrt{1-2rz+r^2}^{\,3}} = \sum_{n=0}^{\infty} nP_n(z)r^{n-1}$$

or

$$\frac{z-r}{\sqrt{1-2rz+r^2}} = (z-r)\sum_{n=0}^{\infty} P_n(z)r^n = (1-2rz+r^2)\sum_{n=0}^{\infty} nP_n(z)r^{n-1}.$$

Comparing the coefficients of r^n leads to

$$(n+1)\,P_{n+1}(z) - z(2n+1)\,P_n(z) + nP_{n-1}(z) = 0 \quad \text{for } n = 0, 1, 2, \ldots \qquad \text{(IV.15)}$$

For $h = 0$ the polynomial P_{-1} appears formally, but this has not been defined. However, since it is multiplied by the factor $n = 0$, this term disappears. This equation (IV.15) is very suitable for the recursive calculation of $P_n(z)$ from $P_0 = 1$.

4. Another relation. Differentiating (IV.11) with respect to z gives

$$\frac{r}{\sqrt{1-2rz+r^2}^{\,3}} = \sum_{n=0}^{\infty} P_n'(z)r^n$$

or

$$\frac{r}{\sqrt{1-2rz+r^2}} = \sum_{n=0}^{\infty} P_n(z)r^n = (1-2rz+r^2)\sum_{n=0}^{\infty} P_n'(z)r.$$

Comparison of the coefficients of r^{n+1} leads to

$$P_{n+1}' - 2zP_n' + P_{n-1}' = P_n \quad (\text{for } n = 0, 1, 2, \ldots). \qquad \text{(IV.16)}$$

In order that this formula shall also hold for $n = 0$, we must put

$$P_{-1}' \equiv 0$$

5. The Legendre differential equation. An equation will now be derived in which P_n appears alone, and neither P_{n-1} nor P_{n+1}.

The plan of the method of elimination becomes clearer if the individual steps are represented in a diagram, Fig. IV.6, in which the terms appearing in any one of the equations are represented by the points joined together.

Equation (IV.15) yields on differentiation

$$(n+1)\,P_{n+1}' - (2n+1)\,P_n - z(2n-1)\,P_n' + nP_{n-1}' = 0. \qquad \text{(IV.17)}$$

From this P_{n+1}' or P_{n-1}' can be eliminated by means of (IV.16):

$$nP_n - zP_n' + P_{n-1}' = 0, \qquad \text{(IV.18)}$$

$$P_{n+1}' - (n+1)\,P_n - zP_n' = 0. \qquad \text{(IV.19)}$$

Here n is replaced by $n-1$, and the result remains true even for $n = 0$ if we

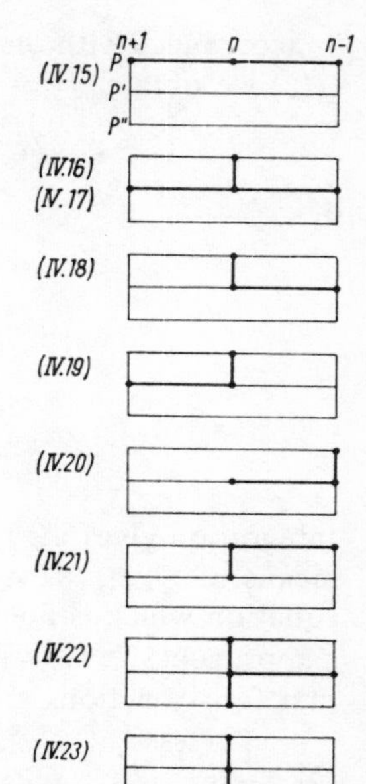

Fig. IV.6. On the derivation of the Legendre differential equation

define P_{-1} as identically zero:

$$P_n' - nP_{n-1} - zP_{n-1}' = 0. \qquad \text{(IV.20)}$$

Next, P_{n-1}' is eliminated by means of (IV.18):

$$znP_n + (1 - z^2)\, P_n' - nP_{n-1} = 0. \qquad \text{(IV.21)}$$

and this equation is differentiated to give

$$(1 - z^2)\, P_n'' + z(n-2)\, P_n' + nP_n - nP_{n-1}' = 0. \qquad \text{(IV.22)}$$

Finally, combination with (IV.18) gives the Legendre differential equation, agreeing with (IV.7):

$$(1 - z^2)\, P_n'' - 2zP_n' + n(n+1)\, P_n = 0. \qquad \text{(IV.23)}$$

3. Legendre functions of the second kind

The Legendre differential equation (IV.23) has singularities at $z = \pm 1$ because at these points the factor multiplying the highest derivative vanishes. We can prescribe the value $P_n(1) = 1$ at $z = 1$, but not the value of P_n' which has, rather, to be found from the differential equation.

We now enquire about further solutions $Q(z)$ of the differential equation and to find them we can make the substitution

$$Q(z) = P_n(z)\, v(z)$$

in accordance with the recursion method of Chapter II, 7: on substituting for $Q(z)$ we obtain

$$(1-z^2)\,2P_n'v' + (1-z^2)\,P_n v'' - 2zP_n v' = 0$$

or

$$\frac{2P_n'}{P_n} + \frac{v''}{v'} - \frac{2z}{1-z^2} = 0,$$

$$\ln P_n^2 + \ln|v'| + \ln|1-z^2| = c_1,$$

$$v' = \frac{c_2}{P_n^2(1-z^2)}.$$

Integration gives $v(z)$, and with a suitable determination of the constants by means of $Q_n(z) = P_n(z)\,v(z)$, we obtain in $(-1, 1)$ a solution of Legendre's equation which is linearly independent of $P_n(x)$; it increases without limit as z approaches ± 1. $Q_n(z)$ is called Legendre's function of the second kind. The first four-functions are:

$$Q_0 = \frac{1}{2}\ln\frac{1+z}{1-z}, \quad Q_1 = P_1Q_0 - 1, \quad Q_2 = P_2Q_0 - \frac{3}{2}z, \quad Q_3 = P_3Q_0 - \frac{5}{2}z^2 + \frac{2}{3}.$$

At the same time we see that the spherical harmonics of the first kind are, apart from constant factors, the only solutions of Legendre's equation which remain finite at $z = \pm 1$, and that the Legendre polynomials $P_n(z)$ are therefore uniquely determined as solutions of the differential equation (IV.23) by the prescription of the single condition $P_n(1) = 1$.

We state without proof that a simple explicit formula can be given for $Q_n(z)$:

$$Q_n(z) = \frac{1}{2}P_n(z)\ln\frac{1+z}{1-z} - W_{n-1}(z) \qquad \text{(IV.24)}$$

$$W_{n-1}(z) = \sum_{j=1}^{n}\frac{1}{j}P_{j-1}(z)\,P_{n-j}(z).$$

(see J. Horn and H. Wittich: *Gewöhnliche Differentialgleichungen,* 6th ed., Berlin 1960, p. 179).

4. Another explicit representation of the Legendre polynomials

The Legendre differential equation (IV.23) can be transformed in the following way (cf. L. A. Pipes, *Applied Mathematics for Engineers and Physicists,* McGraw-Hill, New York, 1958, p. 365). By the Leibniz rule for the differentiation of a product

$$\left[(u\cdot v)^{(n)} = u^{(n)}v + \binom{n}{1}u^{(n-1)}v' + \binom{n}{2}u^{(n-2)}v'' + \cdots\right]$$

we have, for any $(n+2)$ times differentiable function w,

$$[1-z^2)\,w'\,]^{(n+1)}+2n[zw]^{(n+1)}$$
$$=(1-z^2)\,w^{(n+2)}+(n+1)(-2z)\,w^{(n+1)}-(n+1)\,nw^{(n)}$$
$$+\,2nzw^{(n+1)}+2n(n+1)\,w^{(n)}$$
$$(1-z^2)\,P''-2zP'+n(n+1)\,P,$$

where $P=w^n$.

But this last is precisely the differential expression which occurs in Legendre's differential equation. The differential equation is therefore satisified by $P_n=w^{(n)}$ if w satisfies the equation

$$W^{(n+1)}=0 \quad \text{with } W=(1-z^2)\,w'+2nzw.$$

For $w=(z^2-1)^n$ we have $W=0$ and hence also $W^{(n+1)}=0$, and so, for any c,

$$P=c\cdot\frac{\mathrm{d}^n}{\mathrm{d}z^n}\,[(z^2-1)^n]$$

is a solution of the Legendre differential equation; it is, in fact, a polynomial of the nth degree, and must therefore up to a constant factor (i.e., for a suitable c) coincide with $P_n(x)$. The constant c is determined most easily from the highest coefficient, which for P reads

$$c\cdot 2n(2n-1)\ldots(n+1);$$

This must be equal to the value of a_n from (IV.14), and so c is found to be

$$c=\frac{1}{2^n}\,\frac{(2n)!}{n!n!}\,\frac{1}{2n(2n-1)\ldots(n+1)}=\frac{1}{2^n n!}.$$

Hence, finally we have *Rodrigues's formula*

$$P_n(z)=\frac{1}{2^n n!}\,\frac{\mathrm{d}^n}{\mathrm{d}z^n}\,[(z^2-1)^n]. \tag{IV.25}$$

5. ORTHOGONALITY

This property of $P_n(z)$ can easily be deduced either from the formula (IV.25) or from the differential equation. For $k<n$, it follows, by integrating by parts and writing $Z=(z^2-1)^n$, that

$$\frac{1}{c}\int_{-1}^{1}P_n(z)\,z^k\,\mathrm{d}z=\int_{-1}^{1}Z^{(n)}z^k\,\mathrm{d}z=[Z^{(n-1)}z^k]_{-1}^{1}-\int_{-1}^{1}k\cdot Z^{(n-1)}z^{k-1}\,\mathrm{d}z.$$

The integrated part vanishes at the limit points, and the integral on the right can again be integrated by parts; the process can be repeated $(k+1)$ times, until the integral drops out altogether. The integrated parts disappear, because Z certainly has n-fold zeros at $z=\pm 1$, and so the derivatives of Z up to the $(n-1)$th order inclusive are equal to 0 at $z=\pm 1$. $P_n(z)$ is therefore orthogonal to $1, z, z^2, \ldots, z^{n-1}$ and so also to linear combinations of them, i.e., to $P_0, P_1, \ldots, P_{n-1}$. For normalization, the following integral is needed, which again can be transformed by integrating by parts, in which process the integrated parts again vanish:

$$J = \int_{-1}^{1} Z^{(n)} Z^{(n)}\, dz = -\int_{-1}^{1} Z^{(n-1)} Z^{(n+1)}\, dz = \cdots = (-1)^n \int_{-1}^{1} Z Z^{(2n)}\, dz.$$

But now, since Z is of degree $2n$,

$$Z^{(2n)} = (2n)!,$$

and therefore

$$J = (-1)^n (2n)! \int_{-1}^{1} Z\, dz = (2n)! \int_{-1}^{1} (1-z)^n (1+z)^n\, dz.$$

Integrating by parts again n times, we have

$$\int_{-1}^{1} (1-z)^n (1+z)^n\, dz = \frac{n}{n+1} \int_{-1}^{1} (1-z)^{n-1} (1+z)^{n+1}\, dz = \cdots$$

$$= \frac{n(n-1)\ldots 1}{(n+1)(n+2)\ldots 2n} \int_{-1}^{1} (1+z)^{2n}\, dz.$$

Putting $1 + z = u$, this becomes

$$\int_{-1}^{1} (1+z)^{2n}\, dz = \int_{0}^{2} u^{2n}\, du = \frac{2^{2n+1}}{2n+1},$$

and therefore

$$J = (2n)! \frac{n!n!}{(2n)!} \frac{2^{2n+1}}{2n+1} = \frac{(n!)^2}{2n+1} 2^{2n+1}.$$

Hence we have altogether

$$\int_{-1}^{1} P_n(z) P_k(z) = \begin{cases} 0 & \text{for } n \neq k \\ \dfrac{2}{2n+1} & \text{for } n = k \end{cases}$$

Theorem. *With the usual metric as in* (III.85) *the functions* $\sqrt{\dfrac{2n+1}{2}}\, P_n(z)$ *form an orthonormal system in the interval* $-1 \leq z \leq 1$.

§2. BESSEL FUNCTIONS

6. The partial differential equation for vibrations of a membrane

In vibrations of a membrane (say, a circular membrane) the deflection z is a function of position and time. If the quotient, always positive, of the surface stress to density is denoted by c^2, then the wave equation can be shown to be

$$c^2\left(\frac{\partial^2 z}{\partial x^2}+\frac{\partial^2 z}{\partial y^2}\right)=\frac{\partial^2 z}{\partial t^2} \tag{IV.27}$$

where c is the wave velocity, or, more shortly, using the Laplacian operator, (cf. Chapter I, 2, Example 3)

$$c^2\,\Delta z=\ddot{z}.$$

Since there is circular symmetry in this problem, it is sensible to change over to polar co-ordinates:

$$r=\sqrt{x^2+y^2}, \qquad \varphi=\text{arc tan}\,\frac{y}{x};$$

then

$$\left.\begin{aligned} r_x&=\frac{x}{\sqrt{x^2+y^2}}=\frac{x}{r}, & \varphi_x&=\frac{-y}{x^2+y^2}=-\frac{y}{r^2},\\ r_{xx}&=\frac{1}{r}-\frac{x}{r^2}\,r_x=\frac{y^2}{r^3}, & \varphi_{xx}&=2\,\frac{y}{r^3}\,r_x=\frac{2xy}{r^4}. \end{aligned}\right\} \tag{IV.28}$$

By the generalized chain-rule we have

$$z_x=z_r r_x+z_\varphi \varphi_x,$$

$$z_{xx}=z_{rr}r_x^2+2z_{r\phi}r_x\varphi_x+z_r r_{xx}+z_{\varphi\varphi}\varphi_x^2+z_\varphi\varphi_{xx}.$$

Calculating z_y and z_{yy} in the same way, we find that

$$z_{xx}+z_{yy}=z_{rr}\left(\frac{x^2}{r^2}+\frac{y^2}{r^2}\right)+z_r\left(\frac{x^2}{r^3}+\frac{y^2}{r^3}\right)+z_{\varphi\varphi}\left(\frac{x^2}{r^4}+\frac{y^2}{r^4}\right)$$

or

$$\Delta z=z_{rr}+\frac{1}{r}\,z_r+\frac{1}{r^2}\,z_{\varphi\varphi}. \tag{IV.29}$$

7. The Bernoulli substitution for vibrations of a membrane

The natural vibrations under investigation of a membrane thus satisfy the wave equation

$$z_{rr}+\frac{1}{r}\,z_r+\frac{1}{r^2}\,z_{\varphi\varphi}=\frac{1}{c^2}\,z_u \tag{IV.30}$$

with the boundary condition

$$z=0 \quad \text{for } r=R.$$

The Bernoulli product substitution $z=u(r,\varphi)f(t)$ makes it possible to open up the differential equation in such a way that, instead of the one function in three independent variables $z(r,\varphi,t)$, two new functions appear in it, one $u(r,\varphi)$ of two independent variables, and one $f(t)$ of a single independent variable:

$$f \cdot \left(u_{rr} + \frac{1}{r} u_r + \frac{1}{r^2} u_{\varphi\varphi} \right) = \frac{1}{c^2} \ddot{f} u$$

The two functions can now be separated:

$$\frac{1}{u} \left(u_{rr} + \frac{1}{r} u_r + \frac{1}{r^2} u_{\varphi\varphi} \right) = \frac{1}{c^2} \frac{\ddot{f}}{f}. \tag{IV.31}$$

By the same argument as in Chapter III, No. 16, this equation between two functions, one a function of r and φ, and the other of t only, can subsist only if both sides are equal to the same constant. Since vibrations are expected, we call this constant $-k^2$. Thus we have two differential equations, coupled by the parameter k^2,

$$\ddot{f} + k^2 c^2 f = 0$$

and

$$u_{rr} + \frac{1}{r} u_r + \frac{1}{r^2} u_{\varphi\varphi} + k^2 u = 0. \tag{IV.31a}$$

The first equation has as its general solution (with the circular frequency $\omega = kc$)

$$f(t) = C_1 \cos \omega t + C_2 \sin \omega t,$$

and we try to split the second equation with a new product substitution

$$u = v(r) w(\varphi);$$

this gives

$$v'' w + \frac{1}{r} v' w + \frac{1}{r^2} v w'' + k^2 v w = 0.$$

Separation of the variables leads to the same circumstances as above. Again the two sides, because they depend on distinct variables, must equal to a constant, n^2;

$$\frac{r^2}{v} \left(v'' + \frac{1}{r} v' + k^2 v \right) = -\frac{w''}{w} = n^2.$$

Here w must be periodic with period 2π, because φ and $\varphi + 2\pi$ relate to the same position on the circular membrane and must give the same deflection. Of the two differential equations coupled through n

$$w'' + n^2 w = 0$$

and

$$v'' + \frac{1}{r} v' + \left(k^2 - \frac{n^2}{r^2} \right) v = 0 \tag{IV.32}$$

the first has the general solution

$$w(\varphi) = C_3 \sin n\varphi + C_4 \cos n\varphi;$$

n here is an integer.

The second equation is a Bessel differential equation and cannot be solved in closed form. It frequently arises in the treatment of cylinder problems. Its solutions are called Bessel functions (in foreign mathematical literature they are often called cylinder functions). The further treatment of the membrane problem is postponed to No. 14.

8. The generating function

The Bessel functions will now be introduced in quite a different way, namely, by means of a generating function $F(z, t)$, as we did for Legendre functions:

$$\begin{aligned} F(z,t) &= \exp\left[\frac{1}{2} z\left(t - \frac{1}{t}\right)\right] = \exp\left(\frac{1}{2} zt\right) \cdot \exp\left(-\frac{1}{2}\frac{z}{t}\right) \\ &= \left(1 + \frac{1}{1!}\frac{1}{2} zt + \frac{1}{2!}\left(\frac{1}{2} zt\right)^2 + \cdots\right) \times \left(1 - \frac{1}{1!}\frac{1}{2}\frac{z}{t} + \frac{1}{2!}\left(\frac{1}{2}\frac{z}{t}\right)^2 - \cdots\right). \end{aligned} \tag{IV.33}$$

The series converge not only for $t = 0$ but also absolutely for all z and t; the product can therefore be multiplied out and the terms re-arranged. We collect all terms with the same power of t together and call the factor of t^n the Bessel function of the first kind $J_n(z)$:

$$F(z,t) = \exp\left[\frac{1}{2} z\left(t - \frac{1}{t}\right)\right] = \sum_{n=-\infty}^{\infty} J_n(z)\, t^n. \tag{IV.34}$$

In particular,

$$J_0(z) = 1 - \frac{1}{(1!)^2}\left(\frac{z}{2}\right)^2 + \frac{1}{(2!)^2}\left(\frac{z}{2}\right)^4 - \frac{1}{(3!)^2}\left(\frac{z}{2}\right)^6 + - \cdots \tag{IV.35}$$

$$J_1(z) = \frac{1}{1!}\frac{z}{2} - \frac{1}{1!2!}\left(\frac{z}{2}\right)^3 + \frac{1}{2!3!}\left(\frac{z}{2}\right)^5 - + \cdots. \tag{IV.36}$$

And, in general,

$$F(z,t) = \sum_{k=0}^{\infty} \frac{1}{k!}\left(\frac{zt}{2}\right)^k \cdot \sum_{m=0}^{\infty} \frac{1}{m!}\left(\frac{-z}{2t}\right)^m = \sum_{k,m=0}^{\infty} (-1)^m \frac{1}{k!m!}\left(\frac{z}{2}\right)^{k+m} t^{k-m}$$

In the k,m-plane, Fig. IV.7, each term corresponds to a lattice-point. By 'collecting' the points on the dotted lines shown, i.e., for constant $k - m = n$ we obtain $J_n(z)$:

$$F(z,t) = \sum_{n=-\infty}^{\infty} \sum_{m=0}^{\infty} (-1)^m \frac{1}{(m+n)!m!}\left(\frac{z}{2}\right)^{2m+n} t^n = \sum_{n=-\infty}^{\infty} J_n(z) t^n,$$

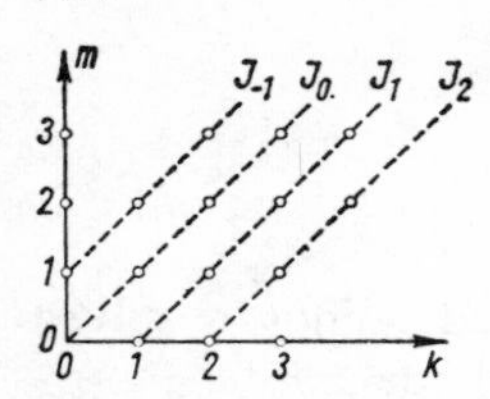

Fig. IV.7. On the power series for Bessel functions of the first kind

i.e.,

$$J_n(z) = \left(\frac{z}{2}\right)^n \sum_{m=0}^{\infty} \frac{(-1)^m}{m!\Gamma(m+n+1)} \left(\frac{z}{2}\right)^{2m}. \tag{IV.37}$$

Here $\Gamma(m+n+1)$ has been written instead of $(m+n)!$, where Γ stands for the gamma function, which, for positive integer values of the argument q has the value

$$\Gamma(q) = (q-1)! = 1 \cdot 2 \ldots (q-1)$$

The reciprocal of the gamma function $1/\Gamma(z)$ is, however, a well-defined holomorphic function for all complex values of z $(z \neq \infty)$, and so (IV.37) defines the Bessel function of the first kind for non-integer indices n too. The series are absolutely convergent for all finite z; they converge 'better' than the series for e^z.

9. Deductions from the generating function

As in No. 2 for Legendre functions, here too we can draw from the generating functions a number of conclusions about Bessel functions.

1. Recursion formulae. Differentiation of (IV.34) with respect to z (termwise differentiation of this Laurent series is justifiable) gives

$$\frac{\partial F(z,t)}{\partial z} = \exp\left[\frac{1}{2} z\left(t - \frac{1}{t}\right)\right] \cdot \frac{1}{2}\left(t - \frac{1}{t}\right) = \frac{1}{2}\left(t - \frac{1}{t}\right) \sum_{n=-\infty}^{\infty} J_n(z)t^n = \sum_{n=-\infty}^{\infty} J_n'(z)t^n.$$

Comparison of the coefficients of t^n on the two sides then yields

$$J_{n-1}(z) - J_{n+1}(z) = 2J_n'(z). \tag{IV.38}$$

Differentiation of (IV.34) with respect to t similarly gives

$$\frac{\partial F(z,t)}{\partial t} = \frac{z}{2}\left(1 + \frac{1}{t^2}\right) \sum_{n=-\infty}^{\infty} J_n(z)t^n = \sum_{n=-\infty}^{\infty} nJ_n(z)t^{n-1}.$$

and comparison of coefficients of t^{n-1} on both sides yields

$$J_{n-1}(z) + J_{n+1}(z) = \frac{2n}{z} \cdot J_n(z). \tag{IV.39}$$

2. A differential equation for $J_n(z)$. (IV.38) and (IV.39) imply

$$J_{n-1} = \frac{n}{z} J_n + J_n', \quad J_{n+1} = \frac{n}{z} J_n - J_n'. \tag{IV.40}$$

We replace n by $n+1$ in the first relation

$$J_n = \frac{n+1}{z} J_{n+1} + J_{n+1}'$$

and express J_{n+1} in terms of J_n by the second relation in (IV.40):

$$J_n''(z) + \frac{1}{z} J_n'(z) + \left(1 - \frac{n^2}{z^2}\right) J_n(z) = 0. \tag{IV.41}$$

3. Boundedness for real z. If t is replaced by $-1/t$ in (IV.34), the generating function remains unchanged; n is then changed on the right to $-n$, and hence it follows that

$$F(z, t) = \exp\left[\frac{1}{2} z\left(t - \frac{1}{t}\right)\right] = \sum_{n=-\infty}^{\infty} J_n(z)(-1)^n t^{-n} = \sum_{n=-\infty}^{\infty} J_{-n}(z)(-1)^n t^n.$$

Comparison with (IV.34) gives

$$J_n(z) = (-1)^n J_{-n}(z). \tag{IV.42}$$

Next t is replaced by $1/t$ in (IV.34):

$$F\left(z, \frac{1}{t}\right) = \exp\left[-\frac{1}{2} z\left(t - \frac{1}{t}\right)\right] = \sum_{m=-\infty}^{\infty} J_m(z) t^{-m}$$

and this series is multiplied by the series (IV.34); the absolutely convergent double series

$$1 = \sum_{m,\, n=-\infty}^{\infty} J_m(z) J_n(z) t^{n-m} \tag{IV.43}$$

is obtained. Comparison of the coefficients of t^p on both sides, taking (IV.42) into account, gives

$$1 = \sum_{n=-\infty}^{\infty} [J_n(z)]^2 = [J_0(z)]^2 + 2\{[J_1(z)]^2 + [J_2(z)]^2 + \cdots\} \tag{IV.44}$$

$$0 = \sum_{n=-\infty}^{\infty} J_n(z) J_{n+p}(z) \qquad (p = 1, 2, \ldots). \tag{IV.45}$$

It follows from (IV.44), in particular, that

$$|J_0(z)| \leq 1, \qquad |J_n(z)| \leq \frac{1}{\sqrt{2}} \tag{IV.46}$$

for real z and $n = 1, 2, \ldots$.

4. Theorems on separation of the zeros. We can also write the equations (IV.46) as

$$z^n J_{n-1} = \frac{\mathrm{d}}{\mathrm{d}z} (z^n J_n), \qquad -z^{-n} J_{n+1} = \frac{\mathrm{d}}{\mathrm{d}z} (z^{-n} J_n). \tag{IV.47}$$

We consider $J_n(z)$ along the positive real axis. If $J_n(z)$ has two zeros z', z'' there, then $z^n J_n$ and $z^{-n} J_n$ also have zeros at these points, and by Rolle's theorem, the derivative vanishes at at least one point between z' and z'', i.e., J_{n-1} and likewise J_{n+1} have at least one zero between z' and z''. This implies at the same time that between every two zeros of J_{n-1} there is at least one zero of J_{n-2} and J_n. J_n being a solution of a regular homogeneous second-order differential equation (if $z \neq 0$), $J_n(z)$ cannot have a double zero ζ for $z \neq 0$, for otherwise we would have $J_n(\zeta) = J_n'(\zeta) = 0$ and so $J_n(z)$ would be identically zero. Between every two simple zeros of J_n lies at least one zero of J_{n-1}, but also between every two zeros of J_{n-1} there is at least one zero of J_n; therefore between every two successive zeros of J_n there is precisely one zero of J_{n-1}, and also precisely one zero of J_{n+1}. For $z > 0$ (and similarly for $z < 0$) the zeros of J_n and those of J_{n-1} separate one another. Fig. IV.8 shows the graphs of the first few functions $J_n(z)$ in the reals.

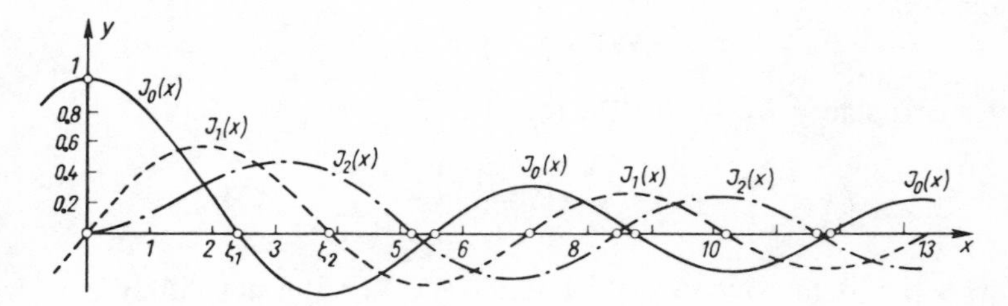

Fig. IV.8. Bessel functions of the first kind

10. An integral representation

If we put

$$t = \mathrm{e}^{\mathrm{i}\varphi}, \qquad \frac{1}{t} = \mathrm{e}^{-\mathrm{i}\varphi}, \qquad t - \frac{1}{t} = 2\mathrm{i} \sin \varphi$$

in (IV.34), then

$$\mathrm{e}^{\mathrm{i}z \sin \varphi} = \sum_{n=-\infty}^{\infty} J_n(z)\, \mathrm{e}^{\mathrm{i}n\varphi}. \tag{IV.48}$$

This series converges uniformly for $0 \leq \varphi \leq 2\pi$, and so the series can be multiplied by $\mathrm{e}^{-\mathrm{i}m}$ and integrated term by term over φ from 0 to 2π. Since

$$\int_0^{2\pi} \mathrm{e}^{\mathrm{i}k\varphi}\, \mathrm{d}\varphi = \begin{cases} 0 & \text{for } k \neq 0 \\ 2\pi & \text{for } k = 0 \end{cases} \tag{IV.49}$$

only the term with $n = m$ remains on the right-hand side, and we have

$$\int_0^{2\pi} e^{i(z \sin\varphi - m\varphi)}\, d\varphi = 2\pi J_m(z) \quad (m = 0, \pm 1, \pm 2, \ldots). \qquad \text{(IV.50)}$$

For real z the right-hand side is real, and so we need take only the real part of the left-hand side:

$$J_m(z) = \frac{1}{2\pi} \int_0^{2\pi} \cos(z \sin\varphi - m\varphi)\, d\varphi. \qquad \text{(IV.51)}$$

From (IV.48) the following series can be derived. Using (IV.42) and the Euler formula we have

$$e^{iz\sin\varphi} = \cos(z \sin\varphi) + i \sin(z \sin\varphi) = \sum_{n=-\infty}^{\infty} J_n(z)\, e^{in\phi}$$

$$= J_0(z) + 2i \sum_{n=1}^{\infty} J_{2n-1}(z)\sin(2n-1)\varphi + 2 \sum_{n=1}^{\infty} J_{2n}(z)\cos 2n\varphi.$$

If for real z we compare the terms with and without i, we obtain the Fourier series

$$\cos(z \sin\varphi) = J_0(z) + 2J_2(z)\cos 2\varphi + 2J_4(z)\cos 4\varphi + \cdots$$
$$\sin(z \sin\varphi) = 2J_1(z)\sin\varphi + 2J_3(z)\sin 3\varphi + \cdots.$$

For $\varphi = \pi/2$ these give

$$\cos z = J_0(z) - 2J_2(z) + 2J_4(z) - + \cdots$$
$$\sin z = 2[J_1(z) - J_3(z) + J_5(z) - + \cdots].$$

Further integral representations can easily be obtained from this point, but the reader must be referred to more extensive works on Bessel functions, such as C. N. Watson, *A Treatise on the Theory of Bessel Functions,* Cambridge Univ. Press, 1966.

11. An example from astronomy; the Kepler equation

Bessel was led to introduce the functions named after him when he was calculating planetary orbits. A planet P, Fig. IV.9, moves in an ellipse (with

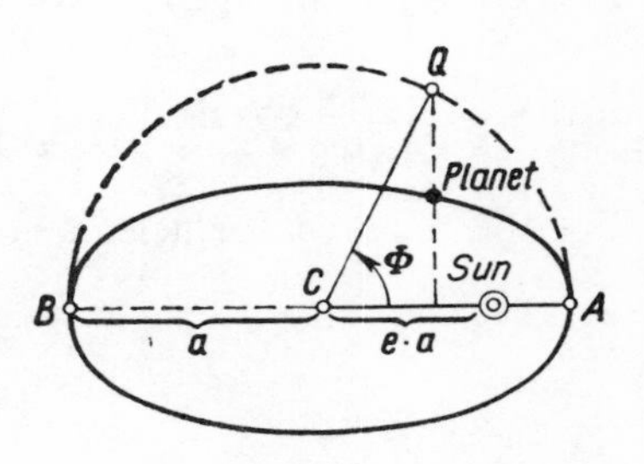

Fig. IV.9. A planetary orbit

semi-major axis a) about the sun S, standing at one of the foci. Using the notation in Fig. IV.9, the following definitions are made:

$$\text{Eccentric anomaly } \Phi = \frac{\text{Area } ACQ}{\text{Half area of ellipse}} \cdot \pi$$

$$\text{Mean anomaly} \quad \mu = \frac{\text{Area } ASP}{\text{Half area of ellipse}} \cdot \pi$$

$$\text{Eccentricity} \quad e = \frac{CS}{CA} = \frac{CS}{a} < 1.$$

By Kepler's second law, μ is proportional to the time t. At the apsides (perihelion and aphelion)

$$\Phi = \mu = 0 \quad \text{and} \quad \Phi = \mu = \pi \quad \text{respectively.}$$

Since Φ is a quantity which fixes the position of P, and μ fixes the time, the relation between Φ and μ is therefore of interest. Since

$$\mu = \frac{\text{Area } ASQ}{\text{Area of semicircle}} \cdot \pi = \frac{ACQ - SCQ}{a^2/2} \cdot \pi = \frac{\frac{1}{2}\Phi a^2 - \frac{1}{2} e a^2 \sin \Phi}{\frac{1}{2} a^2}$$

we have the 'Kepler equation'

$$\mu = \Phi - e \sin \Phi. \tag{IV.52}$$

We now enquire about the inverse function $\Phi(\mu)$. Since $e < 1$, $\Phi(\mu)$ is a single-valued function of μ. $\Phi - \mu = e \sin \Phi$ is periodic in Φ and therefore is also periodic in μ and can be expanded in a trigonometrical series, and because of the symmetry we need only consider the sine terms:

$$\Phi - \mu = \sum_{n=1}^{\infty} A_n \sin n\mu.$$

The A_n are now the required quantities. Differentiation gives

$$\frac{d\Phi}{d\mu} - 1 = \sum_{n=1:}^{\infty} nA_n \cos n\mu.$$

The Fourier coefficients are determined in the known way (cf. Chapter III, 19). In the integral we again introduce Φ instead of μ as the integration variable:

$$nA_n = \frac{1}{\pi} \int_0^{2\pi} \frac{d\Phi}{d\mu} \cos n\mu \, d\mu = \frac{1}{\pi} \int_0^{2\pi} \cos n(\Phi - e \sin \Phi) \, d\Phi \qquad (n = 1, 2, \ldots).$$

For the required coefficients A_n we thus obtain, by (IV.51) the expression

$$A_n = \frac{2}{n} J_n(n \cdot e). \tag{IV.53}$$

12. Bessel functions of the second kind

The Bessel differential equation (IV.41) has the power series (IV.37) as a solution; for integer-valued n this is clearly true from the way in which the equation was obtained, but it is also true if n is not an integer, as can be verified directly by substituting the power series (we omit this simple demonstration). Like $J_n(z)$, the function $J_{-n}(z)$ is also a solution of (IV.41). For n not an integer, J_n and J_{-n} form a fundamental system; J_n and J_{-n} then differe not merely by a constant factor, since, if $n > 0$ say, then $J_n(z)$ vanishes as $z \to 0$ but J_{-n} is unbounded as $z \to 0$. If n is an integer, however, then J_n and J_{-n} do not form a fundamental system, because of (IV.42). As in the case of spherical harmonics in No. 3, we can obtain here also a second solution independent of $J_n(z)$ by making the substitution

$$\Phi(z) = J_n(z) \cdot w(z);$$

(IV.41) then becomes

$$2 J_n' w' + J_n w'' + \frac{1}{z} J_n w' = 0$$

$$\frac{w''}{w'} + \frac{1}{z} + 2 \frac{J_n'}{J_n} = 0$$

$$\ln |w'| + \ln |z| + \ln J_n^2 = \text{const}$$

$$w' = \frac{c}{z J_n^2};$$

Hence

$$J_n(z) \cdot \int_{z_0}^{z} \frac{dt}{t[J_n(t)]^2} \tag{IV.54}$$

is a solution of the Bessel differential equation (IV.41) which is independent of $J_n(z)$; It is called a *Bessel function of the second kind.* Depending on the choice of z_0 and of a further free factor the Bessel functions of the second kind can be normalized in different ways, Various definitions and notations are actually found in the literature, e.g., $N_n(z)$, $Y_n(z)$. But in any case they do not remain bounded as $z \to 0$.

13. More general differential equations giving Bessel functions

If $Z_p(z)$ denotes a Bessel function of the first or second kind of 'order' p, then, as can easily be checked by transforming (IV.41), $y = z^\alpha Z_p(\beta z^\gamma)$ is a solution of the more general equation

$$y'' + \frac{1-2\alpha}{z} y' + \left[(\beta\gamma z^{\gamma-1})^2 + \frac{\alpha^2 - p^2\gamma^2}{z^2}\right] y = 0. \tag{IV.55}$$

Here α, β, γ, p are arbitrary real constants, by a suitable choice of which any particular differential equation that gives Bessel functions can be brought under the form (IV.55) (see Jahnke–Emde–Lösch, *Tables of Functions*, Dover Publications, New York, pp. 146–7). For example,

$$y'' + Ax^m y = 0 \tag{IV.56}$$

has the solution

$$y = \sqrt{x} Z_s\left(2s\sqrt{A}x^{1/(2s)}\right) \quad \text{with } s = \frac{1}{2+m}. \tag{IV.57}$$

For $A = 1$ and $m = 0$, i.e., for $s = 1/2$, (IV.56) has the solution

$$y = c_1 \cos x + c_2 \sin x,$$

i.e., the Bessel functions of order 1/2 can be expressed in terms of $\sqrt{x}$ and trigonometric functions, and are therefore elementary functions. From (IV.37) we obtain, since $\Gamma(\frac{1}{2}) = \sqrt{\pi}$,

$$J_{1/2}(x) = \sqrt{\frac{2}{\pi x}} \sin x; \; J_{-1/2}(x) = \sqrt{\frac{2}{\pi x}} \cos x. \tag{IV.58}$$

Consequently, by (IV.39), all the $J_{m+1/2}(x)$ for $m = 0, \pm 1, \pm 2, \ldots$ are elementary functions.

Example

To find the buckling load P for a homogeneous vertical column of length l, encastré at its lower end, under the influence of its own weight. With the notation as in Fig. IV.10, and

μ = mass per unit length,
g = acceleration cue to gravity,
EJ = bending stiffness,

we have for the bending moment in a deflected position

$$M(x) = \int_0^x \mu g[y(\xi) - y(x)] \, d\xi = EJy''.$$

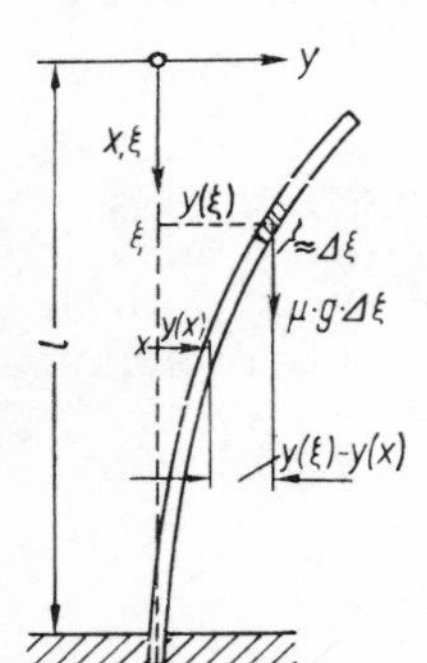

Fig. IV.10. Buckling of a column taking its own weight into account

Differentiation with respect to x using the rule (III.16) gives

$$-\int_0^x \mu g y'(x)\,d\xi = -\mu g x y'(x) = EJy'''.$$

For $u(x) = y'(x)$ we thus have the eigenvalue problem

$$u'' + \frac{\mu g}{EJ} x u = 0 \tag{IV.59}$$

with the boundary conditions

$$\begin{aligned} &u'(0) = y''(0) = 0 \text{ (no moment at the upper end)} \\ &u(l) = y'(l) = 0 \text{ (lower end encastré)} \end{aligned} \tag{IV.60}$$

From (IV.56), (IV.57) we can give the solution of (IV.59) at once

$$u(x) = \sqrt{x} Z_{1/2}\left(\frac{2}{3}\sqrt{\frac{\mu g}{EJ}}\, x^{3/2}\right).$$

Since the order 1/3 is not an integer, the general solution is

$$u(x) = \sqrt{x}\left\{c_1 J_{1/3}\left(\frac{2}{3}\, a x^{3/2}\right) + c_2 J_{-1/3}\left(\frac{2}{3}\, a x^{3/2}\right)\right\} = c_1\varphi_1(x) + c_2\varphi_2(x), \tag{IV.61}$$

$$\text{where } a = \sqrt{\frac{\mu g}{EJ}}.$$

The constants now have to be fitted to the boundary conditions. The power-series expansions begin with the terms

$$\varphi_1(x) = \sqrt{x} J_{1/2}\left(\frac{2}{3}\, a x^{3/2}\right) = \sqrt{x}\,[\alpha_1 x^{1/2} + \alpha_2 x^{7/2} + \cdots],$$

$$\varphi_2(x) = \sqrt{J}_{-1/3}\left(\frac{2}{3}\, a x^{3/2}\right) = \sqrt{x}\,[\beta_1 x^{-1/2} + \beta_2 x^{5/2} + \cdots]$$

with certain constants α_j, β_j which could easily be specified exactly by using (IV.37), but here we need only to use the fact that $\alpha_1 \neq 0$, $\beta_1 \neq 0$. Hence,

$$\varphi_1'(x) = \alpha_1 + 4\alpha_2 x^3 + \ldots, \qquad \varphi_2'(x) = 0 + 3\beta_2 x^2 + \ldots,$$
$$u'(0) = c_1 \cdot \alpha_1 + c_2 \cdot 0 = 0.$$

Therefore $c_1 = 0$, and only the term with the Bessel function $J_{-1/3}$ remains:

$$u(l) = c_2\sqrt{l} J_{-1/3}\left(\frac{2}{3}\, a l^{3/2}\right) = 0. \tag{IV.62}$$

c_2 must be $\neq 0$. We have, then, to determine the zero ξ of

$$J_{-1/3}(\xi) = 0 \tag{IV.63}$$

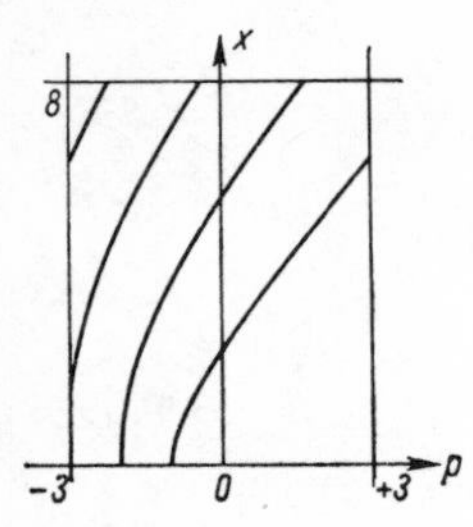

Fig. IV.11. Zeros of the Bessel functions of the first kind

and it is, in fact, the smallest positive root which determines the buckling. In the Jahnke–Emde–Lösch Tables of functions there are graphs of the curves $J_p(x)$ curves in the p,x-plane, from which the zeros can be read off. The curves $J_p x$ appear as shown in Fig. IV.11. From the tables mentioned we find

$$\xi = \frac{2}{3}\sqrt{\frac{\mu g}{EJ}}\, l^{3/2} \approx 1.9. \tag{III.64}$$

ξ can also easily be calculated without knowing the values of Bessel functions, by substituting a power series

$$u(x) = \sum_{\nu=0}^{\infty} a_\nu x^\nu$$

into (IV.59). Since $u'(0) = 0$ we have $a_1 = 0$, and by (IV.59) $a_2 = 0$; putting $a_0 = 1$, we find

$$a_3 = -\frac{a^2}{2\cdot 3}, \qquad a_6 = -\frac{a^2}{5\cdot 6}a_3, \ldots, \qquad a_{3k+1} = a_{3k+2} = 0 \qquad (k = 0, 1, 2, \ldots)$$

or

$$u(x) = 1 - \frac{a^2x^3}{2\cdot 3} + \frac{(a^2x^3)^2}{2\cdot 3\cdot 5\cdot 6} - \frac{(a^2x^3)^3}{2\cdot 3\cdot 5\cdot 6\cdot 8\cdot 9} + - \cdots. \tag{IV.65}$$

The zeros of such rapidly covergent power-series are determined not by breaking off the series but by calculating the value of the series at a few points, and interpolating:

a^2x^3	7.2	7.8	8.4
$u(x)$	0.06042	0.00347	-0.05094

Hence, we obtain

$$a^2l^3 = 7.837\,35, \qquad al^{3/2} = 2.800, \qquad \xi = \frac{2}{3}al^{3/2} = 1.866, \tag{IV.66}$$

and similarly, for the second zero,

$$a^2l^3 = 55.977\,03, \qquad al^{3/2} = 7.482, \qquad \xi = \frac{2}{3}al^{3/2} = 4.988.$$

The equation (IV.65) has infinitely many zeros.

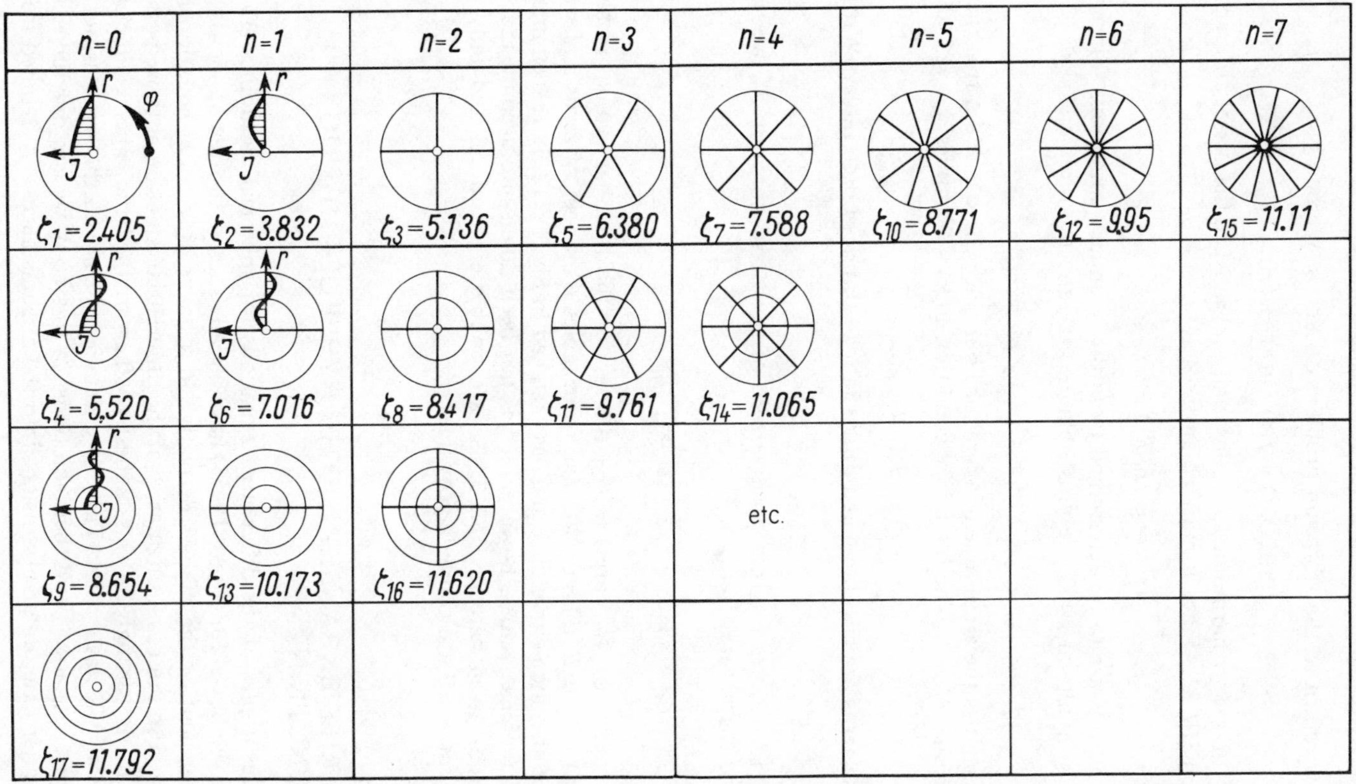

$n=0$	$n=1$	$n=2$	$n=3$	$n=4$	$n=5$	$n=6$	$n=7$
r, φ, J $\zeta_1=2.405$	r, J $\zeta_2=3.832$	$\zeta_3=5.136$	$\zeta_5=6.380$	$\zeta_7=7.588$	$\zeta_{10}=8.771$	$\zeta_{12}=9.95$	$\zeta_{15}=11.11$
r, J $\zeta_4=5.520$	r, J $\zeta_6=7.016$	$\zeta_8=8.417$	$\zeta_{11}=9.761$	$\zeta_{14}=11.065$			
r, J $\zeta_9=8.654$	$\zeta_{13}=10.173$	$\zeta_{16}=11.620$		etc.			
$\zeta_{17}=11.792$							

Fig. IV.12. The first 17 natural vibrations of a circular diaphragm, with the corresponding nodal curves

14. VIBRATION MODES OF A CIRCULAR MEMBRANE

We now return to the vibration problem of No. 7. We had put

$$z(r, \varphi, t) = f(t) \cdot v(r) \cdot w(\varphi),$$

where f and w are harmonic functions, and $v(r)$ satisfied the Bessel differential equation (IV.32). Since n must be an integer,

$$v = C_1 J_n(kr) + C_2 N_n(kr).$$

The constants are to be determined from the boundary conditions. Since N_n becomes infinite at $r = 0$, we must take $C_2 = 0$, and there remains

$$v = C_1 J_n(kr).$$

C_1 determines the amplitude of vibration, but since there is still a free factor we can take $C_1 = 1$. The boundary condition $v(R) = 0$ implies

$$J_n(kR) = 0. \tag{IV.67}$$

These zeros are also tabulated in the Jahnke–Emde–Lösch tables. Writing

$$kR = \zeta_m^{(n)}$$

where $\zeta_m^{(n)}$ denotes the mth positive zero of the Bessel function J_n, we then have for the natural frequency

$$\omega = \frac{c}{R} \zeta_m^{(n)}.$$

The values of $\zeta_m^{(n)}$ are the *eigenvalues* for the boundary-value problem (IV.31a) with $u(R, \varphi) = 0$. We take from tables a number of zeros and write them on the node-curve diagrams, as sketched in Fig. IV.12. In this enumeration of the eigenvalues, however, it has not been taken into account that all the values given here $n = 1, 2, 3, \ldots$ are double eigenvalues, and only the values for $n = 0$ are simple eigenvalues.

§3. SERIES EXPANSIONS; THE HYPERGEOMETRIC FUNCTION

As the examples of the differential equations for spherical harmonics (IV.7) and for the Bessel functions (IV.55) have already shown, it often happens that the coefficient of the highest derivative in a linear differential equation has zeros, and so there is a singularity of the differential equation at those points. If we divide through by this highest coefficient, then the other coefficients in general will have poles. There is a detailed theory on the singularities of differential equations and their solutions; in this book we can only give, for a particular case, a cursory insight into the circumstances. (A detailed treatment can be found, e.g., in Bieberbach: *Theorie der gewöhnlichen Differentialgleichungen,* Berlin. Göttingen, Heidelberg 1953.)

15. THE SERIES SUBSTITUTION; THE INDICIAL EQUATION

The theory is more complete if all quantities are considered to be in the complex domain. We shall therefore write $w(z)$ instead of $y(x)$, and consider a linear homogeneous differential equation of the form

$$L[w] = w^{(n)} + q_1(z)w^{(n-1)} + \cdots + q_n(z)w \equiv \sum_{\nu=0}^{n} q_\nu(z)w^{(n-\nu)} = 0. \qquad \text{(IV.69)}$$

Here $q_0(z) = 1$. Let the other $q_j(z)$ have the form

$$q_j(z) = \frac{p_j(z)}{(z-a)^j} \quad \text{with } p_j(z) = \sum_{k=0}^{\infty} p_{j,k}(z-a)^k. \qquad \text{(IV.70)}$$

We enquire about the behaviour of the solutions in the neighbourhood of a point $z = a$, i.e., where the coefficients $q_j(z)$ can have a pole of order j. It would be convenient to introduce a new independent variable $z^* = z - a$, but we can avoid using this new variable simply by assuming, without loss of generality, that $a = 0$. We make, for w, a substitution

$$w = \sum_{\nu=0}^{\infty} a_\nu z^{\rho+\nu}, \qquad \text{(IV.71)}$$

where ρ is a fixed number still to be determined, and we may assume that $a_0 \neq 0$, say $a_0 = 1$. In the following array we have to multiply out the series standing side by side in each line and add the results:

$p_0(z) = 1$	$w^{(n)} = \rho(\rho-1)\ldots(\rho-n+1)a_0 z^{\rho-n} + (\rho+1)\rho(\rho-1)\ldots(\rho-n+2)a_1 z^{\rho-n+1} + \cdots$
....	....
$p_{n-1,0} + p_{n-1,1}z + \cdots$	$w' \cdot z^{1-n} = \rho a_0 z^{\rho-n} + (\rho+1)a_1 z^{\rho-n+1} + \cdots$
$p_{n,0} + p_{n,1}z + p_{n,2}z^2 + \cdots$	$wz^{-n} = a_0 z^{\rho-n} + a_1 z^{\rho-n+1} + a_2 z^{\rho-n+2} + \cdots$

In order to form $L[w]$ we arrange the result according to powers of z; then the coefficients of $z^{\rho-n}, z^{\rho-n+1}, \ldots$ are each to be put equal to zero. Using the abbreviations

$$\begin{aligned} \varphi_0(\rho) &= \rho(\rho-1)\ldots(\rho-n+1) + p_{1,0}\rho(\rho-1)\ldots(\rho-n+2) + \cdots + p_{n,0} \\ \varphi_k(\rho) &= \sum_{j=1}^{n} p_{j,k}\rho(\rho-1)\ldots(\rho-n+j+1) \quad \text{for } k = 1, 2, \ldots \end{aligned} \qquad \text{(IV.72)}$$

(an 'empty' product is, as usual, to be replaced by 1) we thus obtain the equations:

the factor of $z^{\rho-n}$: $a_0\varphi_0(\rho)=0$

factor of $z^{\rho-n+1}$: $a_1\varphi_0(\rho+1)+a_0\varphi_1(\rho)=0$

factor of $z^{\rho-n+2}$: $a_2\varphi_0(\rho+2)+a_1\varphi_1(\rho+1)+a_0\varphi_2(\rho)=0$

. . . .

the factor of $z^{\rho-n+k}$: $a_k\varphi_0(\rho+k)+a_{k-1}\varphi_1(\rho+k-1)+\cdots+a_0\varphi_k(\rho)=0$

. . . .

Since $a_0 \neq 0$, we must have

$$\varphi_0(\rho)=0; \tag{IV.73}$$

this is called the 'indicial equation'.

16. THE ROOTS OF THE INDICIAL EQUATION

Equation (IV.73) is an algebraic equation of the nth degree in ρ; let its roots be $\rho_1,\ldots,\rho_n$. For a given differential equation we can set up the indicial equation $\varphi_0(\rho)=0$ at the very beginning and distinguish between the two cases:
Case I. No two of the roots $\rho_1,\ldots,\rho_n$ differ by an integer.
Case II. There are at least two roots ρ_j,ρ_k whose difference $\rho_j-\rho_k$ is an integer; multiple zeros fall into Case II as a special case.

In Case I, therefore, it can never happen that $\phi_0(\rho)=0$ and $\varphi_0(\rho+k)=0$ for the same ρ and $k>0$. This means that the $a_1, a_2, \ldots$ can be calculated one after the other from the above equations. The series (IV.71) for $w(z)$ can therefore be constructed, and it can also be shown that the a_k decrease sufficiently rapidly as $k\to\infty$ for the series (IV.71) to converge in a finite circle of convergence (for the proof, see, e.g., E. L. Ince: *Ordinary Differential Equations,* Dover Publications 1956, p. 398). One solution of the differential equation is obtained for each of the roots $\rho_1,\ldots,\rho_n$.

In Case II there is among the roots at least one with the maximum real part; let ρ_1 be such a root. Then $\varphi_0(\rho_1+k)\neq 0$ for all $k>0$, and for this ρ_1 a solution $w_1(z)$ of the form (IV.71) is obtained; the other solutions of the differential equation may have more complicated forms. What form a second solution of the differential equations takes will be shown here only for the particular case $n=2$, though this does include many important differential equations. As in Chapter II, 7, the substitution

$$w(z)=w_1(z)\cdot W(z) \tag{IV.74}$$

transforms the equation

$$w''+\frac{p_1(z)}{z}w'+\frac{p_2(z)}{z^2}w=0 \tag{IV.75}$$

into

$$\frac{W''}{W'}+\frac{2w_1'}{w_1}+\frac{p_1}{z}=0 \tag{IV.76}$$

(we have divided by $w_1 W'$; neither w_1 nor W' can vanish identically, since a solution is being sought which differs from w_1 not merely by a constant; so a z-interval can be picked out in which w_1 and W' are not zero).
(IV.76) gives on integration

$$\ln W' + \ln w_1^2 + \int_{z_0}^{z} \frac{p_1(s)}{s}\, ds = c_1$$

(c_1, and later, c_2 are constants).

If $Q(z)$ is an antiderivative of $p_1(z)/z$, then $W' = c_2 \exp[-Q(z)]\, \dfrac{1}{w_1^2}$.

We can think of the right-hand side as being expanded in a series which, apart from a certain power z^σ, is a Laurent series

$$W' = z^\sigma \cdot \sum_{\nu \| -\infty}^{\infty} b_\nu z^\nu, \tag{IV.77}$$

from which W is found by integration. W itself will then again be of the form (IV.77) plus possibly a term $c \cdot \log z$. Then, by (IV.74) $w(z)$ will contain, apart from powers of z, a further term

$$c \cdot w_1(z) \cdot \ln z.$$

A point $z = a$ in a neighbourhood of which, with the possible exception of the point $z = a$ itself, the coefficients $q_\nu(z)$ of the differential equation (IV.69) are single-valued holomorphic functions and at least one of the functions $q_\nu(z)$ has a singularity at $z = a$, is called a *regular singular point* if there is a number q such that, for every solution w of the differential equation, the expression $|(z-a)^q w|$ tends uniformly to zero as the point $z = a$ is approached radially. If every coefficient $q_j(z)$ is single-valued and holomorphic apart from a finite number of singular points in the entire Gaussian plane, and if the differential equation has a regular singularity at all the points of singularity of the coefficients and at $z = \infty$, then the differential equation is said to *belong to the Fuchsian class.* In other words, a linear differential equation which has no other singularities than regular singular points belongs to the Fuchsian class.

17. EXAMPLES; THE HYPERGEOMETRIC FUNCTION

There is an extensive theory of Fuchsian differential equations (see, e.g., E. L. Ince, *op. cit.*, Chapter XV). Here we give only a few examples.

I.
$$w'' + \frac{2}{z-a} w' = 0$$

has the general solution

$$w = c_1 + \frac{c_2}{z-a};$$

the equation belongs to the Fuchsian class.

II. $w'' + \frac{1}{(z-a)^2} w' = 0$ has, for $a \neq 0$, as the general solution

$$w = c_1 + c_2 \int_0^z \exp\left(\frac{1}{s-a}\right) \mathrm{d}s.$$

The solutions (apart from the constants) have $z = a$ as an essential singularity; the point $z = a$ is not a regular singular point, and the differential equation does not belong to the Fuchsian class.

III. The differential equations for spherical harmonics (IV.7) and for Bessel functions (IV.32) are examples where terms containing logarithms appear.

IV. The 'hypergeometric differential equation'

$$w'' + \frac{(1+a+b)\,z - c}{z(z-1)}\, w' + \frac{ab}{z(z-1)}\, w = 0 \tag{IV.78}$$

belongs to the Fuchsian class. By substituting into it a power series one finds as a solution the 'hypergeometric series' or 'Gauss series' (let c be neither zero nor a negative integer)

$$\begin{aligned} F(a, b, c, z) &= 1 + \frac{a \cdot b}{1 \cdot c} z + \frac{a(a+1)\, b(b+1)}{1 \cdot 2 \cdot c(c+1)} z^2 + \cdots \\ &= \sum_{n=0}^{\infty} \frac{a(a+1)\ldots(a+n-1)\, b(b+1)\ldots(b+n-1)}{1 \cdot 2 \ldots n \cdot c(c+1)\ldots(c+n-1)} z^n \\ &= \sum_{n=0}^{\infty} \frac{\binom{-a}{n}\binom{-b}{n}}{\binom{-c}{n}} (-z)^n, \end{aligned} \tag{IV.79}$$

which converges for $|z| < 1$. If a or b is a negative integer, then the series terminates, and we obtain a polynomial as the solution. The hypergeometric series contains the geometric series as a special case with $a = 1$, $b = c$, and also many other well-known functions, e.g.,

$$\begin{aligned} F\left(\frac{1}{2}, \frac{1}{2}, \frac{3}{2}, z^2\right) &= \frac{1}{z} \arcsin z \\ F\left(\frac{1}{2}, \frac{1}{2}, 1, z^2\right) &= \frac{2}{\pi} K(z) = \frac{2}{\pi} \int_0^{\pi/2} \frac{\mathrm{d}u}{\sqrt{1 - z^2 \sin^2 u}} \\ F\left(\frac{1}{2}, -\frac{1}{2}, 1, z^2\right) &= \frac{2}{\pi} E(z) = \frac{2}{\pi} \int_0^{\pi/2} \sqrt{1 - z^2 \sin^2 u}\, \mathrm{d}u. \end{aligned} \tag{IV.80}$$

Here K and E denote the 'complete elliptic integrals of the first and second kind' (cf. (I.91)). A large number of further particular cases of the hypergeometric function which lead to otherwise known functions, and a numerous collection of numerical and transformational formulae can be found in the book by W. Magnus and F. Oberhettinger, *Formulas and Theorems for the Special Functions of Mathematical Physics,* 3rd ed., Chelsea Publishing Co., New York, 1949.

The hypergeometric series is, in turn, a special case of the more general series

$$ {}_pF_q(\alpha_1, \ldots, \alpha_p; \beta_1, \ldots, \beta_q; z) = \sum_{n=0}^{\infty} \frac{(\alpha_1)_n \ldots (\alpha_p)_n}{(\beta_1)_n \ldots (\beta_q)_n} \frac{z^n}{n!}, \tag{IV.81} $$

where

$$ (a)_n = a(a+1) \ldots (a+n-1); \; (a)_0 = 1. $$

The hypergeometric function can then be written as

$$ F(\alpha, \beta, \gamma, z) = {}_2F_1(\alpha, b; \gamma; z). \tag{IV.82} $$

An important special case of the general series (IV.79) is the confluent hypergeometric function

$$ \Phi(a, c; z) = \lim_{b \to \infty} F\left(a, b, c, \frac{z}{b}\right) = {}_1F_1(a; c; z) $$

$$ = \sum_{n=0}^{\infty} \frac{a(a+1) \ldots (a+n-1)}{c(c+1) \ldots (c+n-1)} \frac{z^n}{n!} = \sum_{n=0}^{\infty} \frac{\binom{-a}{n}}{\binom{-c}{n}} \frac{z^n}{n!}. \tag{V.83} $$

This function Φ is a solution of the 'confluent differential equation' or 'Kummer's differential equation'

$$ zw'' + (c - z)\, w' - aw = 0. \tag{IV.84} $$

If c is an integer, this differential equation has the solution

$$ z^{1-c} \Phi(a - c + 1, 2 - c; z). $$

which is linearly independent of the one mentioned above.

The Kummer differential equation has, in recent times, turned up more and more frequently in application, for example, in eigenvalue problems with the Schrödinger equation (see S. Flügge: *Rechenmethoden der Quantenmechanik,* Berlin, Göttingen, Heidelberg 1947). Many particular transcendental functions are special cases of the confluent function; e.g., Bessel functions, the Laguerre and Hermite polynomials, the incomplete gamma function, the Gaussian error integral, the integral sine, and so on; here we mention only one formula:

$$ J_\nu(z) = \frac{e^{-iz}\left(\frac{z}{2}\right)^\nu}{\Gamma(\nu+1)} \, {}_1F_1(\nu + \tfrac{1}{2}; 2\nu + 1; 2iz). \tag{IV.85} $$

18. THE METHOD OF PERTURBATIONS AND SINGULAR POINTS

Expansion in series constitutes a general method for differential equations which are not soluble in closed form; the required solutions are obtained in the form of an infinite series, and by breaking off the series, approximate solutions can be obtained. Expansions in series have been used several times in this book, e.g., in Chapter I, No. 30, Example 2, in Chapter III, No. 29, Example 2, and in Chapter IV, No. 15. Expansions in Fourier series, and, more generally, in terms of orthogonal systems, have been referred to in Chapter III, Nos. 20 and 21.

Another important application is the 'method of perturbations', where in the given differential equation $L(u, \varepsilon) = 0$, and perhaps also in the boundary conditions, a parameter ε which is regarded as small appears. For $\varepsilon = 0$ (i.e., for the 'unperturbed problem'), suppose the solution $u_0(x)$ is known, and that 'for small ε' the solution $u_\epsilon(x)$ can be expanded as a series in increasing powers of ε

$$u_\varepsilon(x) = \sum_{\nu=0}^{\infty} \varepsilon^\nu v_\nu(x) \tag{IV.86}$$

with $v_0(x) = u_0(x)$; and suppose it is possible, by substituting $u_\varepsilon(x)$ into the differential equation to determine further functions $v_\nu(x)$.

In eigenvalue problems the eigenvalue λ and the eigenfunctions $u(x)$ have to be expanded as power series in ε. This can be illustrated by the example of a self-adjoint problem as in Chapter III, No. 17

$$L[y] + \lambda[p(x) + \varepsilon q(x)]\ y(x) = 0 \quad \text{for } x \in B \tag{IV.87}$$

and the boundary conditions

$$U_\mu[y] = 0, \qquad \mu = 1, \ldots, k \quad \text{on the boundary } \partial B. \tag{IV.88}$$

For $\varepsilon = 0$ let an eigenfunction $y_0(x)$ corresponding to the eigenvalue λ_0 be known. The substitutions

$$y(x) = \sum_{\nu=0}^{\infty} \varepsilon^\nu y_\nu(x), \qquad \lambda = \sum_{\nu=0}^{\infty} \varepsilon^\nu \lambda_\nu \tag{IV.89}$$

are now made into the differential equation and the coefficients are compared for like power of ε. Let $y_\nu(x)$ be comparison functions in the sense of Chapter III, No. 17. We shall not deal with convergence questions here, but only show the calculations for the λ_ν, and $y_\nu(x)$.
From

$$\sum_{\nu=0}^{\infty} \varepsilon^\nu L[y_\nu] + \left(\sum_{\rho=0}^{\infty} \varepsilon^0 \lambda_\rho\right)(p(x) + \varepsilon q(x)) \sum_{\nu=0}^{\infty} \varepsilon^\nu y_\nu(x) = 0 \tag{IV.90}$$

it follows that

as the factor of ε^0: $Ly_0 + \lambda_0\ p(x)\ y_0(x = 0$

as the factor of ε^1: $Ly_1 + \lambda_0 p(x) y_1(x) + \lambda_0 q(x)\ y_0(x) + \lambda_1 p(x)\ y_0(x) = 0$

as the factor of ε^k: $Ly_k + \lambda_0 p(x)\ y_k(x) + \varphi + \lambda_k p(x)\ y_0(x) = 0$ (IV.91) where φ depends on $\lambda_0, \ldots, \lambda_{k-1}, y_0, \ldots, y_{k-1}$, but not on λ_k and not on y_k, and Ly has been written instead of $L[y]$. The first equation is automatically satisfied. Multiplication of the last equation by $y_0(x)$ and integration over the basic domain, using the condition for self-adjointness, gives

$$\int_B (Ly_k)\, y_0\, \mathrm{d}x = \int_B y_k(Ly_0)\, \mathrm{d}x = \int_B [-\lambda_0 p(x)\, y_0 y_k]\, \mathrm{d}x$$

and therefore

$$\int_B (Ly_k + \lambda_0 p(x)\, y_k]\, y_0\, \mathrm{d}x = 0;$$

hence it is possible to calculate λ_k from

$$-\int_B \varphi\, y_0\, \mathrm{d}x = \lambda_k \int_B p(x)[y_0(x)]^2\, \mathrm{d}x,$$

if $\lambda_0, \lambda_1, \ldots, \lambda_{k-1}, y_0, y_1, \ldots, y_{k-1}$ are already known. When λ_k has been found, the comparison function y_k has to be determined from

$$Ly_k + \lambda_0 p y_k + \varphi + \lambda_k p y_0 = 0.$$

It can be shown that this boundary-value problem is always soluble (a detailed presentation can be found, e.g., in L.Collatz: *Eigenwertaufgaben mit technischen Anwendungen,* §24. Here let it suffice to consider an example.

The self-adjoint eigenvalue problem

$$y'' + \lambda(1 + \varepsilon \cos x)\, y = 0, \qquad y(0) = y(\pi) = 0$$

can be interpreted as a buckling-load problem for a column of variable bending stiffness. For $\varepsilon = 0$ the problem is solvable exactly:

$$\lambda_0 = 1,\ y_0(x) = \sin x; \qquad \text{here } L[y] = y'',\ y_0'' = -y_0,$$

The substitution (IV.89), and equation (IV.91) lead to

$$(-\sin x + \varepsilon y_1'' + \varepsilon^2 y_2'' \cdots)$$
$$+ (1 + \varepsilon\lambda_1 + \varepsilon^2\lambda_2 + \cdots)(1 + \varepsilon \cos x)(\sin x + \varepsilon y_1 + \varepsilon^2 y_2 + \cdots) = 0.$$

The factor of ε^0 vanishes:

$$-\sin x + \sin x = 0,$$

the factor of ε^1 gives

$$y_1'' + y_1 + \cos x \sin x + \lambda_1 \sin x = 0.$$

Multiplication by y_0 and integration from 0 to π gives

$$\int_0^\pi (y_1'' + y_1)\, y_0\, \mathrm{d}x = \int_0^\pi (y_0'' y_1 + y_1 y_0)\, \mathrm{d}x = 0,$$

$$\int_0^\pi (\cos x \sin x + \lambda_1 \sin x)\, y_0\, \mathrm{d}x = \int_0^\pi \cos x(\sin x)^2\, \mathrm{d}x + \lambda_1 \int_0^\pi (\sin x)^2\, \mathrm{d}x = 0$$

or

$$\lambda_1 = 0$$

We then calculate $y_1 = \frac{1}{3} \sin x \cos x$ (an additive term $x \cdot \sin x$ could be taken into y_0).

The factor of ε^2 with $\lambda_1 = 0$ gives

$$y_2'' + y_2 + y_1 \cos x + \lambda_2 \sin x = 0;$$

this is multiplied by y_0 and integrated, and we obtain in just the same way

$$0 = \int_0^\pi (y_1 \cos x + \lambda_2 \sin x) \sin x \, dx$$
$$= \int_0^\pi \tfrac{1}{3}(\cos x \sin x)^2 \, dx + \lambda_2 \int_0^\pi (\sin x)^2 \, dx = 0$$

or

$$\lambda_2 = -\frac{1}{12}, \quad \text{and we have approximately}$$
$$\lambda \approx \lambda_0 + \lambda_2 \varepsilon^2 = 1 - \frac{1}{12} \varepsilon^2.$$

19. AN EXAMPLE OF THE SERIES-EXPANSION TECHNIQUE

This technique will be demonstrated (without going into convergence questions) by a typical simple example, namely, for a particular case of the Thomas-Fermi equation for an ionized gas

$$Ty(x) = -y''(x) + \sqrt{\frac{(y(x))^3}{x}} = 0, \quad \text{in an interval } I = (0, a)$$

with the boundary conditions

$$y(0) = 1, \qquad y(a) = 0, \qquad (a > 0).$$

If $a = \infty$, the second boundary condition is to be replaced by

$$\lim_{x \to +\infty} y(x) = 0.$$

Because of the x in the denominator in the differential equation there is a singularity at the point $x = 0$; and if $a = \infty$ there is another singularity at infinity. In order to determine the type of singularity at $x = 0$ we can try a power-series substitution

$$y(x) = 1 + \sum_{\nu=0}^{\infty} a_\nu x^{\rho + \nu\sigma}$$

into the differential equation, and compare the coefficients of the lowest power of x. Without loss of generality it can be assumed that $a_0 \neq 0$, $\rho \neq 0$, $\sigma \neq 0$.

We obtain

$$y^3 = 1 + 3a_0x^\rho + \cdots$$

$$\frac{1}{x}y^3 = x^{-1}(1 + 3a_0x^\rho + \cdots)$$

$$\sqrt{\frac{1}{x}y^3} = \pm x^{-1/2}\left(1 + \frac{3a_0}{2}x^\rho + \cdots\right)$$

$$-y'' = -\rho(\rho-1)a_0x^{\rho-2} + (\rho+\sigma)(\rho+\sigma-1)a_1x^{\rho+\sigma-2} + \cdots$$
$$= A_0x^{\rho-2} + A_1x^{\rho+\sigma-2} + \cdots$$

The terms of the lowest order must vanish; hence either

Case I. $A_0x^{\varrho-2} = 0$

or

Case II. $A_0x^{\rho-2} \pm x^{-1/2} = 0$

must occur. Case I requires that

$$\rho(\rho-1)a_0 = 0 \quad \text{or} \quad \rho = 1.$$

Case II implies that

$$x^{\rho-2} = x^{-1/2}, \quad \rho = \frac{3}{2}; \qquad \rho(\rho-1)a_0 = \pm\frac{4}{3}.$$

From the next lowest terms

$$\pm\frac{3a_0}{2}x^{\rho-1/2} + A_1x^{\rho+\sigma-2} = 0 \quad \text{with } A_1 = (\rho+\sigma)(\rho+\sigma-1)a_1;$$

it then follows that

$$\sigma - 2 = -\frac{1}{2}, \qquad \sigma = \frac{3}{2} = \rho, \qquad A_1 = \mp\frac{3a_0}{2} = -6a_1, \qquad a_1 = \frac{1}{3}.$$

By comparing the coefficients of higher powers we can calculate further a_ν and thus obtain a fully determinate power-series

$$1 + \sum_{\nu=0}^{\infty} a^\nu x^{(\nu-1)\rho}.$$

Case I. $\rho = 1$, and so we must have

$$\pm x^{-1/2} + \sigma(1+\sigma)a_1x^{\sigma-1} = 0$$

or

$$\sigma = \frac{1}{2}, \; a_1 = \mp\frac{4}{3};$$

the series therefore reads

$$y(x) = 1 + a_0 x \mp \frac{4}{3} x^{3/2} + \cdots$$

Here then the factor a_0 is still free; we can therefore work with Case I, because in Case II there is no constant still available in order to be able to satisfy the second boundary condition.

In the case of an infinite interval with $a = \infty$ we are interested in the behaviour at infinity, and we can try a series substitution

$$y = \sum_{\nu=0}^{\infty} a_\nu x^{-(\rho+\nu\sigma)} \quad \text{with } a_0 \neq 0,\ \sigma \neq 0,\ \rho > 0.$$

$$y^3 = a_0^3\, x^{-3\rho} + \cdots$$

$$\sqrt{\frac{y^3}{x}} = \pm\, a_0^{3/2}\, x^{-(3\rho/2)-1/2} + \cdots$$

$$-y'' = -\, a_0 \rho(\rho+1)\, x^{-\rho-2} + \cdots$$

Comparison of coefficients yields

$$-\rho - 2 = -\frac{3}{2}\rho - \frac{1}{2} \quad \text{or } \rho = 3$$

$$-\, a_0 \cdot 3 \cdot 4 = \pm\, a_0^{3/2} \quad \text{or } a_0 = 144,$$

and hence

$$y(x) = 144\, x^{-3} + \cdots$$

As already mentioned, with these formal substitutions the question of convergence still remains unanswered.

V

An excursus into partial differential equations

The general form of a partial differential equation is

$$F(x_1, x_2, \ldots, x_n, u, u_1, \ldots, u_n, u_{11}, u_{12}, \ldots, u_{1n}, \ldots, u_{nn}, \ldots) = 0, \qquad \text{(V.1)}$$

where the subscripts to u are used to denote partial derivatives, so that, for example,

$$u_j = \frac{\partial u}{\partial x_j}; \qquad u_{jk} = \frac{\partial^2 u}{\partial x_j \partial x_k}. \qquad \text{(V.2)}$$

We have already met some particular examples of partial differential equations in Chapter III, §5.

§1. GENERAL SOLUTIONS OF LINEAR PARTIAL DIFFERENTIAL EQUATIONS WITH CONSTANT COEFFICIENTS

1. SIMPLE LINEAR PARTIAL DIFFERENTIAL EQUATIONS

Example 1

The partial differential equation for a function $u(x, y)$

$$u_x = 0$$

states that u does not depend on x. Its general solution is therefore

$$u = w(y),$$

where $w(y)$ denotes an *arbitrary* function of y.

Geometrically, the solution $u = w(y)$ represents a general cylinder with a base curve in the y, z-plane, Fig. V.1.

Comparison with the corresponding ordinary differential equation

$$y' = 0$$

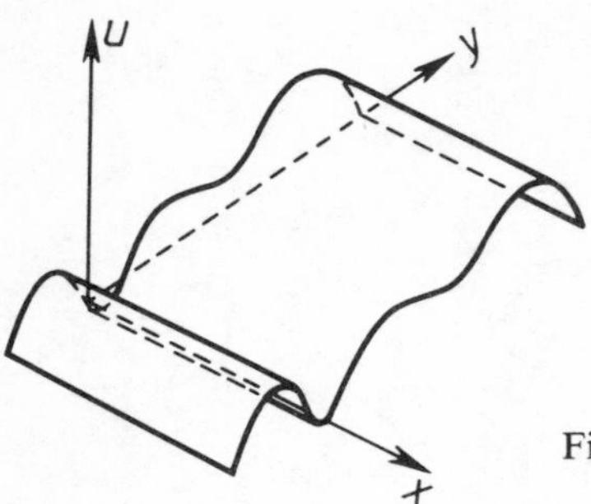

Fig. V.1. A cylinder suface as a solution of $u_x = 0$

with its solution

$$y = c$$

shows already in this simple case that partial differential equations have an incomparably greater manifold of solutions than ordinary differential equations, since arbitrary functions appear instead of arbitrary constants; (the n parameters of the solution of an ordinary differential equation of order n fix only n points of a curve, while an arbitrary function in the solution of a partial differential equation even of the first order is fixed by infinitely many points).

Example 2

The partial differential equation of the *first* order

$$Au_x + Bu_y = 0 \tag{V.3}$$

is *linear* and *homogeneous* and has *constant* coefficients A, B. We try, by a linear transformation of the co-ordinates,

$$\xi = \alpha x + \beta y, \qquad \eta = \gamma x + \delta y$$

to transform the differential equation into an equation which has derivatives with respect to only one of the variables, and which can therefore be solved as an ordinary differential equation. To do this, let the constants α, β, γ, δ at first be free. By the generalized chain-rule

$$u_x = \frac{\partial u}{\partial \xi}\frac{\partial \xi}{\partial x} + \frac{\partial u}{\partial \eta}\frac{\partial \eta}{\partial x} = u_\xi \alpha + u_\eta \gamma, \qquad u_\nu = u_\xi \beta + u_y\, \delta,$$

the differential equation becomes

$$u_\xi(A\alpha + B\beta) + u_\eta(A\gamma + B\,\delta) = 0.$$

If the constants are now chosen so that $\gamma + B\,\delta = 0$, for example, we can choose $\gamma = B$, $\delta = -A$, then the differential equation is reduced to $u_\xi = 0$ with the general solution $u = w(\eta)$. Therefore the solution of (V.3) is

$$u = w(Bx - Ay), \tag{V.4}$$

where w is an arbitrary differentiable function of its argument.

Geometrically considered, the circumstances are the same as in the first example, but relative now to the co-ordinate system u, ξ, η.

Example 3

The general solution of the *homogeneous*, *linear* differential equation of the *second* order

$$u_{xy} = 0$$

can be seen immediately. It is the *sum* of two arbitrary differentiable functions

$$u = w_1(x) + w_2(y).$$

The manifold of all surfaces which can be specified in this form is, conversely, characterized by the above differential equation. The question whether a function $u(x, y)$ of two variables can be split into two terms, one a function of x only and the other a function of y only, is determined by our differential equation. Geometrically, the solution represents a 'surface obtained by a translation', Fig. V.2, which we can visualize as being produced in the following way. A rigid metal template with its lower edge cut into the shape of the curve $w_2(y)$ is held parallel to the y-axis and is then moved in the direction of the x-axis over the surface of sand in a sand-box so that each point of the w_2-curve moves, not in a straight line, as in Fig. V.2, but along a curved track given by $w_1(x)$.

2. The Wave Equation and Laplace's Equation

The *wave equation* for different numbers of variables has already appeared in Chapter III. For two variables it has the form

$$c^2 u_{xx} - u_{yy} = 0 \tag{V.5}$$

where c denotes a velocity of propagation, as will soon be shown.

As in Example 2, a linear transformation is useful (in applications, y often denotes time, and so we shall write t instead of y):

$$\xi = \alpha x + \beta t, \qquad \eta = \gamma x + \delta t.$$

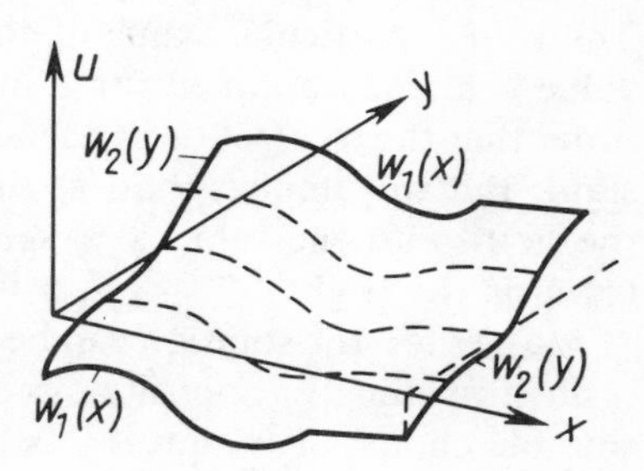

Fig. V.2. A solution surface of the partial differential equation $u_{xy} = 0$

The derivatives are

$$u_x = u_\xi \alpha + u_\eta \gamma,$$

$$u_{xx} = \alpha[u_{\xi\xi}\alpha + u_{\xi\eta}\gamma] + \gamma[u_{\xi\eta}\alpha + u_{\eta\eta}\gamma],$$

$$u_{tt} = \beta[u_{\xi\xi}\beta + u_{\xi\eta}\,\delta] + \delta(u_{\eta\xi}\beta + u_{\eta\eta}\,\delta].$$

The differential equation is thus transformed into

$$u_{\xi\xi}[c^2\alpha^2 - \beta^2] + 2u_{\xi\eta}[c^2\alpha\gamma - \beta\,\delta] + u_{\eta\eta}[c^2\gamma^2 - \delta^2] = 0.$$

If we choose the constants so that the outside terms disappear.

$$c^2\alpha^2 - \beta^2 = 0, \qquad \text{i.e.,} \qquad \beta = c\alpha,$$

and

$$c^2\gamma^2 - \delta^2 = 0, \qquad \text{i.e.,} \qquad \delta = -c\gamma,$$

then only

$$u_{\xi\eta} = 0$$

remains as the differential equation. In the equations linking α and β and linking γ and δ, a negative sign must occur in one of them, for otherwise the transformation relations for ξ and η would be proportional to one another, and then the middle term of the differential equation would also vanish. As in Example 2, we obtain as the solution

$$u = w_1(\xi) + w_2(\eta)$$

with

$$\xi = \alpha(x + ct), \qquad \eta = \gamma(x - ct),$$

or, if we take the α and γ into the functions themselves, we have

$$u = w_3(x + ct) + w_4(x - ct). \tag{V.6}$$

Every sum of two twice-continuously differentiable functions which have respectively as their arguments the linear expressions $x + ct$ and $x - ct$ will therefore satisfy the differential equation (V.5).

We can picture the situation by considering the functions w_3 and w_4 separately, $u = u_3 + u_4$.

At the initial time $t = 0$ the function $u_3 = w_3(x + ct)$ reduces to $u_3 = w_3(x)$. Think of a particular value of the function $u_3 = w_3(x)$; at a later time t this value is already attained for a smaller x; x must in fact, be reduced by ct in order that the total value of the argument within the bracket shall remain constant. But this implies that, as time goes on, the function w_3 travels towards the 'left' with the velocity c; see Fig. V.3. Similarly the function w_4 travels towards the 'right'.

Altogether, the solution can be interpreted as the superposition of two wave-trains travelling in opposite directions in, say, an oscillatory process. With a suitable choice of w_3 and w_4 'standing waves' can also occur.

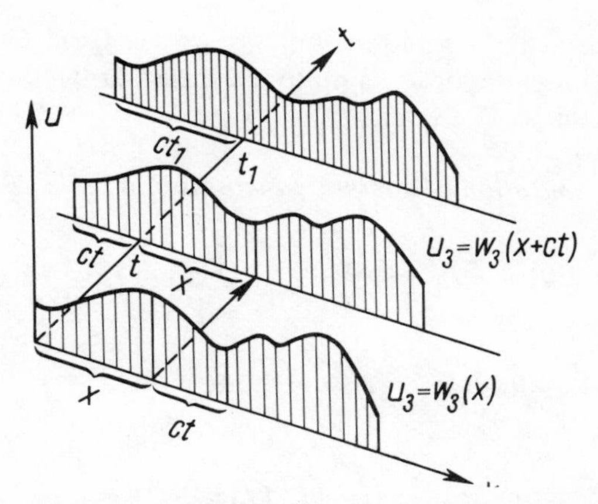

Fig. V.3. Travelling wave trains as solutions of the wave equation

If we put $c^2 = -1$, i.e., $c = i$, in the wave equation, then it is transformed into Laplace's equation

$$u_{xx} + u_{yy} = \Delta u = 0, \tag{V.7}$$

the study of which constitutes an independent branch of mathematics, viz., potential theory. It plays a fundamental rôle in other branches, too; for example, in vector analysis, in the theory of functions, in conformal mappings, and for many fields of applications. By comparison with the wave equation, we have the general solution

$$u = w_1(x + \mathrm{i}y) + w_2(x - \mathrm{i}y). \tag{V.7a}$$

All (twice) continuously differentiable functions of the complex arguments $x + \mathrm{i}y$ and $x - iy$ are thus solutions of Laplace's equation; so likewise the real and imaginary parts of such functions, and also linear combinations of them. Functions of the complex argument $x + iy$ are called 'analytic or holomorphic functions', provided that they have a derivative at every point of the region considered in the complex plane.

Further, since the imaginary part of any analytic function can also be regarded as the real part of another analytic function

$$\operatorname{Im} f(x + iy) = \operatorname{Re}\left(\frac{f(x + iy)}{i}\right)$$

we can condense the manifold of all solutions, the so-called potential functions, to

$$u = \operatorname{Re} f(x + \mathrm{i}y), \tag{V.8}$$

where f is again to be understood as an analytic function. With ordinary differential equations, the order of the equation is equal to the number of free parameters in the general solution; but, as we see from (V.8), with partial differential equations, the order is no longer necessarily equal to the number of arbitrary functions.

Examples of the separation of analytic functions into real and imaginary parts: Suppose

$$f(x+\mathrm{i}y)=(x+\mathrm{i}y)^3=x^3+\mathrm{i}3x^2y-3xy^2-\mathrm{i}y^3.$$

Both

$$\mathrm{Re}\, f(x+\mathrm{i}y)=u(x,y)=x^3-3xy^2$$

and

$$\mathrm{Im}\; f(x+\mathrm{i}y)=v(x,y)=3x^2y-y^3$$

are each solutions of Laplace's equation. The real solutions of Laplace's equation are also called *harmonic functions*; $u(x, y)$ and $v(x, y)$ are called *conjugate harmonic functions*.

Often the use of polar co-ordinates is advisable:

$$x=r\cos\varphi,\qquad y=r\sin\varphi.$$

For the complex argument we then have

$$x+\mathrm{i}y=r(\cos\varphi+\mathrm{i}\sin\varphi)=r\,\mathrm{e}^{\mathrm{i}\varphi}.$$

So, for example

$$(x+\mathrm{i}y)^n=r^n\,\mathrm{e}^{\mathrm{i}n\varphi}=\underbrace{r^n\cos n\varphi}_{u(r,\varphi)}+\mathrm{i}\underbrace{r^n\sin n\varphi}_{v(r,\varphi)}.$$

Here again

$$u(r,\varphi)=r^n\cos n\varphi\quad\text{and}\quad v(r,\varphi)=r^n\sin n\varphi$$

are harmonic functions.

3. NON-LINEAR DIFFERENTIAL EQUATIONS. THE BREAKING OF A WAVE

In recent times non-linear differential equations have gained ever increasing importance. With them many new phenomena have been observed. Here we present a simple physical example which leads to a non-linear equations, known as the Burgers equation. While the motivation presented in Part I of this number is to be regarded as merely heuristic, in Part II we investigate mathematically a simplified form (V.9) of the Burgers equation.

I. We consider quite generally a one-dimensional flow along the x-axis, for example, the flow of a liquid, a flow of traffic, or the like. In the interval from x_1 to x_2 an 'amount' q_1 flows into the interval at x_1 and an amount q_2 flows out of it at x_2; let the 'density', i.e., the amount per unit length, be $u(x, t)$.

Then the equation

$$\frac{d}{dt}\int_{x_1}^{x_2} u\,dx + q_1 + q_2 = 0$$

holds for the 'amount-balance'. On passage to the limit $x_2 \to x_1$ we obtain the 'conservation law'

$$\frac{\partial u}{\partial t} + \frac{\partial q}{\partial x} = 0.$$

Now an assumption about the connection between q and u is required. The simplest assumption would be $q = f(u)$; here, however, we shall also introduce a correction term $\nu \cdot u_x$. This is plausible if we consider the case of a traffic flow. Let u correspond to the number of cars per unit length of road; if the density of cars increases, i.e., if $u_x > 0$, a driver will reduce his speed, and will increase it if $u_x < 0$. It is reasonable, therefore, to make the substitution

$$q = f(u) - \nu \frac{\partial u}{\partial x}$$

where ν is a small, given, positive constant. Substitution of q into the conservation law then gives

$$u_t + f'(u) \cdot u_x - \nu u_{xx} = 0.$$

The derivative

$$f'(u) = \frac{df}{du}$$

is again a function of u, say $c(u)$. If we were to put $c = \text{const}$, we would have a linear differential equation, which would correspond to a linear model. Here, to obtain a very simple non-linear model, we shall put $c(u) = u$, and so the Burgers equation arises:

$$u_t + uu_x - \nu u_{xx} = 0. \tag{V.8}$$

Of course, very many other models have been studied; one of the best known of them is described by the Korteweg, de Vries (KdV) equation

$$u_t + uu_x + u_{xxx} = 0.$$

Another motivation for this last equation starts from the well-known Navier–Stokes equation, the derivation of which can be found in many textbooks (for example, in Courant–Hilbert: *Methods of Mathematical Physics*, vol. II, Interscience Publishers, 1962). For a 'Newtonian fluid' the time derivative $\dot{\gamma}$ of the shear angle γ depends linearly on the shear stress τ: $\tau = \eta \cdot \gamma$, where the constant η is the 'viscosity'. For non-Newtonian fluids, η is still a function of τ.

For a three-dimensional flow in the x, y, z-space, let $\vec{v} = (u, v, w)$ be the velocity vector, ρ be the density, p the pressure, and $\vec{f} = (f_x, f_y, f_z)$ the external force per unit volume. In vector notation, using the symbolic vector ∇ for differentiation

$$\nabla = \left(\frac{\partial}{\partial x}, \frac{\partial}{\partial y}, \frac{\partial}{\partial z}\right)$$

the Navier–Stokes equations for a Newtonian fluid of constant density read

$$\rho\left((\vec{v}\cdot\nabla)\vec{v}+\frac{\partial\vec{v}}{\partial t}\right) = -\nabla p+\vec{f}+\eta\,\nabla\vec{v}.$$

Additionally there is the equation of continuity

$$\frac{\partial u}{\partial x}+\frac{\partial v}{\partial y}+\frac{\partial w}{\partial z}=0.$$

For the sake of clarity we write out the first component of this vector equation:

$$(\vec{v}\,\mathrm{grad}\,u+u_t)=\rho(uu_x+vu_y+wu_z+u_t)=-p_x+f_x+\eta\,\Delta u,$$

where Δ denotes $u_{xx}+u_{yy}+u_{zz}$. Introducing the force vector

$$\vec{K}=(K_x,K_y,K_z)=-\nabla p+\vec{f},$$

we have in the case of a one-dimensional space ($v=w=0$) the equation

$$\rho(uu_x+u_t)=K_x+\eta u_{xx}.$$

For $K_x=0$ and $\eta=\rho\cdot\nu$ we again have the equation (V.8) exactly.

II. We now investigate mathematically the 'simplified Burgers equation'

$$u_t+c(u)u_x=0$$

which arises from the conservation law without the correcting term ($\nu=0$), or, rather more generally, the equation

$$u_t+c(u,t)u_x=0. \tag{V.9}$$

To understand what follows the motivations given above in Part I are not needed. Instead, we may think of $u(x,t)$ as the height of water in a channel at the point x and at time t. For $c(u)=c$, a constant, the general solution $u=w(x-ct)$, where w is an arbitrary function, represents a wave-train travelling to the right, as we already know from section V. 3, equation (V.6).

Now we consider first in the x,t-plane an arbitrary curve K given by $x=x(t)$; let this curve cut the x-axis in $x(0)=\xi$, and let $u=u(x,t)$ be a particular solution of the differential equation (V.9). Along the curve K the function u takes the value

$$u(x(t),t)=U(t),\quad \text{say;}$$

then along K we have

$$\frac{\mathrm{d}U}{\mathrm{d}t}=\frac{\partial u}{\partial x}\frac{\mathrm{d}x}{\mathrm{d}t}+\frac{\partial u}{\partial t}=u_x\frac{\mathrm{d}x}{\mathrm{d}t}+u_t.$$

Now let the curve K be chosen so that along it

$$\frac{dx}{dt} = c(u, t).$$

Since $u(x, t)$ satisfies the equation (V.9), it follows that $dU/dt = 0$, i.e., $u(x, t)$ is constant along K:

$$u = u(x(0), 0) = u_0.$$

Thus we get for the co-ordinates $x(t)$, $u(t)$ of the curve K the system of ordinary differential equations

$$\frac{dx}{dt} = c(u, t), \qquad x(0) = \xi,$$

$$\frac{du}{dt} = 0, \qquad u(0) = u_0.$$

Provided the given function $c(u, t)$ is bounded and satisfies a Lipschitz condition, it follows from the general existence and uniqueness theorem of Chapter I, No. 20 that by prescribing ξ and u_0 a continuous curve K is uniquely determined.

If the water level at the time $t = 0$ is given by a function $f(\xi)$, i.e., if $u(\xi, 0) = f(\xi)$, then through each point of the initial water level the corresponding curve $K = K[\xi]$ can be drawn, and so the water level can be calculated for times $t > 0$. Whether these curves from a 'smooth' surface we shall now investigate for the case when $c = c(u)$ depends only on u and not on t. In this case it follows from $u = \text{const} = f(\xi)$ with $F(\xi) = x(f(\xi))$ as the equation for K that

$$x = \xi + tF(\xi), \qquad u = f(\xi).$$

K is therefore a straight line.

Example

Let $c(u) = u$. Let the water level at the time $t = 0$ be given by

$$u(\xi, 0) = f(\xi) = \frac{1}{1 + \xi^2};$$

the 'characteristic' going through

$$x = \xi, \qquad u = \frac{1}{1 + \xi^2}$$

has the equation

$$x = \xi + \frac{t}{1 + \xi^2}, \qquad u = \frac{1}{1 + \xi^2}.$$

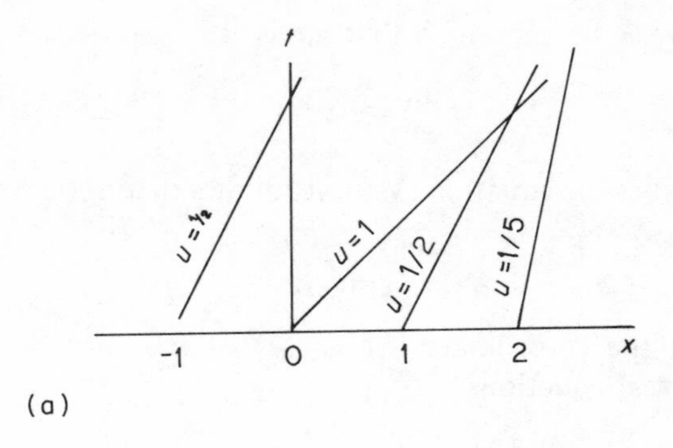

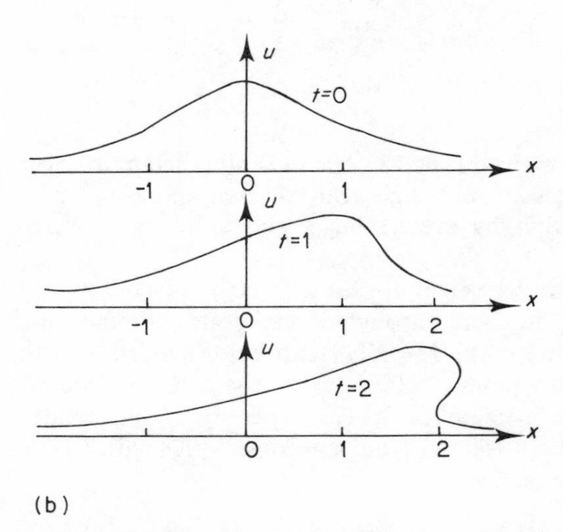

Fig. V.4. Breaking of a wave in the case of the simplified Burgers equation

Figure V.4a shows the projections of some of the characteristics in the x, t-plane, and Fig. V.4b shows some 'snapshots' of the curve $u(x)$ for $t = 0, 1, 2$. Since at $t = 2$ for $x = 2$ the values $u = 1/2$ and $u = 1$ appear, a breaking of the wave must have occurred somewhat earlier than $t = 2$. (When the wave breaks, of course, the differential equation loses its validity.)

In general it is true that: if two projections of the K curves for different values of u intersect in the x, t-plane, then a breaking of the wave must have already occurred earlier. This certainly occurs if the straight lines which are the projections of the K curves become 'steeper' with increasing ξ. This is expressed more clearly in the following

Theorem

Let a wave process described by the function $u(x, t)$ be given by the initial-

value problem for the simplified Burgers equation

$$u_t + c(u)u_x = 0, \qquad (-\infty < x < \infty, t = 0), \qquad u(x,0) = f(x),$$

where $c(u)$ and $f(x)$ are continuous functions; $f(x)$ describes the initial state. If there is an x_0 such that the function $F(x) = c(f(x))$ decreases strictly monotonely for $x > x_0$, then the wave will break.

The example can also serve as a model for the breaking of ocean waves.

§2. INITIAL-VALUE PROBLEMS AND BOUNDARY-VALUE PROBLEMS

4. The three fundamental types of a quasi-linear partial differential equation of the second order

The methods of the previous section will now be applied to the general differential equation of the second order

$$Tu = au_{xx} + 2bu_{xy} + cu_{yy} + \psi_1(x, y, u, u_x, u_y) = 0. \tag{V.9}$$

Here a, b, c are given constants, not all zero, and ψ_1 and later ψ_2 are continuous functions, continuously partially differentiable with respect to all their arguments. We again introduce new variables ξ, η (but keeping now the variable y instead of the t introduced previously)

$$\xi = \alpha x + \beta y, \qquad \eta = \gamma x + \delta y. \tag{V.10}$$

where the determinant $\Phi = \alpha\,\delta - \beta\gamma$ is not to vanish (for otherwise ξ and η would be proportional to one another). Just as before we calculate u_{xx}, u_{yy}, and also

$$u_{xy} = \alpha[u_{\xi\xi}\beta + u_{\xi\eta}\,\delta] + \gamma[u_{\xi\eta}\beta + u_{\eta\eta}\,\delta].$$

Substitution of these expressions into (V.9) gives

$$Tu = Au_{\xi\xi} + 2Bu_{\xi\eta} + Cu_{\eta\eta} + \psi_2(\xi, \eta, u, u_\xi, u_\eta) = 0$$

with

$$\begin{cases} A = a\alpha^2 + 2b\alpha\beta + c\beta^2 \\ B = a\alpha\gamma + b(\alpha\,\delta + \beta\gamma) + c\beta\,\delta \\ C = a\gamma^2 + 2b\gamma\,\delta + c\,\delta^2. \end{cases}$$

We calculate directly that

$$AC - B^2 = (ac - b^2)\Phi^2 \tag{V.11}$$

The determinant $d = ac - b^2$ has become

$$D = AC - B^2 = d\Phi^2$$

Under the transformation, therefore, the determinant preserves its sign and D will be zero if and only if d was equal to zero.

A classification into types based on the sign or vanishing of the determinant d therefore suggests itself; in analogy with the conic sections, the ellipse, parabola, and hyperbola, in plane analytical geometry, for which too the value of a determinant classifies the corresponding quadratic form, we say that in equation (V.9) the differential expression Tu is

$$\left.\begin{array}{ll}\text{elliptic, if} & d = ac - b^2 > 0 \\ \text{parabolic if} & d = ac - b^2 = 0 \\ \text{hyperbolic if} & d = ac - b^2 < 0\end{array}\right\} \tag{V.12}$$

Now it can always be arranged that in the transformed equation no term with the mixed partial derivative $u_{\xi\eta}$ appears, i.e., that $B = 0$. In the case $a = c = 0$ (and therefore $b \neq 0$) we can choose

$$\alpha = \beta = \delta = 1, \qquad \gamma = -1, \qquad (\Phi = 2 \neq 0),$$

and if not both the constants a and c vanish (if, say, $c \neq 0$), then we can take $\alpha = \delta = 1$, $\gamma = 0$, $\beta = -(b/c)$, $(\Phi = 1 \neq 0)$. Having arranged that $B = 0$, we can also arrange, by suitable scale factors in the directions of the ξ and η co-ordinates, to make the coefficients of the pure second derivatives $u_{\xi\xi}$ and $u_{\eta\eta}$, if they are not zero, equal to ± 1. We thus obtain, writing x and y again instead of ξ and η, the 'normal forms'

$$\begin{array}{ll}\text{Elliptic equation} & Tu = u_{xx} + u_{yy} + \psi(x, y, u, u_x, u_y) = 0 \\ \text{Parabolic equation} & Tu = u_{xx} \qquad\quad + \psi(x, y, u, u_x, u_y) = 0 \\ \text{Hyperbolic equation} & Tu = u_{xx} - u_{yy} + \psi(x, y, u, u_x, u_y) = 0\end{array} \tag{V.13}$$

Example

Laplace's equation	$u_{xx} + u_{yy} = 0$	Elliptic
Heat-conduction equation	$u_{xx} - u_y = 0$	Parabolic
Wave equation	$u_{xx} - u_{yy} = 0$	Hyperbolic

If a, b, c in the equation (V.9) are not constants, but given continuous functions of x and y, then, for the reduction to normal form, equation (V.10) has to be replaced by

$$\xi = \xi(x, y), \qquad \eta = \eta(x, y)$$

with the functions $\xi(x, y)$ and $\eta(x, y)$ still to be determined (a detailed exposition can be found in Courant and Hilbert, II, p. 154 et seq.). It then depends, in general, on the point (x_0, y_0) to what type the differential equation belongs at the point (x_0, y_0).

Examples 1

The Tricomi equation

$$yu_{xx} + u_{yy} = 0 \tag{V.14}$$

which models the transition from subsonic to supersonic flow in aerodynamics, is elliptic in the half-plane $y > 0$, hyperbolic in the half-plane $y < 0$, and parabolic for $y = 0$. The line $y = 0$ is called the 'parabolic boundary line'.

If the coefficients a, b, c depend not only on x and y but also on u, u_x, and u_y, then in general one can only still say that a particular solution $u(x, y)$ of the differential equation at a particular point (x_0, y_0) behaves elliptically, hyperbolically, or parabolically.

2. In aerodynamics, for a steady, compressible, two-dimensional flow the differential equation

$$(\gamma^2 - u_x^2)u_{xx} - 2u_x u_y u_{xy} + (\gamma^2 - u_y^2)u_{yy} = 0 \qquad \text{(V.15)}$$

in obtained, where γ is the velocity of sound, and u_x, u_y the components of the velocity vector v. Here the determinant has the value

$$(\gamma^2 - u_x^2)(\gamma^2 - u_y^2) - u_x^2 u_y^2 = \gamma^2(\gamma^2 - v^2).$$

In the 'sub-critical domain' ($|v| < \gamma$) the behaviour is elliptic, in the 'super-critical domain' ($|v| > \gamma$) hyperbolic, and at the passage through the critical velocity ($|v| = \gamma$, the 'sound barrier') parabolic.

There are, however, differential equations which belong to a fixed type for all solutions in the whole domain of definition of the solutions.

3. If a wire whose axis is bent into a closed curve C is dipped into a soap solution and then removed, a soap film surface is formed having the curve C as its boundary. Under certain simple assumptions, the thin film forms a 'minimal surface', which satisfies the differential equation

$$(1 + u_y^2)u_{xx} - 2u_x u_y u_{xy} + (1 + u_x^2)u_{yy} = 0 \qquad \text{(V.16)}$$

Here the differential equation is simply elliptic, because, for all solutions $u(x, y)$ the determinant

$$\begin{vmatrix} 1 + u_y^2 & u_x u_y \\ u_x u_y & 1 + u_x^2 \end{vmatrix} = 1 + u_x^2 + u_y^2 > 0$$

is always positive.

The differential equation (V.16) is not linear, but is still quasi-linear. A differential equation for a function u is said to be 'almost linear' if, regarded as a function, it is linear in all the derivatives of u which appear in it; it is said to be 'quasi-linear' if it is linear in all the highest-order derivatives which appear in it.

The classification into types adopted, as in (V.12), according to the sign of the determinant d may at first appear artificial and merely formal, but in the further theory of quasi-linear differential equations of the second order it turns out to be fundamental and determinative for essential properties of the solutions; moreover, it has an analogue in the description of processes and states occurring in nature, in that, often elliptic differential equations describe

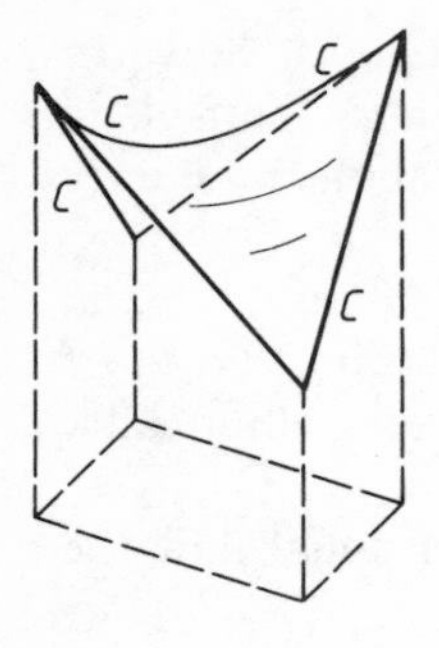

Fig. V.5. Minimal surface through a closed curve C (the four edges of a tetrahedron)

steady-state phenomena, but parabolic and hyperbolic equations, on the other hand, often describe diffusion processes and states.

5. Range of influence and domain of continuation

As with ordinary differential equations, we distinguish also with partial differential equations between

1. initial-value problems
2. boundary-value problems, and
3. eigenvalue problems.

Here, however, there is also a combination of 1 and 2, the 'mixed initial and boundary value problem'; we give an example of this at the end of this No.

A particularly simple example of an initial-value problem is provided by the differential equation (V.5) for a vibrating string. Let $u(x, t)$ be the deflection of the string at the point x at time t. At the time $t = 0$ let the initial deflection be given by $u(x, 0) = f(x)$ and the initial velocity by $u_t(x, 0) = g(x)$, the functions f and g being, say, continuous. The arbitrary functions w_3 and w_4 in the general solution (V.6) must now be fitted to satisfy the initial conditions

$$\left.\begin{aligned} u(x, 0) &= f(x) = w_3(x) + w_4(x) \\ u_t(x, 0) &= g(x) = cw_3'(x) - cw_4'(x) \end{aligned}\right\} \tag{V.17}$$

or

$$G(x) = \int_0^x g(s)\, ds = cw_3(x) - cw_4(x) - \gamma$$

where

$$\gamma = c(w_3(0) - w_4(0)).$$

We now have the two equations

$$w_3(x) = \frac{1}{2} f(x) + \frac{1}{2c} [G(x) + \gamma], \qquad w_4(x) = \frac{1}{2} f(x) - \frac{1}{2c} [G(x) + \gamma].$$

for determining w_3 and w_4. So the solution of this initial-value problem is

$$\left.\begin{aligned} u(x,t) &= w_3(x+ct) + w_4(x-ct) \\ &= \frac{1}{2}\,[f(x+ct) + f(x-ct)] + \frac{1}{2c}\int_{x-ct}^{x+ct} g(s)\,\mathrm{d}s. \end{aligned}\right\} \qquad \text{(V.18)}$$

We see that only the initial values from the interval

$$J = [x_1 - ct_1, x_1 + ct_1]$$

have an influence on the value of the function u at a point $(x - t)$; see Fig. V.6. The rhombus R in Fig. V.6

$$R = \{(x,t), |x - x_1| + c|t| \leq c|t_1|\}$$

is called the 'domain of continuation' of the interval J, and the domain of points (x, t) at which the values of the function u can be influenced by the initial values of J is called the 'range of influence' F of the interval J; see Fig. V.6.

$$F = \{(x,t), |x - x_1| - c|t| \leq c|t_1|\}.$$

The equation (V.18) further shows that for continuous changes of the initial data f and g the solution $u(x, t)$ likewise changes continuously. We also see from this equation that discontinuities in the derivatives of the initial data ('wave fronts', 'shock waves') are propagated along the straight lines

$$x \pm ct = \text{const},$$

which are known as the 'characteristics' (cf. Fig. V.3).

It may further be mentioned that, in applications, other boundary conditions may also occur; for example, with a vibrating string of finite length l,

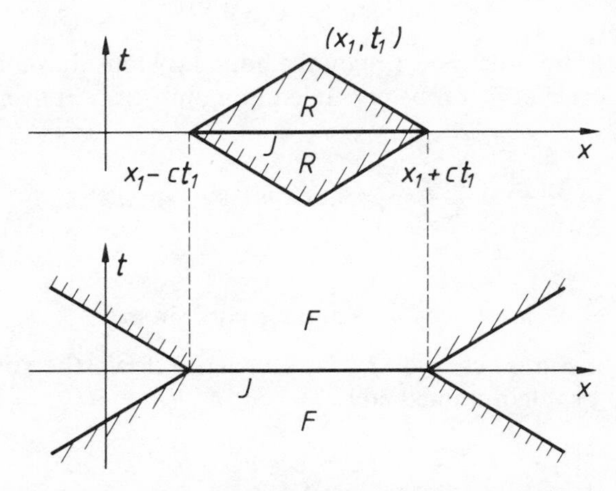

Fig. V.6. The domain of continuation R and the range of influence F of an interval

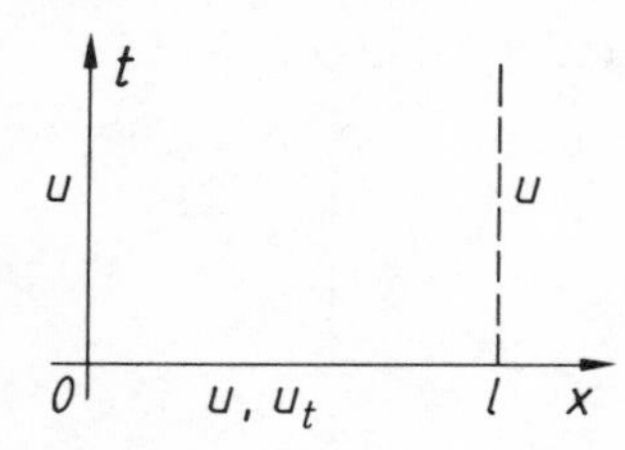

Fig. V.7. An initial and boundary value problem with the wave equation

which extends over the interval, say, $0 \le x \le l$, the deflections along the two half-lines $t \ge 0$, $x = 0$ and $t \ge 0$, $x = l$ are given, and the values of u and u_t are given along the interval $0 \le x \le l$, $t = 0$; see Fig. V.7. One then has an 'initial and boundary value problem' on hand.

6. Solution of a Boundary-Value Problem for a Circular Domain

Suppose a cylindrical-shaped body of radius $R = 1$ is given, which is insulated on all sides against heat flow and heat radiation. A certain temperature distribution is impressed on it round its circular boundary. We may treat the problem as a plane problem in an x,y-plane. The question asked is, what is the temperature distribution in the interior of the body? This temperature field satisfies Laplace's equation

$$u_{xx} + u_{yy} \equiv \Delta u = 0.$$

So the boundary-value problem may be formulated thus: *What harmonic function assumes on the boundary of the body prescribed values, which are, say, continuous along the boundary?*

The function $u(x, y)$ or, in polar co-ordinates $u^*(r, \varphi)$, has thus to satisfy Laplace's equation and the boundary condition

$$u^*(1, \varphi) = g(\varphi).$$

Here the temperature function $g(\varphi)$ on the boundary must have the period 2π. It may be assumed that it can be expanded in a uniformly convergent Fourier series:

$$g(\varphi) = \frac{a_0}{2} + \sum_{n=1}^{\infty} (a_n \cos n\varphi + b_n \sin n\varphi). \tag{V.19}$$

Since

$$u = r^n \cos n\varphi \quad \text{and} \quad v = r^n \sin n\varphi$$

are harmonic functions (cf. No. 2), we can write down the solution of the boundary-value problem immediately:

$$u^*(r, \varphi) = \frac{a_0}{2} + \sum_{n=1}^{\infty} (a_n r^n \cos n\varphi + b_n r^n \sin n\varphi). \tag{V.20}$$

This series converges uniformly in the whole disc, and therefore represents a continuous function; for $r \to 1$, the series goes over into the series for $g(\varphi)$, and so the prescribed boundary values are assumed. In the interior of the disc, for $r < 1$, the series satisfies Laplace's equation, because for constant φ the series is a power series in r, which certainly converges for $r \leq 1$, and therefore it can be differentiated twice term-wise with respect to r. For constant $r < 1$ the series converges uniformly in φ, and also the series arising from differentiation twice term-wise with respect to φ converges uniformly in φ, since $\sum_{n=1}^{\infty} C \cdot n^2 r^n$ (where C is a constant) is a convergent majorant for $r < 1$ and so the series may also be differentiated twice term-wise with respect to φ. In the series, therefore, the formation of the Laplacian operator may be interchanged with the summation. If the Fourier constants a_n and b_n are now determined from the given function $g(\varphi)$ by means of the well-known formulae in Fourier analysis

$$\begin{Bmatrix} a_n \\ b_n \end{Bmatrix} = \frac{1}{\pi} \int_{-\pi}^{\pi} g(\varphi) \begin{Bmatrix} \cos n\varphi \\ \sin n\varphi \end{Bmatrix} \mathrm{d}\varphi,$$

the boundary-value problem will be solved.

The function $u(r, \varphi)$ (if we again omit the asterisk) can also be written in closed form:

$$u(r, \varphi) = \int_{-\pi}^{\pi} G(r, \varphi, \psi) g(\psi) \, \mathrm{d}\psi,$$

where $G(r, \varphi, \psi)$ denotes the Green's function

$$\pi \cdot G(r, \varphi, \psi) = \frac{1}{2} + \sum_{n=1}^{\infty} r^n [\cos(n\varphi)\cos(n\psi) + \sin(n\varphi)\sin(n\psi)].$$

The expression in the square brackets is precisely, by the addition theorem for cosines) equal to

$$\cos(n\alpha) = \mathrm{Re}\, \mathrm{e}^{\mathrm{i}n\alpha} = \text{real part of } \mathrm{e}^{\mathrm{i}n\alpha}.$$

where $\alpha = \varphi - \psi$. Hence the sum can be evaluated from the geometric series:

$$\pi \cdot G(r, \varphi, \psi) = \frac{1}{2} + \left(\sum_{n=0}^{\infty} r^n \cos(n\alpha) - 1 \right) = -\frac{1}{2} + \mathrm{Re} Z,$$

where

$$Z = \sum_{n=0}^{\infty} r^n \mathrm{e}^{\mathrm{i}n\alpha} = \frac{1}{1 - r\,\mathrm{e}^{\mathrm{i}\alpha}} = \frac{1}{1 - r(\cos\alpha + \mathrm{i}\sin\alpha)} = \frac{1 - r\cos\alpha + \mathrm{i}r\sin\alpha}{N}$$

and

$$N = (1 - r\cos\alpha)^2 + r^2(\sin\alpha)^2 = 1 - 2r\cos\alpha + r^2.$$

Hence

$$\mathrm{Re}\, Z = \frac{1 - r\cos\alpha}{N}$$

and

$$\pi G = \frac{1}{2N}[-N + 2 - 2r\cos\alpha] = \frac{1}{2N}(1 - r^2)$$

The solution of the boundary-value problem can therefore be written in the form of the so-called 'Poisson integral'

$$u(r, \varphi) = \frac{1 - r^2}{2\pi}\int_{-\pi}^{\pi}\frac{g(\psi)\,d\psi}{1 - 2r\cos(\varphi - \psi) + r^2} = \int_{-\pi}^{\pi} G(r, \varphi, \psi)g(\psi)\,d\psi \tag{V.21}$$

with the Green's function

$$G(r, \varphi, \psi) = \frac{1 - r^2}{2\pi(1 - 2r\cos(\varphi - \psi) + r^2)}. \tag{V.22}$$

Important consequences can be drawn from this formula. For example, we obtain from (V.21) for $r = 0$

$$u(0, \varphi) = \frac{1}{2\pi}\int_{-\pi}^{\pi} g(\psi)\,d\psi. \tag{V.23}$$

This is the so-called 'mean-value theorem of potential theory' (in the plane). It states that the value of a harmonic function at the centre of a circle K is equal to the mean-value of the harmonic function on the circumference of the circle. If the boundary values $g(\varphi)$ are continuous in the interval $I = \{\varphi;\ -\pi \le \varphi \le \pi\}$ then it follows, in particular, that

$$\min_I g(\psi) \le u(0, \varphi) \le \max_I g(\psi) \tag{V.24}$$

which is a particular form of the 'boundary-maximum principle', which will be derived later in general form in No. 8.

The method which has been used here, of first finding the solutions, and subsequently proving their validity, is frequently used in solving boundary-value problems.

7. Example: A Temperature Distribution

As a concrete example, suppose it is given that

$$g(\varphi) = \begin{cases} A & \text{for } |\varphi| < \alpha \\ 0 & \text{for } |\varphi| > \alpha \end{cases} \quad \text{and} \quad |\varphi| < \pi, 0 < \alpha < \pi,$$

(Fig. V.8). (The circular curves of constant temperature have been drawn as broken curves.) The Fourier coefficients are now easily calculated to be

$$a_0 = \frac{1}{\pi}\int_{-\pi}^{+\pi} g(\varphi)\,d\varphi = \frac{2}{\pi}A\alpha;$$

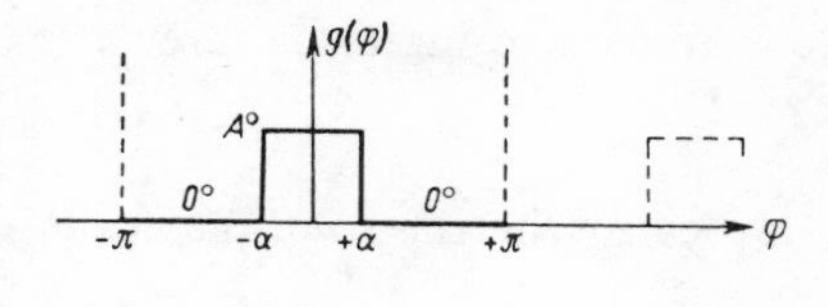

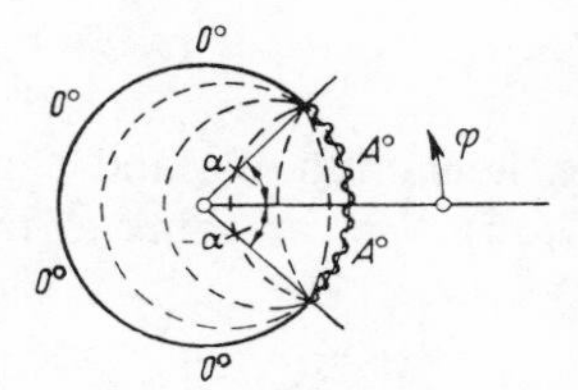

Fig. V.8. Temperature distribution on surface of cylinder given, temperature distribution in interior required

$$a_n = \frac{1}{\pi}\int_{-\pi}^{\pi} g(\varphi)\cos n\varphi\, d\varphi = \frac{A}{\pi}\int_{-\alpha}^{\alpha}\cos n\varphi\, d\varphi$$

$$= \frac{2A}{\pi n}\sin n\alpha \qquad (n = 1, 2, \ldots),$$

$$b_n = \frac{A}{\pi}\int_{-\alpha}^{\alpha}\sin n\varphi\, d\varphi = 0.$$

So the boundary-value function in the form of a Fourier series reads

$$g(\varphi) = \frac{A}{\pi}\left[\alpha + \sum_{n=1}^{\infty}\frac{2}{n}\sin n\alpha\cos n\varphi\right].$$

This series does not converge uniformly in φ, because the boundary function has a jump discontinuity, and it is therefore not yet certain that (V.20) provides the solution of the problem. From what has so far been done, it cannot even be stated whether the problem posed has a solution at all. However, we can use (V.20) as a heuristic device, and establish subsequently whether the solution obtained in accordance with (V.20) actually satisfies all the requirements. This we shall now do. Equation (V.20) gives

$$u^*(r,\varphi) = \frac{A}{\pi}\left[\alpha + \sum_{n=1}^{\infty}\frac{2}{n}r^n\sin n\alpha\cos n\varphi\right].$$

Summation of the series in closed form. An evaluation of this series can again be conveniently achieved by way of the *complex* representation.

We have

$$\sin n\alpha\cos n\varphi = \frac{1}{2}\sin n(\alpha+\varphi) + \frac{1}{2}\sin n(\alpha-\varphi)$$

$$= \frac{1}{2}\,\mathrm{Im}\, e^{in(\alpha+\varphi)} + \frac{1}{2}\,\mathrm{Im}\, e^{in(\alpha-\varphi)}.$$

Therefore, using the abbreviations

$$re^{i(\alpha+\varphi)} = q, \qquad re^{i(\alpha-\varphi)} = p,$$

we can write the solution in the form

$$u^*(r,\varphi) = \frac{A\alpha}{\pi} + \frac{A}{\pi}\operatorname{Im}\left\{\sum_{n=1}^{\infty}\frac{q^n}{n} + \sum_{n=1}^{\infty}\frac{p^n}{n}\right\}.$$

Since the moduli of q and p are less than 1, the series converge (in the complex field too). We can think of these series as being obtained from the geometric series

$$\sum_{n=0}^{\infty} q^n = 1 + q + q^2 + q^3 \ldots = \frac{1}{1-q} \quad \text{with } |q| < 1$$

by integration

$$\sum_{n=1}^{\infty}\frac{q^n}{n} = q + \frac{q^2}{2} + \frac{q^3}{3} + \frac{q^4}{4} + \ldots = -\ln(1-q) \quad \text{for } |q| < 1.$$

So we obtain for the sums the closed expression

$$u^*(r,\varphi) = \frac{A\alpha}{\pi} - \frac{A}{\pi}\operatorname{Im}\ln\overbrace{(1-q)\cdot(1-p)}^{\tilde{u}}.$$

Here we have to form the natural logarithm of a complex number $\tilde{z} = \tilde{x} + i\tilde{y}$ For this,

$$\ln\tilde{u} = \ln(\tilde{x} + i\tilde{y}) = \ln[\,|\tilde{u}|\,e^{i(\varphi+2k\pi)}] = \ln|\tilde{u}| + i(\tilde{\varphi} + 2k\pi)$$

where

$$\tilde{\varphi} = \arctan\frac{\tilde{y}}{\tilde{x}} \quad \text{and} \quad k = 0, \pm 1, \pm 2, \ldots$$

The value for $k = 0$ and $-\pi < \tilde{\varphi} \leq \pi$ is called the *principal value* of the logarithm. Since the addition of $2k\pi i$ has no effect on the result, we obtain

$$u^*(r,\varphi) = \frac{A}{\pi}(\alpha - \tilde{\varphi}).$$

Example

First let $\alpha = \pi/2$. Then

$$u^*(r,\varphi) = \frac{A}{2} - \frac{A}{\pi}\operatorname{Im}\ln\left[\left(1 - r\exp\left(i\left(\frac{\pi}{2}+\varphi\right)\right)\right)\left(1 - r\exp\left(i\left(\frac{\pi}{2}-\varphi\right)\right)\right)\right].$$

Since

$$e^{i\pi} = -1, e^{i\pi/2} = i \quad \text{and} \quad e^{i\varphi} + e^{-i\varphi} = 2\cos\varphi$$

it follows that

$$u^*(r,\varphi) = \frac{A}{2} - \frac{A}{\pi}\operatorname{Im}\ln(1 - ir\,2\cos\varphi - r^2). \tag{V.25}$$

For $\varphi = \pi/2$ the logarithm has a real positive argument

$$u^*\left(r, \frac{\pi}{2}\right) = \frac{A}{2}.$$

The temperature set up there is the arithmetic mean.

For $\varphi = 0$, we find

$$u^*(r, 0) = \frac{A}{2} - \frac{A}{\pi} \operatorname{Jm} \ln(1 - 2ir - r^2) = \frac{A}{2} - \frac{2A}{\pi} \operatorname{Im} \ln(1 - ir)$$

$$= \frac{A}{2} - \frac{2A}{\pi} \operatorname{Jm} \ln(\sqrt{1 + r^2} \cdot e^{i \arctan r})$$

$$= \frac{A}{2} - \frac{2A}{\pi} \operatorname{Jm} (\ln \sqrt{1 + r^2} - i \arctan r)$$

$$= \frac{A}{2} + \frac{2A}{\pi} \arctan r.$$

At the cross-section considered ($\varphi = 0$) the temperature function behaves as shown in Fig. V.9.

For an arbitrary point x, y inside the unit circle $|r| < 1$ we likewise obtain by picking out the imaginary part from (V.25)

$$u(x, y) = u^*(r, \varphi) = A\left(\frac{1}{2} - \frac{1}{\pi} \operatorname{Jm} \ln(1 - 2ix - x^2 - y^2)\right)$$

$$= A\left(\frac{1}{2} + \frac{1}{\pi} \arctan \frac{2x}{1 - x^2 - y^2}\right).$$

We can now easily establish that this function does satisfy the differential equation and the boundary conditions, and hence that it really is a solution of the problem. That the differential equation is satisfied follows immediately from (V.7a), since

$$\ln(1 - 2ix - x^2 - y^2) = \ln(1 - i(x + iy)) + \ln(1 - i(x - iy)).$$

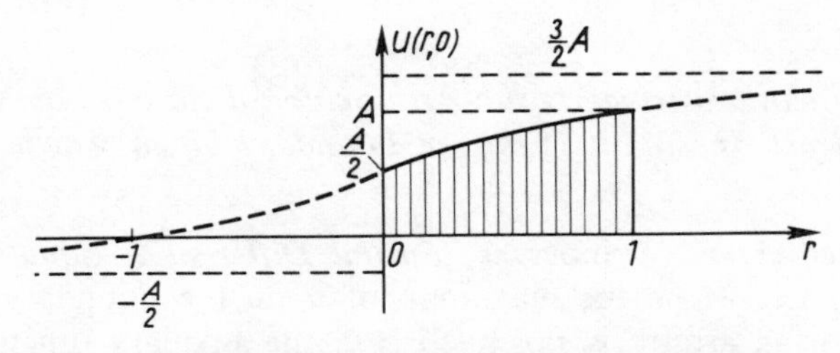

Fig. V.9. The temperature variation along the diameter $\varphi = 0$

For an arbitrary α from $0 < \alpha < \pi$ we shall merely state the result:

$$u = A\frac{\alpha}{\pi} + \frac{A}{\pi}\operatorname{arc\,tan}\frac{2x\sin\alpha - r^2\sin 2\alpha}{1 - 2x\cos\alpha + r^2\cos 2\alpha}.$$

The curves of constant temperature u are therefore circles (unless they degenerate into straight lines).

§3. THE BOUNDARY-MAXIMUM THEOREM AND MONOTONICITY

8. The boundary-maximum theorem in potential theory in the plane and in three-dimensional space

Let u be a solution, defined in an open, bounded, connected region B of the x,y-plane (or of the x,y,z-space), of the differential equation

$$-\Delta u(x, y) = h(x, y) < 0 \quad \text{for all } (x, y) \in B \tag{V.26}$$

where h is a given, continuous, negative function in B. If ∂B is the boundary of B and $\bar{B} = B \cup \partial B$ is the closure of B, let u be continuous in $\bar{B}$. It then follows that u assumes a maximum in $\bar{B}$, at a point P, say, in $\bar{B}$. But then P cannot be an interior point of $\bar{B}$, for if u had a differentiable maximum at P, i.e., if we had

$$\frac{\partial u}{\partial x} + \frac{\partial u}{\partial y} = 0, \qquad \frac{\partial^2 u}{\partial x^2} \leq 0, \qquad \frac{\partial^2 u}{\partial y^2} \leq 0, \tag{V.27}$$

at P, then would $-\Delta u \geq 0$, contrary to the hypothesis that $-\Delta u = h < 0$. (in space, $\partial u/\partial z = 0$, $\partial^2 u/\partial z^2 \leq 0$ would have to be included in (V.27)).

The function u therefore assumes its maximum *only* on the boundary ∂B.

A function u in B is said to be 'sub-harmonic' (resp. 'super-harmonic') if

$$-\Delta u = h^*(x, y) \leq 0 \text{ (resp. } -\Delta u = h^*(x, y) \geq 0) \quad \text{in } B. \tag{V.29}$$

(In space, we would have to write $h^*(x, y, z)$.)

The following theorem now holds:

Theorem

For every function u which is sub-harmonic in a closed, bounded domain B there is at least one point P on the boundary ∂B at which u assumes its maximum.

Proof. (After H. F. Weinberger, *Partial Differential Equations*, Blaisdell 1965, p. 55) Let M be the maximum of u on the boundary ∂B. We now introduce, for an arbitrary, positive $\varepsilon > 0$, the auxiliary function $v(x, y)$

$$v(x, y) = u(x, y) + \varepsilon(x^2 + y^2) = u(x, y) + \varepsilon r^2 \quad \text{with } r^2 = x^2 + y^2 \tag{V.29}$$

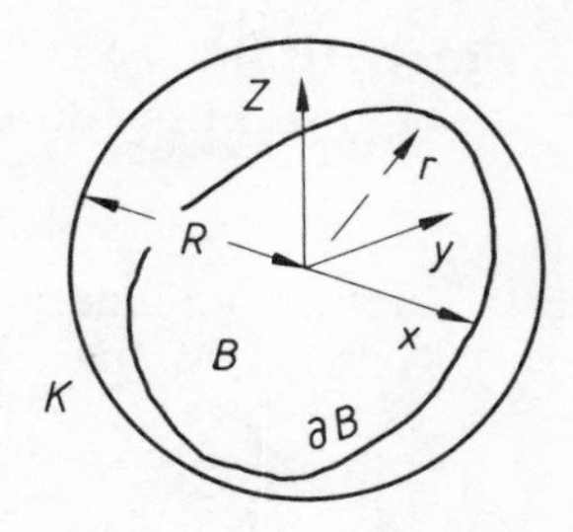

Fig. V.10. For the proof of the maximum theorem

(or in the x,y,z-space,

$$v(x, y, z) = u(x, y, z) + \varepsilon r^2 \quad \text{with } r^2 = x^2 + y^2 + z^2).$$

After an easy calculation, it then follows that

$$-\Delta v = -\Delta u - \varepsilon\, \Delta(x^2 + y^2) = h^*(x, y) - 4\varepsilon < 0 \text{ in } B \qquad \text{(V.30)}$$

(or, in space, $-\Delta v = h^* - 6\varepsilon < 0$ in B). v thus satisfies the condition (V.26) and assumes its maximum value on the boundary ∂B.

Now let R be the radius of a circle (resp. a sphere, in space), Fig. V.10, which contains all the points of $\bar{B}$. Then from (V.29) it follows that

$$\max_B v = \max_{\partial B} v \leq \max_{\partial B} u + R^2 = m + \varepsilon R^2. \qquad \text{(V.31)}$$

Letting $\varepsilon \to 0$ we obtain from this our assertion that

$$\max_{\bar{B}} v = \lim_{\varepsilon \to 0} \max_{\bar{B}} v \leq M \text{ in } B. \qquad \text{(V.32)}$$

Similarly, it follows, for every function $u(x, y)$ which is super-harmonic in a closed region b, that there is a point P on the boundary ∂B at which u assumes its minimum value.

Every function which is 'harmonic in B. i.e., for which $\Delta u = 0$, is both sub-harmonic and super-harmonic, and therefore it assumes its minimum and its maximum on the boundary. If such a function is constant on the boundary, $u = M$ on ∂B, then it must also be constantly equal to M in the interior. In particular, if a harmonic function $u = 0$ on ∂B, then u must be identically zero in the whole of B. Hence follows the

Uniqueness theorem

For a function $u(x, y)$ in an open, bounded, connected region B with boundary ∂B let the following boundary-value problem be posed:

$$\begin{cases} -\Delta u = h(x, y) & \text{in } B \\ u = k(x, y) & \text{on } B \end{cases} \qquad \text{(V.33)}$$

where $h(x, y)$ and $k(x,\ y)$ are given continuous functions. This boundary-value problem, if it has a solution at all, has precisely one solution $u(x, y)$. (In space, we would have to write $u(x, y, z)$, $h(x, y, z)$, and $k(x, y, z)$.)

Proof. If the boundary-value problem has two solutions $u_1(x, y)$ and $u_2(x, y)$, then the difference

$$v(x, y) = u_1(x, y) - u_2(x, y)$$

satisfies the boundary-value problem

$$\begin{cases} \Delta v = \Delta u_1 - \Delta u_2 = h - h = 0 & \text{in } B \\ v = u_1 - u_2 = k - k = 0 & \text{on } \partial B. \end{cases} \tag{V.34}$$

v is therefore a harmonic function in B, and vanishes identically on the boundary ∂B, and therefore, by the preceding result, it vanishes identically in B. Hence

$$u_1(x, y) \equiv u_2(x, y),$$

i.e., if a solution exists, then it is uniquely determined (for another proof, see No. 14).

9. Continuous dependence of the solution of the boundary-value problem on the data

A further important consequence is the

Theorem (Continuous dependence on the data).

If the boundary-value problem (V.33) *has a solution* (x,y), *then* u *depends continuously on the given data* $h(x,y)$ *and* $k(x,y)$. (*In space, we would correspondingly have to write* $u(x,y,z)$, $h(x,y,z)$, *and* $k(x,y,z)$.)

What is meant by continuous dependence on the data will be seen at once from the proof.

Proof. Let

$$H = \sup_B |h(x, y)|, \qquad K = \max_{\partial B} |k(x,y)|; \tag{V.35}$$

then in the interior

$$-\Delta\left(u + \frac{1}{4}H(x^2 + y^2)\right) = h(x, y) - H \leq 0 \quad \text{in } B \tag{V.36}$$

(in space, we would, correspondingly with

$$H = \sup_B h(x, y, z),$$

have to write

$$\frac{1}{6}H(x^2 + y^2 + z^2)$$

instead of

$$\frac{1}{4}H(x^2+y^2,$$

and also in (V.39) the factor 1/6 would appear instead of 1/4), and also

$$u+\frac{1}{4}H(x^2+y^2)\leq K+\frac{1}{4}HR^2=M \quad \text{on the boundary } \partial B. \tag{V.37}$$

The function

$$v=u+\frac{1}{4}H(x^2+y^2)$$

is therefore sub-harmonic and takes its maximum on the boundary. This maximum is $\leq M$, and so

$$v\leq M \text{ in } B.$$

Hence we have for u the upper bound

$$u=v-\frac{1}{4}H(x^2+y^2)\leq K+\frac{1}{4}HR^2-\frac{1}{4}H(x^2+y^2)\leq K+\frac{1}{4}HR^2. \tag{V.38}$$

If we now replace u by $-u$, i.e., h and k by $-h$ and $-k$, then we have the same bound for $-u$

$$-u\leq K+\frac{1}{4}HR^2$$

and hence altogether for u the estimate

$$|u|\leq K+\frac{1}{4}HR^2=\max_{\partial B}|k|+\frac{1}{4}R^2\sup_B|h|. \tag{V.39}$$

This is also a useful estimate, for numerical purposes, of the solution u of the boundary-value problem (V.33).

The desired proposition on continuity now follows immediately. Let $u^*(x, y)$ be the solution of the boundary-value problem

$$\left.\begin{aligned} -\Delta u^* &= h^*(x,y) \quad \text{in } B\\ u^* &= k^*(x,y) \quad \text{on } \partial B;\end{aligned}\right\} \tag{V.40}$$

then for the difference $\varphi=u-u^*$ we have

$$\left.\begin{aligned} -\Delta\varphi &= h-h^* \quad \text{in } B\\ \varphi &= k-k^* \quad \text{on } \partial B.\end{aligned}\right\} \tag{V.41}$$

By (V.39) it follows from this that

$$|\varphi|=|u-u^*|\leq\max_{\partial B}|k-k^*|+\frac{1}{4}R^2\sup_B|h-h^*|; \tag{V.42}$$

this means, roughly speaking, that if h and k are changed, but only sufficiently slightly, then the solution u also changes 'only slightly'. The precise form of this assertion about continuity is given by the formula (V.42): given any $\varepsilon > 0$, we can arrange that $|u - u^*| < \varepsilon$, and, indeed, uniformly in B, by making the right-hand side of (V.42) sufficiently small.

10. Monotonicity theorems. Optimization and approximation

Another consequence of the boundary-maximum theorem, which is very important for numerical methods, is the

Monotonicity Theorem I

Let B be an arbitrary, open, bounded, connected region with boundary ∂B. Then for every solution $u(x, y)$ (or $u(x, y, z)$ in space) of the inequalities

$$\left.\begin{aligned} -\Delta u &\leq 0 \quad \text{in } B \\ u &\leq 0 \quad \text{on } \partial B \end{aligned}\right\} \tag{V.43}$$

it is true that

$$u \leq 0 \quad \text{in } B. \tag{V.44}$$

Proof. Since u is sub-harmonic, u assumes its maximum on the boundary ∂B: but u is not positive there, and so u is also not positive in the whole region B.

Rather differently expressed, this states that:

Monotonicity Theorem II

The operator

$$Tu = \begin{cases} -\Delta u & \text{in } B \\ u & \text{on } \partial B \end{cases} \tag{V.45}$$

is of 'monotonic type', i.e., for two functions $v(x,y)$ and *$w(x,y)$ (or, in space, $v(x,y,z)$ and $w(x,y,z)$) which are twice continuously differentiable in an open, bounded, connected region B and are continuous functions in $\bar{B}$, it is true that*

$$Tv \leq Tw \quad \text{implies } v \leq w \text{ in } \bar{B}. \tag{V.46}$$

By this again it is meant that

$$\left\{\begin{aligned} -\Delta v \leq -\Delta w \quad &\text{in } B \\ v \leq w \quad &\text{on } \partial B \end{aligned}\right\} \quad \text{implies } v \leq w \text{ in } \bar{B}, \tag{V.47}$$

The proof follows immediately from the first monotonicity theorem if we put $u = v - w$.

Application to optimization and approximation

An important approximation process for setting up an approximate solution

v for the required solution u of the boundary-value problem (V.33) consists in the following.

We approximate to u by means of an expression v of the form

$$u \approx v = \varphi_0 + \sum_{\nu=1}^{p} a_\nu \varphi_\nu(x, y),$$

where the $\varphi_\nu(x, y)$ are fixed chosen functions which satisfy the differential equations

$$\Delta\varphi_0 = -h, \qquad \Delta\varphi_\nu = 0 \qquad (v = 1, \ldots, p)$$

and the a_ν are at first unknown real constants. For every choice of the a_ν, the function v, and therefore also the error $\varepsilon = v - u$, satisfies Laplace's equation $\Delta\varepsilon = 0$. For arbitrarily chosen values of the a_ν, we know the error ε for the corresponding function v on the boundary ∂B and so we can determine a number δ such that

$$-\delta \leq \varepsilon(P) \leq \delta, \tag{V.48}$$

where P runs through the set of points of the boundary ∂B. By the monotonicity theorem, the error estimate (V.48) for the chosen function v also holds for all interior points P of B, or

$$|\varepsilon(P)| = |v(P) - u(p)| \leq \delta \quad \text{for } P \in B. \tag{V.49}$$

Next, in order to obtain the most favourable error estimate, we try to choose the still free constants a_ν so that the bound δ shall turn out to be as small as possible. We thus have a 'linear optimization' problem on hand: δ is the target function to be minimized

$$\delta = \min$$

under the infinitely many side-conditions (or 'restrictions')

$$-\delta \leq \varphi_0(P)t \sum_{\nu=1}^{p} a_\nu \varphi_\nu(P) - k(P) \leq \delta \quad \text{for all } P \in \partial B. \tag{V.50}$$

The $(p + 1)$ variables of the optimization problem are the a_ν and δ.

In a numerical treatment the problem is 'discretized' by choosing a finite number of points $P_\rho(\rho = 1, \ldots, q)$ and writing down the side conditions only for $P = P_\rho$. We then have a finite optimization, for the treatment of which various algorithms have been developed (e.g., the 'simplex algorithm', cf. Collatz and Wetterling [71], and also Glashoff and Gustafson, *Einführung in die lineare Optimierung*, Darmstadt 1978, U. Grothkopf, Diplomarbeit Univ. Hamburg 1980). In this way the solutions of many boundary-value problems have been calculated numerically on computers.

In this example of a 'boundary-value problem of the first kind' the boundary-maximum theorem can also be used instead of the monotonicity theorem, and a Chebyshev approximation problem can be formulated as follows by means of the maximum norm of a continuous function φ defined on the boundary

$$\|\varphi\| = \max_{P \in \partial B} |\phi(P)|.$$

$$\|v - u\| = \left\| \varphi_0 + \sum_{\nu=1}^{p} a_\nu \varphi_\nu - u \right\| = \max_{P \in \partial B} \left| \varphi_0(P) + \sum_{\nu=1}^{p} a_\nu \varphi_\nu(P) - u(P) \right| \qquad \text{(V.51)}$$

is to be made as small as possible.

For the problem (V.33) the boundary-maximum theorem and the monotonicity theorem are equivalent. In more complicated cases, however, the range of the monotonicity theorem is very much greater than that of the boundary-maximum theorem. For non-linear boundary-value problems only the monotonicity theorem can be used directly (without special tricks) for obtaining two-sided estimates of the required solutions, and even for linear boundary-value problems the boundary-maximum theorem cannot be used directly for this purpose; for example, it is already unusable for boundary-value problems of the second and third kinds, in which the value of the normal derivative or of a combination of the function values and the values of the normal derivative are prescribed on parts of the boundary. For non-linear problems the monotonicity theorem often provides the only possibility for a usable estimation of the error in approximation solutions (cf. Collatz, Günther and Sprekeln, *Z. angew. Math. Mech.* 56 (1976), 1–11).

11. A NUMERICAL EXAMPLE; A TORSION PROBLEM

A homogeneous prismatic beam whose cross-section B is simply connected (i.e., has no holes in it) is subjected to torsional stress by a torque M (see Fig. V.11). In its unstressed state, the cross-section lies in an x,y-plane. Under certain simplifying assumptions (see, e.g., S. Timoshenko and J. N. Goodier, *Theory of Elasticity*, McGraw-Hill, 3rd ed., 1970, p. 309) the torsion stresses are proportional to the derivatives $\partial\Phi/\partial x$, $\partial\Phi/\partial y$ of a stress function $\Phi(x, y)$, where Φ satisfies Laplace's differential equation in B and has the boundary value 0:

$$\Delta\Phi = -1 \quad \text{in } B, \qquad \Phi = 0 \quad \text{on } \partial B. \qquad \text{(V.52)}$$

Let B be the domain $|x| < 1$, $|y| < \frac{1}{2}$. By putting

$$u = \Phi + \frac{x^2 + y^2}{4}$$

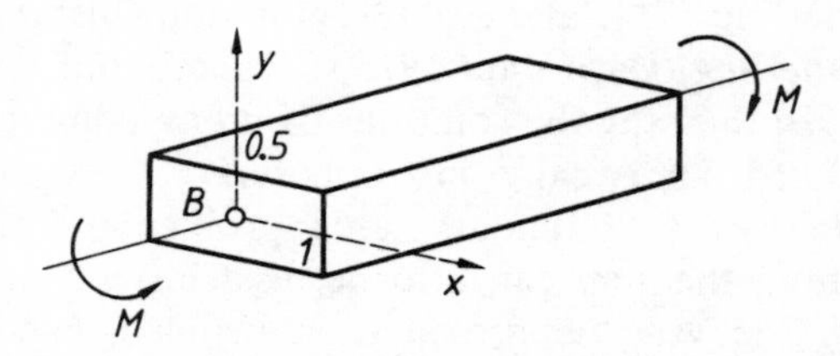

Fig. V.11. Torsion problem

the inhomogeneity is transferred to the boundary condition:

$$\Delta u = 0 \quad \text{in } B, \qquad u = \frac{x^2 + y^2}{4} \quad \text{on } \partial B. \tag{V.53}$$

We can now approximate to u by an expression of the form v

$$v = \sum_{\nu=1}^{p} a_\nu \Phi_\nu(x, y) \quad \text{with} \quad \Phi_\nu(x, y) = \operatorname{Re}(x + \mathrm{i}y)^{2(\nu-1)} \quad (v = 1, \ldots, p)$$

and so

$$\Phi_1 = 1, \qquad \Phi_2 = x^2 - y^2, \quad \Phi_3 = x^4 - 6x^2y^2 + y^4;$$

for $p = 1$ we have to approximate to the boundary value $\frac{1}{4}(x^2 + y^2)$ as well as possible by a constant a_1; the best value is $a_1 = 3/16$, and then

$$\left| u - \frac{3}{16} \right| = \left| \frac{1}{4}(x^2 + y^2) - \frac{3}{16} \right| \le \frac{1}{8} \quad \text{on the whole boundary } \partial B.$$

The boundary value can be bracketed appreciably more closely by making the substitution $a_1 + a_2(x^3 - y^2)$; with $a_1 = a_2 = 3/20$ we have

$$\left| u - \frac{3}{20}(1 + x^2 - y^2) \right| \le \frac{1}{20} \quad \text{on } B.$$

By taking in further functions Φ_ν we can easily increase the accuracy. The table below (for the computation of which I have to thank Mr. Uwe Grothkopf) shows, for $p = 1, 2, 3, 4, 5, 6$, the accuracy of δ achieved, and the value of $a_1 = v(0, 0)$ as the approximation value for the value of u at the centre of the rectangle (the numbers are rounded values):

p	δ	$a_1 = v(0,0)$
1	0.123	0.188
2	0.050	0.150
3	0.008 62	0.112
4	0.002 070	0.113 0
5	0.001 318	0.113 99
6	0.000 372 0	0.113 857

We have obtained, for $p = 6$, with a modest amount of computing work, the inclusion

$$| u(0, 0) - 0.113\,857 | \le 0.000\,372,$$

while with other methods, such as the difference method, finite elements, or the Ritz method, such a guaranteed error-estimate could scarcely be obtained, or only after a long computing time.

§4. WELL-POSED, AND FREE, BOUNDARY-VALUE PROBLEMS

12. WELL-POSED AND NOT WELL-POSED PROBLEMS

Suppose the solution u of a differential equation

$$Tu = 0 \quad \text{in a region } B \tag{V.54}$$

is required, which satisfies the initial or boundary conditions

$$Ru = r \quad \text{on the boundary } \partial B \text{ of the region } B \tag{V.55}$$

(B and ∂B as in No. 7). The problem is said to be 'well-posed' if

1. there is a solution u to the problem,
2. u is uniquely determined, and
3. the solution u depends continuously on r.

We consider a few examples.

(a) A particular initial-value problem with a hyperbolic equation which is well posed.

The initial-value problem set up in No. 4 for the wave equation with prescribed continuous intial values (V.17)

$$u(x, 0) = f(x), \qquad u_t(x, 0) = g(x)$$

is well-posed, because a unique solution is given by (V.18) and this solution depends continuously on the initial data. The uniqueness of the solution follows from the fact that (V.6) is the general solution of (V.5) and the functions w_3 and w_4 are determined (up to a constant) by the initial conditions.

(b) The following initial-value problem with an elliptic equation is not well-posed. With Laplace's equation (V.7) let the same initial values be prescribed as in the previous example (a). Consider in particular the sequence of initial-value problems with the initial values

$$u(x, 0) = 0, \qquad u_y(x, 0) = \tfrac{1}{2} \cos (nx) \quad (\text{for } n = 1, 2, \ldots)$$

(here y is written instead of t as in the previous example). These intial-value problems have the solutions

$$u_n(x, y) = \frac{\cos (n, x) \sinh (ny)}{n^2}.$$

The initial values of u and u_y converge uniformly to 0. $u \equiv 0$ is the solution of (V.7) with the initial values $u(x, 0) = u_y(x, 0) = 0$. But the solutions $u_n(x, y)$ of the initial-value problems considered do not converge to the null function for $y = 0$, indeed, they do not converge at all. There is no continuous dependence on the initial data.

(c) A particular boundary-value problem with a hyperbolic equation which is not well-posed.

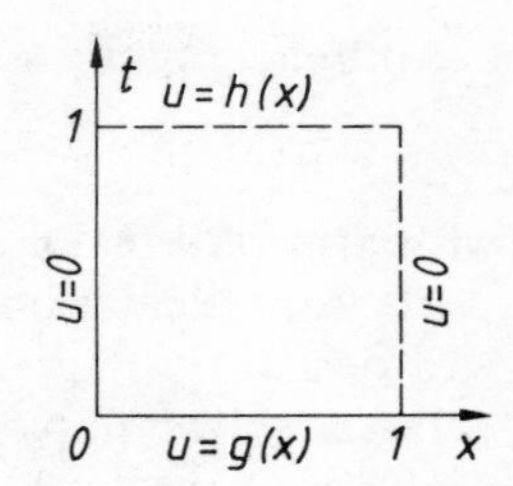

Fig. V.12. A bondary-value problem with the wave equation which is not well-posed

For the wave equation with $c = 1$, let the boundary values be, as in Fig. V.12,

$$u(x, 0) = g(x), \qquad u(x, 1) = h(x), \qquad u(0, y) = u(1, y) = 0.$$

To make matters simple we have chosen $g(x) = 0$. The solution, if it exists, is of the form (V.6).

$u(x, 0) = 0$	requires (1)	$w_3(x) = -w_4(x)$	for $0 < x < 1$
$u(0, y) = 0$	requires (2)	$w_3(y) = -w_4(-y)$	for $0 < y < 1$
$u(1, y) = 0$	requires (3)	$w_3(1 + y) = -w_4(1 - y)$	for $0 < y < 1$
$u(x, 1) = h(x)$	requires (4)	$h(x) = \varphi(x)$	for $0 < x < 1$

with

$$\psi(x) = w_3(1 + x) + w_4(x - 1)$$

Now

$$\begin{aligned} w_4(x-1) &= -w_3(1-x) && \text{by (2)} \\ &= w_4(1-x) && \text{by (1)} \\ &= -w_3(1+x) && \text{by (3)} \end{aligned}$$

(see Fig. V.13), and so

$$\psi(x) = w_3(1 + x) + w_4(x - 1) \equiv 0 \quad \text{for } 0 < x < 1.$$

If, now, $h(x) \neq 0$ in $0 < x < 1$ is prescribed, then there is no solution of the boundary-value problem. But if $h(x) \equiv 0$ is prescribed, then the continuity condition is infringed, because, with a 'small' perturbation ($h(x)$ 'small' but not $\neq 0$), a solution no longer exists. The uniqueness is also infringed; for, if the boundary value $u = 0$ is prescribed on the whole boundary, then there

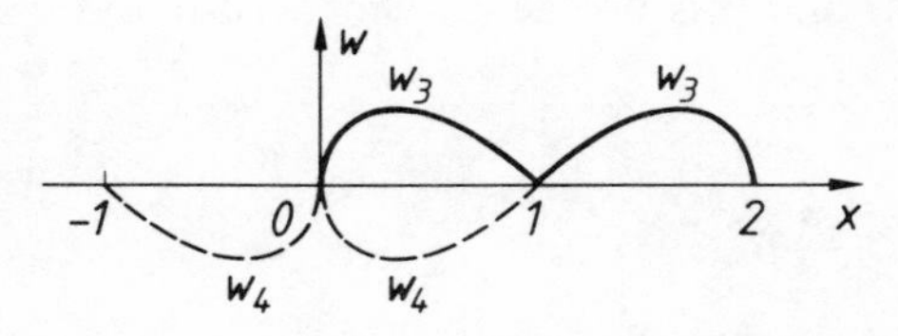

Fig. V.13. For the construction of the auxiliary function

are infinitely many solutions

$$u = c \sin(k\pi x) \sin(k\pi y)$$

with arbitrary real c and $k = 1, 2, \ldots$.

(d) A particular boundary-value problem with an elliptic equation which is well-posed.

It was shown in No. 5 that in potential theory the so-called boundary-value problem of the 1st kind for a circular region with prescribed boundary conditions $g(\varphi)$ has a solution which depends continuously on the boundary values, provided we restrict ourselves to boundary distributions which can be expanded in uniformly convergent Fourier series (this assumption, which was used in the proof, can be somewhat weakened); cf. also No. 8. We then further need only the uniqueness of the solution in order to say that the problem is well-posed. This uniqueness was proved in No. 7.

13. Further examples of problems which are not well-posed

Normally boundary-value problems with elliptic equations, and initial-value problems or mixed initial and boundary-value problems with hyperbolic and parabolic equations, are regarded as being well-posed problems. In recent times, however, problems which are not well-posed have acquired increasing importance. A few simple instances may be given.

1. Conclusions about the past with parabolic equations. We observe the present state of a medium, and we want to know about the state at earlier times. For example, we observe a temperature distribution, or a density distribution, or a concentration distribution, and we would like to draw inferences about the distribution at an earlier time.

2. We observe the state of a medium in a bounded region of space, and we would like to draw conclusions about the state in a neighbourhood outside this region. As an example, suppose observations are made of the height u of the tide along a coastal strip from A to B round a bay of the ocean. If n is the direction of the normal, we have that $\partial u/\partial n = 0$ along the length of the arc (since no water goes into the interior of the land). Let F be the ocean region between the coast from A to B and the secant from A *to* B. In F let the equation $\Delta u = 0$ hold. From these data we would like to determine the height of tide u along the secant from A to B, because it would be somewhat difficult actually to measure u there. This is an initial-value problem for an elliptic differential equation, and this problem is not well-posed.

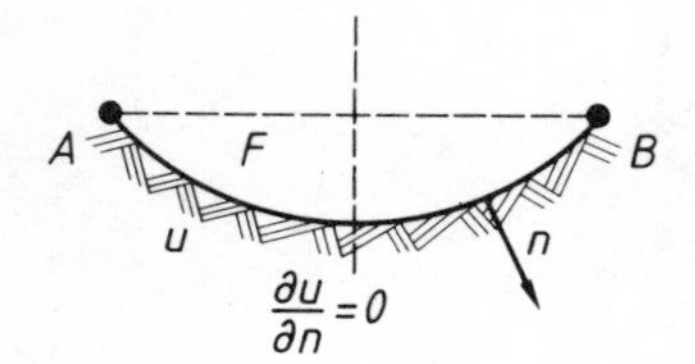

Fig. V.14. Height of tide in an ocean bay

3. Inverse differential equation problems. One would like to deduce from measurements what the coefficients of the differential equation are. For example, the equation for heat-conduction

$$\frac{\partial u}{\partial t}=\frac{\partial}{\partial x}\left(k(x)\frac{\partial u}{\partial x}\right) \tag{V.56}$$

holds for a bar; from measurements, the temperature distribution in the bar, $u(x, t_0)$ and $u(x, t_1)$, is known for two different times t_1 and t_2. From them it is desired to determine $k(x)$.

14. EXAMPLES OF FREE BOUNDARY-VALUE PROBLEMS

Again, an entirely different type of problem is concerned with free boundaries. Here the region considered consists of two (or even several) domains B_1, B_2. It is known that in B_1 a differential equation $T_1u=0$ holds, and in B_2 another differential equation $T_2u=0$ holds; but the boundary Γ between the two domains (the 'interface') is not known. We know only that certain transition conditions shall hold on Γ.

Several examples may be mentioned.

1. The Stephan problem. As a simple case of a Stephan problem, let us consider (as an idealized problem) a sea which, in the x,y-plane, covers the half-plane $x>0$, while $x<0$ is land, see Fig. V.15. At time t let the strip S: $0<x<s(t)$, y arbitrary, be free from ice (water), and let the region $x>s(t)$ be covered with ice. Let $u(x,t)$ be the temperature of the water on the straight lines $x,y\in S$ (y arbitrary) at time t. Heat flows from the land into the water in accordance with the equation

$$\frac{\partial u}{\partial x}=-g(t) \quad \text{for } x=0, t>0, \tag{V.57}$$

where $g(t)$ is to be regarded as given. Further, let the initial condition be known:

$$u(x,0)=f(x) \quad \text{for } 0\leq x\leq s_0=s(0). \tag{V.58}$$

We want to know the melting curve C (Fig. V.16):

$$x=s(t).$$

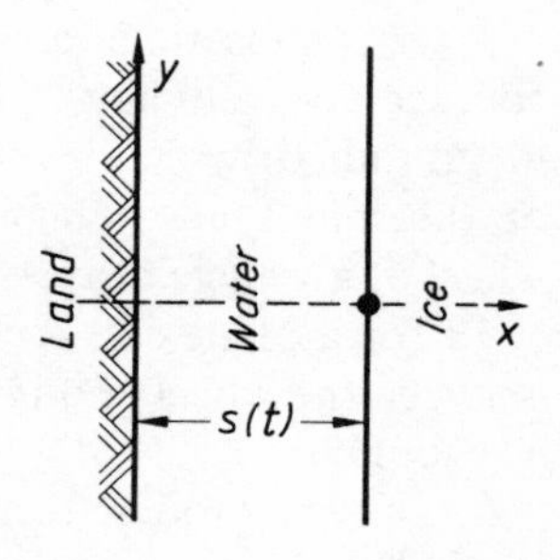

Fig. V.15. A Stephan problem

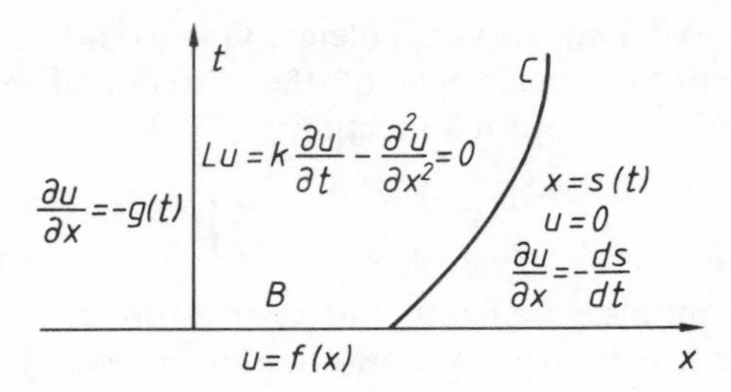

Fig. V.16. The melting curve as a free boundary

In the water region

$$B: \qquad \{0 < x < x(t), y > 0\}$$

u satisfies the equation of heat conduction (with a given constant k)

$$Lu = \frac{\partial u}{\partial t} - \frac{\partial^2 u}{\partial x^2} = 0 \quad \text{in } B. \tag{V.59}$$

In the ice a different differential equation holds. At the melting-curve C we have, as the transition condition,

$$u = 0 \quad \text{and} \quad \frac{\partial u}{\partial x} = -\frac{\mathrm{d}s}{\mathrm{d}t} \quad \text{(the heat-conduction balance).} \tag{V.60}$$

C is a free boundary. The problem is made more difficult, compared with classical problems, by the fact that we do not yet know where the differential equation (V.59) holds. The part C of the boundary is being sought, on which, however, we do have the two boundary conditions (V.60). Similar problems arise in solidification processes (e.g., with molten steel).

2. Free boundaries between regions of elastic behaviour and plastic behaviour occur in the continuum mechanics of deformable bodies. The differential equations in the elastic region and in the plastic region are different from one another, but the boundary surface Γ between the two regions is not known in advance, see Fig. V.17.

3. Free boundaries occur as free water surfaces on the sea with waves, or ships' wakes, in flow through porous media, etc.: see, e.g., K. Wieghardt: Theoretische Strömungslehre, Stuttgart, Teubner, 1965.

4. If two elastic bodies are in contact with one another under pressure, the contact does not take place at a single point but over a surface area, the magnitude of which is not known in advance, and the stresses arising depend on this magnitude. This is the so-called 'contact problem', If, for example, a wheel of a vehicle bears on a beam support, as shown diagrammatically in Fig. V.18 for a plane section, the beam will undergo a deformation, and contact between the wheel and beam will extend from P_1 to P_2. The points P_1 and P_2 are sought.

5. In recent times free boundary-value problems have arisen and have been investigated in many other applications, for flows through porous media, con-

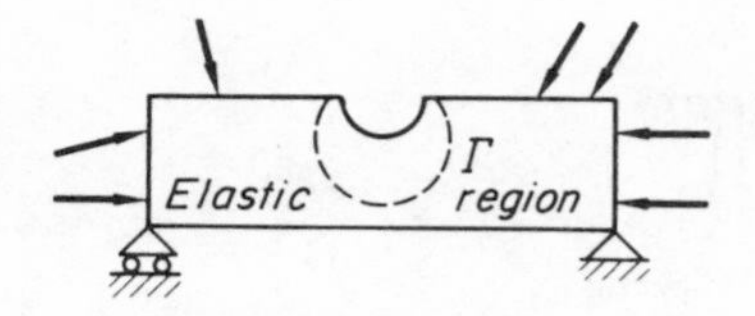

Fig. V.17. Free boundary between the regions of elastic and plastic behaviour

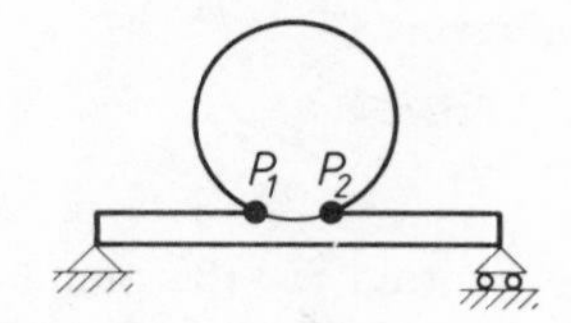

Fig. V.18. The contact problem

centration distributions, and the like; see, e.g., D. G. Wilson, A. D. Solomon and P. T. Boggs, *Moving-boundary Problems*, Acad. Press, 1978, p. 329.

§5. RELATIONS WITH THE CALCULUS OF VARIATIONS AND THE FINITE-ELEMENT METHOD

As with ordinary differential equations, so also with partial differential equations, the relations with the calculus of variations can be made use of for many purposes. Here, however, we have space only to indicate this by means of some very simple examples.

15. The variational problem for the boundary-value problem of the first kind in potential theory

In order not to interrupt the later argument, we mention first as an important tool the so-called 'first Green's formula'.

Without going into topological details, let us assume, for simplicity, that B is an open, bounded region in the x,y-plane (or in x,y,z-space) for which Gauss's integral theorem holds, and that B is bounded by a closed, piecewise smooth boundary curve ∂B free from double points (or by a piecewise smooth boundary surface ∂B). Let n denote the outward normal on ∂B. For two functions v, w belonging to $C^2(B)$ (i.e., functions which have continuous partial derivatives up to the second order inclusive) and which are continuous on the boundary ∂B and 'almost everywhere' differentiable on the boundary, the first Gauss formula (see, e.g., Courant–Hilbert 62, p. 252) holds:

$$\int_B v\,\Delta w\,\mathrm{d}\tau + \int_B (\operatorname{grad} v, \operatorname{grad} w)\,\mathrm{d}\tau - \int_{\partial B} v\,\frac{\partial w}{\partial n}\,\mathrm{d}f = 0. \qquad (V.61)$$

Here $\mathrm{d}\tau$ is a volume-element, $\mathrm{d}f$ a line-element (or a surface-element) and, as usual,

$$\left.\begin{aligned}(\operatorname{grad} v, \operatorname{grad} w) &= \frac{\partial v}{\partial x}\frac{\partial w}{\partial x} + \frac{\partial v}{\partial y}\frac{\partial w}{\partial y} + \frac{\partial v}{\partial z}\frac{\partial w}{\partial z}\\ (\operatorname{grad} v)^2 &= \left(\frac{\partial v}{\partial x}\right)^2 + \left(\frac{\partial v}{\partial y}\right)^2 + \left(\frac{\partial v}{\partial z}\right)^2,\end{aligned}\right\} \qquad (V.62)$$

where the terms

$$\frac{\partial v}{\partial z}\frac{\partial w}{\partial z}, \left(\frac{\partial v}{\partial z}\right)^2$$

are to be omitted for the plane case.

The boundary-value problem. We consider the boundary-value problem of the first kind in potential theory. With the same notation as above, let a function u in a region B for which Gauss's integral theorem holds satisfy the differential equation

$$-\Delta u = q \quad \text{in } B \tag{V.63}$$

and the boundary condition

$$u = p \quad \text{on } \partial B, \tag{V.64}$$

where q and p are given continuous functions of position (as in (V.33)). It is assumed that this boundary-value problem has a solution $u = \hat{u}$. Now let functions φ be called 'admissible' if they satisfy the same conditions as the functions v, w did in the first Green's formula, and if, in addition, they assume the prescribed boundary values p on ∂B. We put $\varphi = \hat{u} + \psi$, where $\psi = 0$ on ∂B, and we then enquire about the minimum of the integral

$$I[\varphi] = \int_B [(\text{grad } \varphi)^2 - 2\varphi q] \, d\tau = \min, \tag{V.65}$$

when φ traverses the set of admissible functions. If we substitute here

$$(\text{grad } \varphi)^2 = (\text{grad}(\hat{v} + \psi))^2 = (\text{grad } \hat{u})^2 + 2(\text{grad } \hat{u}, \text{grad } \psi) + (\text{grad } \psi)^2$$

we can use the Green's formula (V.61)

$$\int_B (\text{grad } \hat{u}, \text{grad } \psi) \, d\tau = \int_{\partial B} \psi \frac{\partial \hat{u}}{\partial n} \, df - \int_B \psi \, \Delta \hat{u} \, d\tau = \int_B q\psi \, d\tau; \tag{V.66}$$

where $\psi = 0$ on ∂B and $-\Delta \hat{u} = q$ has been used. Hence (V.65) gives

$$I[\varphi] = \int_B \{(\text{grad } \hat{u})^2 - 2\hat{u}q + (\text{grad } \psi)^2\} \, d\tau \tag{V.67}$$

or

$$I[\varphi] = I[\hat{u}] + \int_B (\text{grad } \psi)^2 \, d\tau. \tag{V.68}$$

In particular, $I[\varphi] \geq I[\hat{u}]$ for all admissible functions φ, and the equality sign holds if an only if $\text{grad } \psi \equiv 0$ or $\psi = \text{const}$. Since $\psi = 0$ on the boundary, this occurs if and only if $\psi \equiv 0$. Hence the following

Theorem

Let B be a region for which the Gauss integral theorem holds; let its boundary ∂B be piecewise smooth. Suppose the boundary-value problem (V.63), (V.64) *with the given continuous functions p and q has a solution* $u = \hat{u}$. *Then* $\hat{u}$ *gives the integral* $I[\varphi]$ *in* (V.65) *its minimum value as* φ *traverses the set of all admissible functions* (*i.e., functions which are differentiable sufficiently often and which satisfy the boundary condition* (V.64). *Moreover,* $\hat{u}$ *is the unique solution of the boundary-value problem* (V.63), (V.64).

The last of these assertions is obtained directly from the first Green's formula (V.61) if we put $v = w$:

$$\int_B v\, \Delta v\, d\tau + \int_B (\text{grad } v)^2\, d\tau - \int_{\partial B} v \frac{\partial v}{\partial n}\, df = 0. \tag{V.69}$$

Since

$$\Delta v = 0 \text{ in } B \quad \text{and} \quad v = 0 \quad \text{on } \partial B, \tag{V.70}$$

it follows that grad $v \equiv 0$, hence $v = \text{const}$, and since $v = 0$ on ∂B, we have $v \equiv 0$ in B. If, now, the boundary-value problem (V.63), (V.64) has two solutions u_1 and u_2, then the difference $v = u_1 - u_2$ satisfies the boundary-value problem (V.70); therefore $v \equiv 0$, i.e., $u_1 \equiv u_2$. This means that the solution of the boundary-value problem, if it exists at all, is unique.

16. THE RITZ METHOD

As with ordinary differential equations in Chapter II, §7, we can use the connection between boundary-value problems and variational problems to find an approximation solution φ for the solution $u(x)$ of a boundary-value problem.

We substitute

$$\varphi \approx w = w_0 + \sum_{\nu=1}^{s} a_\nu w_\nu \tag{V.71}$$

in the expression $I[\varphi]$ of (V.65), and determine the free, real parameters a_ν so as to make $I[\varphi]$ as small as possible, i.e., the a_ν are to be found from the equations

$$\frac{\partial I[\varphi]}{\partial a_\nu} = 0 \qquad (v = 1, \ldots, s). \tag{V.72}$$

where the w_ν are to satisfy the boundary conditions

$$\left.\begin{cases} w_0 = p \\ w_\sigma = 0 \end{cases} (\sigma = 1, \ldots, s)\right\} \quad \text{on } \partial B.$$

The success of the method depends strongly on the choice of the functions w_ν to be substituted.

Example: The torsion problem with a beam of rectangular cross-section

In the example, discussed in No. 10, of the torsion of a homogeneous, prismatic beam with cross-section B, a stress-function u is to be determined which satisfies the following boundary-value problem:

$$-\Delta u = 1 \quad \text{in } B, \qquad u = 0 \quad \text{on } \partial B. \tag{V.73}$$

A corresponding variational problem reads, as in (V.65),

$$I[\varphi] = \int_B [\varphi_x^2 + \varphi_y^2 - 2\varphi]\, dx\, dy = \min \quad \text{for } \varphi = 0 \text{ on } \partial B. \tag{V.74}$$

As in No. 10, we take B to be the domain $|x| < 1, |y| < \frac{1}{2}$, as a rectangle with sides of length 10 and 20 cm, and 10 cm is used as the unit of length. We choose as the approximation substitution

$$u \approx \varphi = a_1 \cos\left(\frac{\pi x}{2}\right) \cos(\pi y) + a_2 \cos\left(\frac{3\pi x}{2}\right) \cos(\pi y),$$

with the function φ satisfying the boundary conditions. The parameters a_ν are to be chosen so that the integral in (V.74) shall be as small as possible. After forming φ_x and φ_y and working out the integral, we obtain

$$I[\varphi] = \frac{\pi^2}{8}(5a_1^2 + 13a_2^2) - \frac{2}{3\pi^2}(24a_1 - 8a_2) = \min.$$

The equations (V.72) here read

$$\frac{5\pi^2}{4} a_1 - \frac{16}{\pi^2} = 0, \qquad \frac{13\pi^2}{4} a_2 + \frac{16}{3\pi^2} = 0$$

and give

$$a_1 = \frac{64}{5\pi^4} \approx 0.1314, \qquad a_2 = -\frac{64}{39\pi^4} \approx -0.0168$$

The value $u(0, 0)$ at the centre of the rectangle is approximated by

$$\varphi(0, 0) = \frac{64.34}{195\pi^4} \approx 0.11456.$$

Better values can be obtained by taking further terms into φ, such as $\cos(\pi x/2)\cos(3\pi y)$, for example.

17. The Method of Finite Elements

In the method of finite elements a generalization of the connection between boundary-value problems and variational problems is used; the generalization lies in the fact that the minimal property of the expression (V.65) is proved under weaker hypotheses regarding the differentiability of the admissible functions. The basic domain B is divided into a number of sub-domains $B_1, \ldots, B_m$

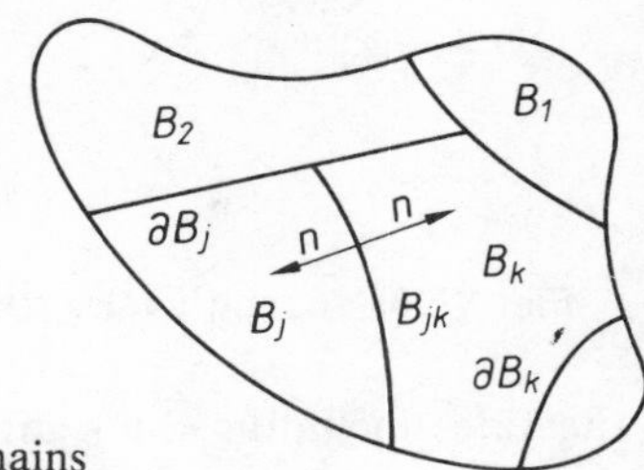

Fig. V.19. Splitting a domain into sub-domains

and their boundaries, as in Fig. V.19. Let B_{jk} be the intersection of the boundaries ∂B_j and ∂B_k. A function φ is said to be 'admissible' if it is continuous in $B \cup \partial B$ and twice continuously differentiable in the interior of each sub-domain B_ν ($\nu = 1, \ldots, m$). Let φ belong to the class '$C_0^2(B)$', where $C_s^r(B)$ denotes the class of functions f which are r times continuously differentiable in B and for which there is a decomposition of the type described above into sub-domains B_j such that f is s times continuously differentiable in each sub-domain B_j. Suppose now that one transformation shown in (V.66) is carried out for each sub-domain B_j and that the equations are added for $j = 1, 2, \ldots, m$:

$$I[\varphi] = \int_B (\operatorname{grad} u)^2 \, d\tau + \int_B (\operatorname{grad} \psi)^2 \, d\tau - \int_B 2\psi q \, d\tau - 2\int_B \psi \, \Delta \hat{u} \, d\tau + 2 \int_{\sum_j \partial Bj} \psi \frac{\partial \hat{u}}{\partial n} \, dJ. \tag{V.75}$$

The last sum contains boundary integrals over the common boundaries (the 'frontiers') B_{jk} between B_j and B_k; on B_{jk} the limit values of ψ from each side are the same, since ψ is continous, and the limit values of $\partial\hat{u}/\partial n$ from each side differ only by the factor -1, since the outward normal for the sub-domain B_j is directly opposite to that for the sub-domain B_k. In the sum the contributions from $\psi(\partial\hat{u}/\partial n)$ from each side cancel out, and since $\psi = 0$ on ∂B, the total boundary integral vanishes. We thus again obtain (V.67) and (V.68), and the variational principle holds also for the more general class of admissible functions φ.

The domain B can be split up in the way most suitable for the particular problem on hand, and this flexibility is one of the strong points of the finite-element method. In plane problems the domain B is often split up into squares, rectangles, or triangles, and we use as the admissible functions piecewise linear or bilinear functions.

A right-angled triangle D

Let the sides including the right-angle have lengths h and k. Let the values of φ at the three vertices be a, b, c; see Fig. V.20. We take the x- and y-axes along

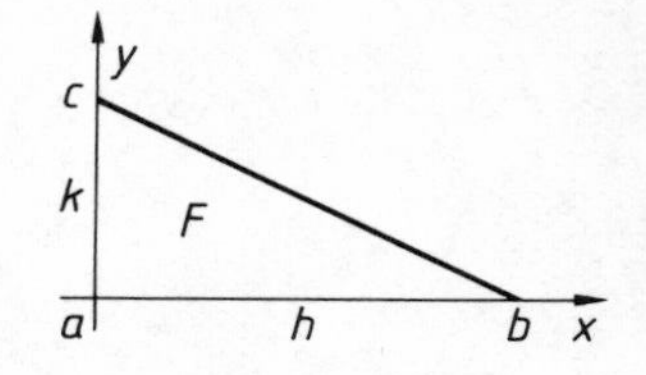

Fig. V.20. A right-angles triangle as a finite element

the sides including the right-angle. Then

$$\varphi = a + (b-a)\frac{x}{h} + (c-a)\frac{y}{k}.$$

The triangle D has the area $F = hk/2$, and we easily calculate

$$\left.\begin{aligned}
\int_D (\text{grad}\,\varphi)^2\,\mathrm{d}f &= \int_D [\varphi_x^2 + \varphi_y^2]\,\mathrm{d}f = F\left[\left(\frac{b-a}{h}\right)^2 + \left(\frac{c-a}{k}\right)^2\right] \\
\int_D \varphi\,\mathrm{d}f &= \frac{F}{3}(a+b+c) \\
\int_D \varphi^2\,\mathrm{d}f &= \frac{F}{6}(a^2+b^2+c^2+ab+bc+ca) \\
\int_D \varphi^3\,\mathrm{d}f &= \frac{F}{10}(a^3+b^3+c^3+a^2(b+c)+b^2(c+a) \\
&\qquad + c^2(a+b)+abc).
\end{aligned}\right\} \quad (\text{V.76})$$

A rectangle R

Let the sides be of length h and k, and let the values of the bilinear function φ at the four corners be a, b, c, d, as in Fig. V.21, and let two sides be along the x- and y-axes. The function φ here is

$$\varphi = a + \frac{b-a}{h}x + \frac{d-a}{k}y + \frac{a+c-b-d}{hk}xy.$$

We easily calculate

$$\left.\begin{aligned}
\int_R \phi\,\mathrm{d}f &= \frac{hk}{4}(a+b+c+d) \\
\int_R (\text{grad}\,\varphi)^2\,\mathrm{d}f &= \frac{1}{3}\left\{\frac{h}{k}[(d-a)^2+(c-b)^2+(d-a)(c-b)]\right. \\
&\qquad \left. + \frac{k}{h}[(b-a)^2+(c-d)^2+(b-a)(c-d)]\right\}.
\end{aligned}\right\} \quad (\text{V.77})$$

The last expression is not much simpler even for a square:

$$\int_Q (\text{grad}\,\varphi)^2\,\mathrm{d}f = \frac{1}{3}\{2(a^2+b^2+c^2+d^2) - 2(ac+bd) - (a+c)(b+d)\}.$$

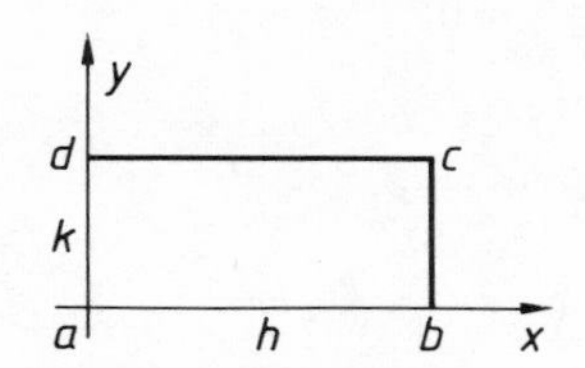

Fig. V.21. A rectangle as a finite element

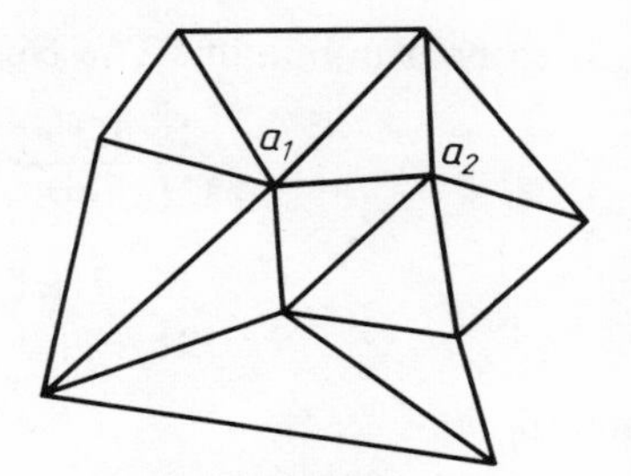

Fig. V.22. Triangulation of a polygon

If the domain B is a polygon, Fig. V.22, we can 'triangulate' it completely, and introduce the function values at all the 'interior' vertices of the triangles as unknown parameters $a_1, a_2, \ldots$ (in Fig. V.20 we use a, b, c); we can then express $I[\varphi]$, at least approximately, as a function of the $a_1, a_2, \ldots$. The equations (V.72) then serve to determine the a_ν and thus we have calculated the approximate values of the function u at the vertices of the triangles.

18. An example

We again consider the torsion problem (V.73) for the bar of rectangular cross-section B as in No. 15. First a triangulation is made as in Fig. V.23; because of the symmetry only two values appear as the unknown function values, and these are denoted by a and b. Here $h = k = 1/2$, and $F = 1/8$, in each case; the shaded triangle D in Fig. V.23 makes a contribution, in accordance with (V.76), of amount

$$F\left[(2(a-b))^2 + (2b)^2 + \frac{a+b}{3}\right];$$

to the integral (V.76); and all 16 triangles of Fig. V.23 together yield

$$I[\varphi] = F\left[16(2a^2 - 2ab + 4b^2) - \frac{16a}{3} - 8b\right]$$

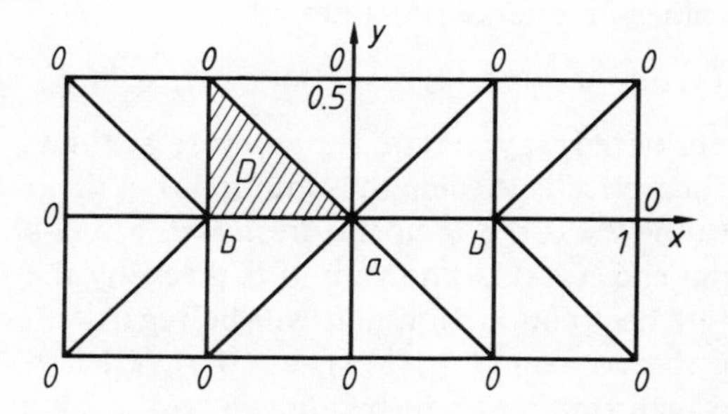

Fig. V.23. Triangulation of a rectangle

which is to be a minimum. The equations (V.72) here are

$$\frac{1}{32F}\frac{\partial I[\varphi]}{\partial a} = 2a - b - \frac{1}{6} = 0$$

$$\frac{1}{32F}\frac{\partial I[\varphi]}{\partial b} = -a + 4b - \frac{1}{4} = 0.$$

We obtain

$$a = \frac{11}{84} \approx 0.13095; \qquad b = \frac{8}{84}.$$

These values are still very rough, but they can, of course, be easily improved by making the mesh finer; here we wanted merely to illustrate the method.

§6. THE LAPLACE TRANSFORMATION AND FOURIER TRANSFORMATION WITH PARTIAL DIFFERENTIAL EQUATIONS

In Chapter II, §9 the Laplace transformation was used to solve ordinary differential equations with constant coefficients. This transformation can, however, also be applied to linear partial differential equations with constant coefficients and with certain domains going to infinity, in order to transform a partial differential equation with m independent variables into an, in general, simpler differential equation with only $m-1$ independent variables. In particular, for a partial differential with two independent variables, we are led to an ordinary differential equation. This will be illustrated here by some simple examples.

19. A PARABOLIC EQUATION (THE EQUATION OF HEAT CONDUCTION)

Suppose the following initial-value problem is presented for the heat conduction equation or diffusion equation (as in No. 3) for a function $u(x, t)$:

$$u_t - k^2 u_{xx} = 0 \tag{V.78}$$

with the initial values or boundary values

$$u(x, 0) = 0, \qquad u(0, t) = f(t) \quad \text{for } t \geq 0,\ x \geq 0. \tag{V.79}$$

$u(x, t)$ can be interpreted, say, as the temperature at time t at the point x of a very long rod (theoretically the length is infinite). Let the rod cover the x-axis and have at the time $t=0$ the temperature $u=0$. Suppose the variation of temperature at the end $x=0$ is known and is given by the function $f(t)$. k^2 is the coefficient of heat conduction and will be regarded as a constant. The temperature u in the domain $B = \{(x, t), x > 0, t > 0\}$ is to be calculated.

The Laplace transformation is to be applied to the equation (V.78) in the time direction, that is, keeping the position x fixed. This means that the differential equation is multiplied by e^{-st} and integrated with respect to time from $t=0$ to $t=\infty$ to obtain

$$\int_0^\infty e^{-st} u_t \, dt - k^2 \int_0^\infty e^{-st} u_{xx}(x,t) \, dt = 0. \tag{V.80}$$

In order that the integrals shall exist at least for large s the following assumption has to be made. There are positive constants m and p such that

$$\left.\begin{cases} |u(x,t)| \le M \, e^{p(x+t)} \\ |u_t(x,t)| \le M \, e^{p(x+t)} \end{cases}\right\} \text{for } x>0, t>0. \tag{V.81}$$

It is not, in general, easy to verify this assumption, since the function $u(x,t)$ is not yet known in the domain B; nevertheless, this assumption seems reasonable on physical grounds when dissipative processes are being dealt with. Under this assumption (x,t) and $u_t(x,t)$ with x held fixed are admissible functions relative to t, and the Laplace transform $U(x,s)$ of $u(x,t)$ can be formed:

$$U(x,s) = \mathscr{L}[u(x,t)] = \int_0^\infty e^{-st} u(x,t) \, dt. \tag{V.82}$$

By (II.127) we have for the derivative $u_t(x,t)$

$$\begin{aligned} \mathscr{L}[u_t(x,t)] &= s\mathscr{L}[u] - u(x,0) = sU(x,s) - u(x,0) \\ &= sU(x,s), \end{aligned} \tag{V.83}$$

by the initial condition $u(x,0)=0$.

Since the second integral appearing in (V.80) is uniformly convergent for large s, the integration and the differentiation with respect to x can be interchanged. If we denote the derivative with respect to x by a prime

$$U' = \frac{\partial}{\partial x} U,$$

then, using (V.83), the equation (V.80) for $U(x,s)$ becomes

$$sU(x,s) = k^2 U''(x,s)$$

which is a second-order ordinary differential equation relative to x with the coefficients s, k^2 independent of x. It has the solution

$$U(x,s) = c_1 \, e^{x\sqrt{s}/k} + c_2 \, e^{-x\sqrt{s}/k}, \tag{V.84}$$

where the constants c_1, c_2 are still to be determined.

From (V.82) we have, for $x=0$,

$$U(0,s) = \mathscr{L}[u(0,t)] = \mathscr{L}[f(t)] = F(s)$$

where $F(s)$ denotes the Laplace transform of $f(t)$. From the growth assumptions (V.81) it follows that the right-hand side of (V.84) must not increase exponentially as $s \to \infty$, and so we must have $c_1 = 0$. Then (V.84) yields, for

$x = 0$, the relation $c_2 = F(s)$. Hence for the image function we have found the solution

$$U(x, s) = F(s)\, e^{-x\sqrt{s/k}}. \tag{V.85}$$

If we use, as in Chapter II, §9, the symbol $\mathscr{L}^{-1}$ for the inverse transformation, then we have for the required solution $u(x, t)$ of the initial-value problem the formal representation

$$u(x, t) = \mathscr{L}^{-1}[U(x, s)] = \mathscr{L}^{-1}[e^{-x\sqrt{s/k}}\mathscr{L}[f(t)]]. \tag{V.86}$$

In actual cases one may be able to evaluate the symbols $\mathscr{L}$, $\mathscr{L}^{-1}$ explicitly or numerically; the process can, however, be very laborious.

20. The Laplace transformation with the wave equation

With the wave equation for a function $u(x, t)$

$$u_{tt} - a^2 u_{xx} = 0 \tag{V.87}$$

with the initial conditions and boundary conditions

$$\left.\begin{array}{l} u(0, t) = f(t) \\ u(x, t) \text{ bounded as } x \to \infty \\ u(x, 0) = 0 \\ u_t(x, 0) = 0 \end{array}\right\} \tag{V.88}$$

(we formally put $f(t) = 0$ for $t < 0$) the Laplace transformation with respect to t gives, writing $\mathscr{L}[u] = U$,

$$s^2\mathscr{L}[u(x, t)] - su(x, 0) - u_t(x, 0)$$

$$= s^2\mathscr{L}[u(x, t)] = a^2\mathscr{L}[u_{xx}] = a^2\frac{d^2}{dx^2}\mathscr{L}[u(x, t)] = a^2\frac{d^2}{dx^2}U(x, s).$$

Here, as in No. 18, the differentiation and integration for U have been interchanged; this is permissible because of the uniform convergence for large s. Thus we have for $U(x, s)$ the ordinary differential equation

$$\frac{d^2U(x, s)}{dx^2} - \frac{s^2}{a^2}U(x, s) = 0 \tag{V.89}$$

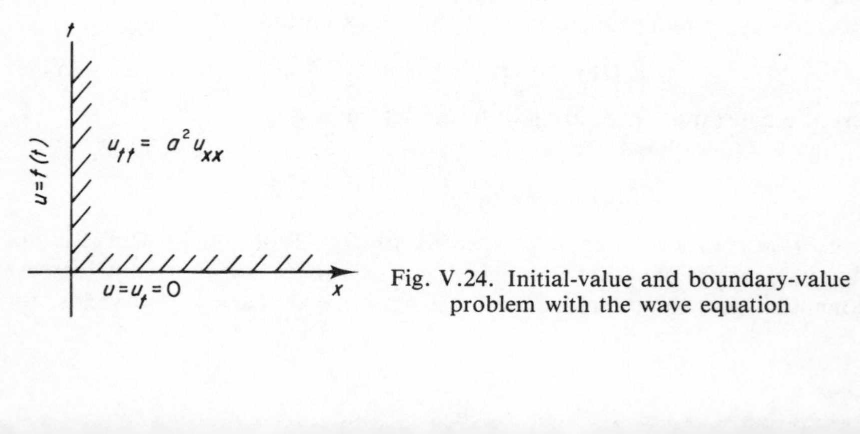

Fig. V.24. Initial-value and boundary-value problem with the wave equation

in which s and a are to be treated as constants. This equation has the solution

$$U(x, s) = \mathscr{L}[u(x, t)] = A(s)\,\mathrm{e}^{-sx/a} + B(s)\,\mathrm{e}^{sx/a};$$

$u(x, t)$ is finite as $x \to \infty$; this implies that

$$B(s) = 0.$$

For $x = 0$ we have

$$\mathscr{L}[u(0, t)] = \mathscr{L}[f(t)] = A(s).$$

The Laplace transformation therefore gives altogether

$$\mathscr{L}[u(x, t)] = \mathscr{L}[f(t)]\,\mathrm{e}^{-sx/a}.$$

This has precisely the form of the shift theorem formula (II.133) (the x, T there are to be replaced by t, x/a). Hence

$$\mathscr{L}[u(x, t)] = \mathscr{L}[f(t - x/a)]$$

Since $f(t) = 0$ for $t < 0$ we can use Heaviside's unit function

$$H(t) = \begin{cases} 1 & \text{for } t > 0 \\ 0 & \text{for } t < 0 \end{cases}$$

and write the solution in the form

$$u(x, t) = f(t - x/a)H(t - x/a). \qquad \text{(V.91)}$$

21. The reciprocity formulae for the Fourier transformation

As well as the Laplace transformation various other integral transformations, such as the Mellin transformation, the Hankel transformation, the Z-transformation, etc., have often proved to be useful and successful for certain classes of problems; perhaps the best known of these transformations is the Fourier transformation.

Definition

Let $f(x)$ be a piecewise smooth function defined for $x \in (-\infty, \infty)$ and such that the integral

$$\int_{-\infty}^{\infty} |f(x)|\,\mathrm{d}x$$

exists. With $f(x)$ is associated as its *Fourier transform* the function

$$\mathscr{F}[f] = g(\omega) = \frac{1}{\sqrt{2\pi}} \int_{-\infty}^{\infty} f(x)\,\mathrm{e}^{-\mathrm{i}\omega x}\,\mathrm{d}x. \qquad \text{(V.92)}$$

This transformation has an inverse transformation constructed symmetrically to it:

$$\mathscr{F}^{-1}[g] = f(x) = \frac{1}{\sqrt{2\pi}} \int_{-\infty}^{\infty} g(\omega)\, e^{i\omega x}\, dx. \tag{V.93}$$

This inversion formula will not be proved here, but only developed heuristically; a mathematical proof can be found in E. C. Titchmarsh: *Introduction to the theory of Fourier integrals*, Oxford, 1948.

The transformation (V.92) is also sometimes defined without the factor $\sqrt{2\pi}$. In applications x often is taken to be the time, and then the function $f(x)$ is said to lie in the 'time domain', and $g(\omega)$ is its image function or 'spectral function' or its 'continuous spectrum'.

If the function $f(x)$ is even, i.e., if $f(-x) = f(x)$, then

$$\int_{-\infty}^{\infty} f(x)\, e^{-i\omega x}\, dx = 2 \int_0^{\infty} f(x) \cos(\omega x)\, dx$$

and we obtain the symmetric formulae of the 'cosine transformation'

$$\begin{aligned} g(\omega) &= \sqrt{\frac{2}{\pi}} \int_0^{\infty} f(x) \cos(\omega x)\, dx \\ f(x) &= \sqrt{\frac{2}{\pi}} \int_0^{\infty} g(\omega) \cos(\omega x)\, d\omega \end{aligned} \tag{V.94}$$

Similarly, for odd functions $f(x) = -f(-x)$ there is the 'sine transformation'

$$\begin{aligned} g(\omega) &= \sqrt{\frac{2}{\pi}} \int_0^{\infty} f(x) \sin(\omega x)\, dx \\ f(x) &= \sqrt{\frac{2}{\pi}} \int_0^{\infty} g(\omega) \sin(\omega x)\, d\omega \end{aligned} \tag{V.95}$$

A heuristic line of reasoning leading to the Fourier formulae

For any piecewise smooth function $f(x)$ which has a real period $2k$ the Fourier coefficients a_n of the expansion

$$f(x) \approx \sum_{n=-\infty}^{\infty} a_n\, e^{in\pi x/k}$$

are given by the formula

$$a_n = \frac{1}{2k} \int_{-k}^{k} f(v)\, e^{-in\pi v/k}\, dv.$$

Hence

$$f(x) = \frac{1}{2k} \sum_{n=-\infty}^{\infty} e^{in\pi x/k} \int_{-k}^{k} e^{-in\pi v/k}\, dv.$$

Now let $f(x)$ be a non-periodic, piecewise smooth function defined in

$(-\infty, \infty)$, and such that the integral

$$\int_{-\infty}^{\infty} |f(x)| \, \mathrm{d}x$$

exists. Let $f_k(x)$ be the function obtained from $f(x)$ be 'periodizing' it, i.e.,

$$f_k(x) = \begin{cases} f(x) & \text{for } |x| < k \\ f_k(x+2k) & \text{for all real } x; \end{cases}$$

see Fig. V.25. Taking $\pi/k = \eta$, $2k = 2\pi/\eta$, we then have

$$f_k(x) = \frac{1}{2\pi} \sum_{n=-\infty}^{\infty} \eta \int_{-k}^{k} f(v) \, \mathrm{e}^{-\mathrm{i}n\eta(v-x)} \, \mathrm{d}v$$

Now let the passages to the limit $k \to \infty, \eta \to 0, n\eta \to \omega$,

$$\sum_{n=-\infty}^{\infty} \eta\varphi(n\eta) \to \int_{-\infty}^{\infty} \varphi(\omega) \, \mathrm{d}\omega$$

be carried out, where ω is a fixed point (Fig. V.25), and φ satisfies the same conditions as $f(x)$. This limit process, which requires a detailed mathematical proof, yields in the present case

$$f(x) = \frac{1}{\sqrt{2\pi}} \int_{-\infty}^{\omega} g(\omega) \, \mathrm{e}^{\mathrm{i}\omega x} \, \mathrm{d}\omega$$

if we merely introduce as an abbreviation

$$g(\omega) = \frac{1}{\sqrt{2\pi}} \int_{-\infty}^{\infty} f(v) \, \mathrm{e}^{-\mathrm{i}\omega v} \, \mathrm{d}v.$$

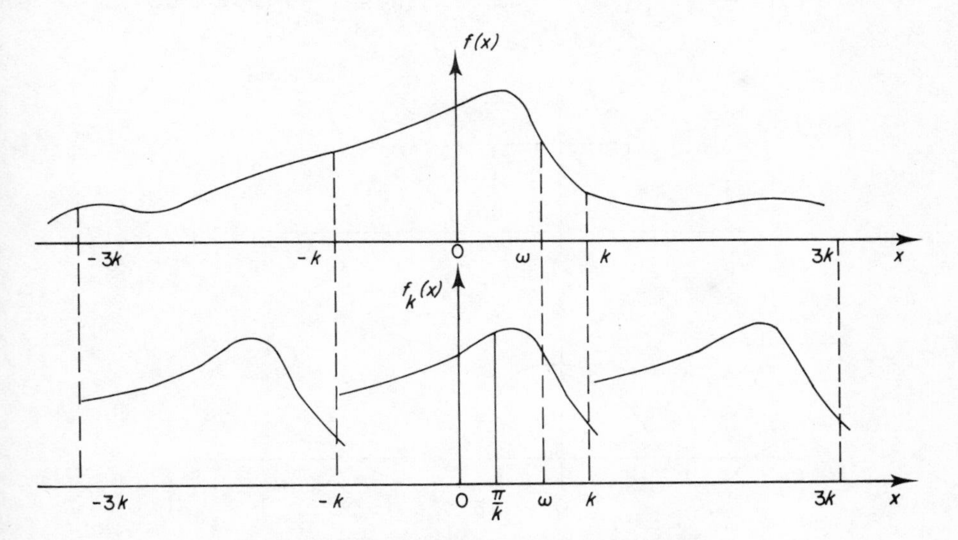

Fig. V.25. 'Periodization' of the function $f(x)$

Example

The impulse function

$$f(x)=\begin{cases} h & \text{for } |x|\le a \\ 0 & \text{otherwise, Fig. V.26} \end{cases} \tag{V.96}$$

has the spectral function

$$g(\omega)=\frac{1}{\sqrt{2\pi}}h\int_{-a}^{a}\mathrm{e}^{-\mathrm{i}\omega x}\,\mathrm{d}x=\sqrt{\frac{2}{\pi}}h\int_{0}^{a}\cos(\omega x)\,\mathrm{d}x$$

$$=\sqrt{\frac{2}{\pi}}h\frac{\sin(\omega a)}{\omega}.$$

A particularly nice symmetric example may be mentioned here without proof: for the even function

$$f(x)=\mathrm{e}^{-x^2/2}$$

the cosine transformation gives

$$g(\omega)=\sqrt{\frac{2}{\pi}}\int_{0}^{\infty}f(x)\cos(\omega x)\,\mathrm{d}x=\mathrm{e}^{-\omega^2/2}$$

By putting $g(\omega)=F(\omega)\cdot\sqrt{2\pi}$, the Fourier transformation (V.92), (V.93) can be brought into the often used form

$$F(\omega)=\frac{1}{2\pi}\int_{-\infty}^{\infty}f(x)\,\mathrm{e}^{-\mathrm{i}\omega x}\,\mathrm{d}x;\qquad f(x)=\int_{-\infty}^{\infty}F(\omega)\,\mathrm{e}^{\mathrm{i}\omega x}\,\mathrm{d}\omega.$$

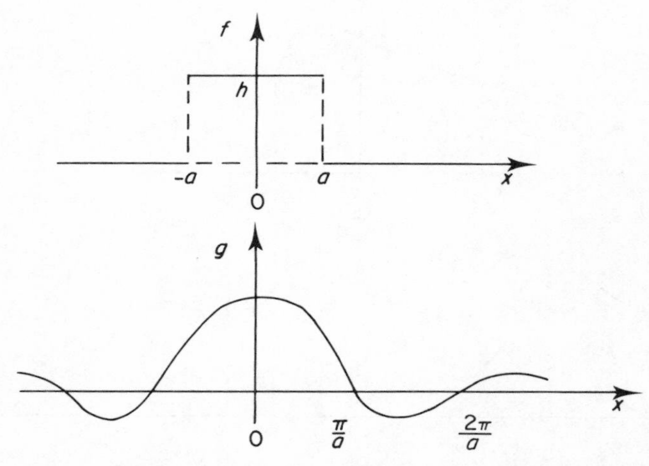

Fig. V.26. The impulse function and its spectral function

Particular examples.

$f(x) = \int_{-\infty}^{\infty} F(\omega)\,\mathrm{e}^{\mathrm{i}\omega x}\,\mathrm{d}x$	$F(\omega) = \frac{1}{2\pi}\int_{-\infty}^{\infty} f(\ )\,\mathrm{e}^{\mathrm{i}\omega x}\,\mathrm{d}\omega$
$\begin{cases} h & \text{for } \lvert x\rvert < a, \\ 0 & \text{for } \lvert x\rvert > a. \end{cases}$	$\frac{h}{\pi}\frac{\sin\omega}{\omega}$
$2\Gamma(1-\nu)\sin\left(\frac{\nu\pi}{2}\right)\frac{1}{\lvert x\rvert^{1-\nu}}$	$\frac{1}{\lvert\omega\rvert^{\nu}}\quad 0 < \operatorname{Re}\nu < 1$
$\mathrm{i}\,\frac{\mathrm{e}^{\mathrm{i}a(\mu+x)} - \mathrm{e}^{\mathrm{i}b(\mu+x)}}{\mu + x}$	$\begin{cases} \mathrm{e}^{\mathrm{i}\mu\omega} & \text{for } a < \omega < b \\ 0 & \text{for } \omega < a \text{ and } \omega > b \end{cases}$
$\frac{\pi}{c}\,\mathrm{e}^{-c\lvert x\rvert}$	$\frac{1}{\omega^2 + c^2}$
$\sqrt{\frac{\pi}{c}}\,\mathrm{e}^{-x^2/(4c)}$	$\mathrm{e}^{-c\omega^2}\quad \operatorname{Re} c > 0, \operatorname{Re}\sqrt{c} > 0$
$\pi J_0(cx)$	$\begin{cases} \frac{1}{\sqrt{c^2-\omega^2}} & \text{for } \lvert\omega\rvert < c. \\ 0 & \text{for } \lvert\omega\rvert > c. \end{cases}$
$\frac{\pi}{2}\left(\frac{1}{\sqrt{\lvert x-c\rvert}} + \frac{1}{\sqrt{\lvert x+c\rvert}}\right)$	$\frac{\cos(c\omega)}{\sqrt{\lvert\omega\rvert}}$
$\begin{cases} \mathrm{i}\,(\operatorname{sgn} x)\,\frac{\pi}{2} & \text{for } \lvert x\rvert < 2c \\ 0 & \text{for } \lvert x\rvert > 2c \end{cases}$	$\frac{\sin^2(c\omega)}{\omega}$
$2\mathrm{i}^n\sqrt{\frac{\pi}{2x}}\,J_{n+1/2}(x)$	$\begin{cases} P_n(\omega) & \text{for } \lvert\omega\rvert < 1 \\ 0 & \text{for } \lvert\omega\rvert > 1 \end{cases}$ where P_n is the Legendre polynomial.

22. THE FOURIER TRANSFORMATION WITH THE HEAT-CONDUCTION EQUATION

As a classical example we apply the Fourier transformation to solve the initial-value problem with the heat-conduction equation for a function $u(x, t)$:

$$Lu \equiv u_t - k^2 u_{xx} = 0 \quad \text{in } B = \{\lvert x\rvert < \infty, t > 0\},$$

with the initial values

$$u(x, 0) = f(x) \quad \text{for } t = 0, \qquad \lvert x\rvert < \infty.$$

Let $f(x)$ satisfy the condition required in the definition of the Fourier transformation, and assume that, for the solution

$$\lim_{t \to \infty} u(x, t) = 0 \text{ for every fixed } x.$$

The function

$$\varphi = e^{isx - k^2 s^2 t}$$

solves the differential equation $L\varphi = 0$ for every fixed s; then

$$\psi(x, t) = \frac{1}{\sqrt{2\pi}} \int_{-\infty}^{\infty} e^{isx - k^2 s^2} g(s) \, ds \tag{V.97}$$

is, for any piecewise smooth, integrable function $g(x)$ such that

$$\int_{-\infty}^{\infty} |g| \, ds < \infty$$

also a solution of the differential equation $L\psi = 0$. For $t = 0$ we require that $\psi(x, 0) = f(x)$:

$$f(x) = \frac{1}{\sqrt{2\pi}} \int_{-\infty}^{\infty} e^{isx} g(s) \, ds.$$

But by (V.93) this is of the form $\mathscr{F}^{-1}[g]$ (with s instead of ω); so by (V.92) the inversion $g(\omega)$ can be written down immediately and substituted in (V.82):

$$u(x, t) = \psi(x, t) = \mathscr{F}^{-1}[e^{-k^2 s^2 t} g(s)]$$

or

$$u(x, t) = \mathscr{F}^{-1}[e^{-k^2 s^2 t} \mathscr{F}[f]].$$

VI

Appendix. Some methods of approximation and further examples for practice

§1. SOME METHODS OF APPROXIMATION TO THE SOLUTION OF ORDINARY DIFFERENTIAL EQUATIONS

Here we shall briefly mention some methods of approximation, although we cannot go fully into their derivation and justification. For more detailed presentations, the reader may consult the books of Collatz, Griegorieff, Werner-Schabach, and others,

1. PRELIMINARY REMARKS AND SOME ROUGH METHODS OF APPROXIMATION

We can approximate, either numerically or graphically, to the solution of a given explicit differential equation of the *first* order $y' = f(x, y)$ for a given initial value (x_0, y_0) in the following way. The slope at the point (x_0, y_0) is determined, and a bit of the tangent so obtained to the solution curve is drawn, or a second point (x_1, y_1) is calculated, or, in general, the $(n + 1)$ the point is calculated from

$$y_{n+1} = y_n + hj_n \quad \text{with} \quad f_n = f(x_n, y_n). \tag{VI.1}$$

The error can, especially with this quite rough approximation formula, already increase considerably even after a few steps. The basic rules which apply are:

for bigger calculations for integration over larger intervals

1. choose small steps h;
2. calculate as accurately as possible (use more accurate formulae, computers)

for smaller calculations

1. large step-widths h permissible
2. slide-rule accuracy often suffices. Any rough method can then be used.

It is often sensible to vary the step-width h according to the curvature, i.e., for greater curvature choose a smaller h, see Fig. VI.1.

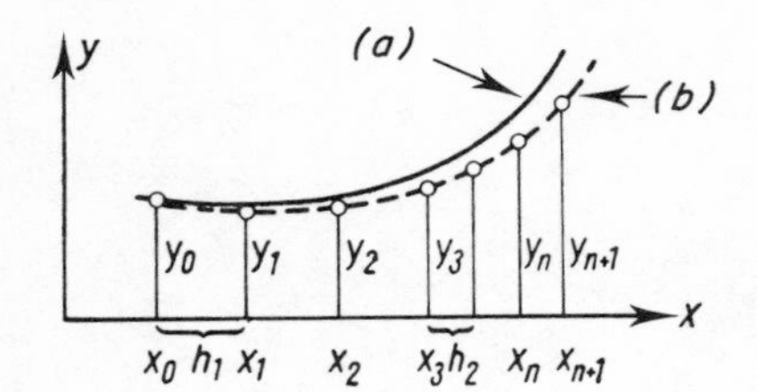

Fig. VI.1. On the approximate integration of differential equations, (a) exact solution; (b) course of the approximations

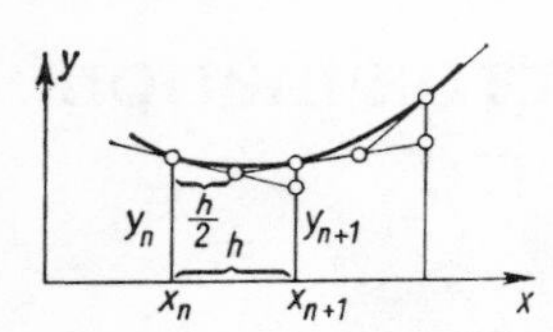

Fig. VI.1a. Improvement by using intermediate points

For points where the slope is very steep, the method is, in general, unsuitable (in such cases, it is often convenient to interchange the x- and y-axes).

The method can be improved by applying at each step the value of the slope calculated for the point $x_n + h$ at the point $x_n + h/2$ instead (see Fig. VI.1a), or by working from the formula

$$y_{n+1} = y_n + hf\left(x_n + \frac{h}{2},\ y_n + \frac{1}{2}hf_n\right). \tag{VI.2}$$

2. The Runge and Kutta Method of Approximate Numerical Integration

For the initial-value problem $y' = f(x, y)$, $y(x_0) = y_0$ approximate values y_n at the points $x_n = x_0 + nh$ $(n = 1, 2, \ldots; h = \text{step-width})$ are to be calculated. In each case the next value y_{n+1} at the point x_{n+1} is to be calculated, from the function values y_n already known, by means of the formulae:

$$\left.\begin{aligned} k_1 &= hf(x_n,\ y_n),\\ k_2 &= hf\left(x_n + \frac{h}{2},\ y_n + \frac{k_1}{2}\right),\\ k_3 &= hf\left(x_n + \frac{h}{2},\ y_n + \frac{k_2}{2}\right),\\ k_4 &= hf(x_n + h,\ y_n + k_3). \end{aligned}\right\} \tag{VI.3}$$

The new value of the function is then

$$y_{n+1} = y_n + k, \quad \text{where } k = \frac{k_1 + 2k_2 + 2k_3 + k_4}{6}. \tag{VI.4}$$

The numerical calculation can conveniently be laid out in tabular form, thus:

x	y	$hf(x, y)$
x_n	y_n	k_1
$x_n + \frac{h}{2}$	$y_n + \frac{k_1}{2}$	k_2
$x_n + \frac{h}{2}$	$y_n + \frac{k_2}{2}$	k_3
$x_n + h$	$y_n + k_3$	k_4
x_{n+1}	$y_{n+1} = y_n + k$	

As a criterion for a suitable choice of the step-width h, it can be taken that the corrections k_2 and k_3 should agree to the first two decimal places. If this condition is not satisfied, a smaller h should be chosen. For the differential equation $y' = f(x)$, this method gives the same result as *Simpson's* rule. Compared with the method of central differences, to be described in No. 3, which is particularly suitable when the function varies smoothly, the Runge–Kutta method has the advantage that it leads to a rapid calculation of the result, even with less smooth solution functions.

Example

For illustration a simple example has been taken, the exact solution of which is known and for which the accuracy of the method can therefore be judged immediately. The initial-value problem

$$y' = -2xy^2, \qquad y(-1) = \frac{1}{2} \tag{VI.5}$$

has the exact solution $y = 1/(1 + x^2)$. Let $h = 0.2$ be chosen as the step-width. The calculation, carried out row by row in the above tabular form, is probably understandable without further explanation; a third column for y^2 has merely been added for convenience in calculating the function values, and a fifth column for calculating the k values; in the fourth column values of $k_j/2$ instead of k_j are calculated. The error is shown in a table included in No. 3.

The Runge-Kutta method; $h = 0.2$

x	y	y^2	$\frac{h}{2} f(x, y) = -0.2\,xy^2$	$3k$ and k
−1	0.5	0.25	0.05	
−0.9	0.55	0.302 5	0.054 45	0.329 237
"	0.554 45	0.307 415	0.055 335	0.109 746
−0.8	0.610 670	0.372 918	0.059 667	
"	0.609 746	0.371 790	0.059 486	
−0.7	0.669 232	0.447 871	0.062 702	0.376 569
"	0.672 448	0.452 186	0.063 306	0.125 523
−0.6	0.736 358	0.542 223	0.065 067	
"	0.735 269	0.540 621	0.064 875	
−0.5	0.800 144	0.640 230	0.064 023	0.380 282
"	0.799 292	0.638 868	0.063 887	0.126 761
−0.4	0.863 043	0.744 843	0.059 587	
−0.4	0.862 030			

3. THE METHOD OF CENTRAL DIFFERENCES

If a part of the solution curve is already known with the values y_ν at the points x_ν for $\nu = 0, 1, \ldots, n$, then, with a step-width h, we can determine the next value

$$y_{n+1} \text{ at the point } x_{n+1}$$

by using the *method of central differences* (see below regarding the determination of the initial values.)

The first difference is defined by

$$\nabla (hf_n)v = v\,\nabla\,[hf(x_n, y_n)]\,v = v\,\nabla_n v = vhf_n v - vhf_{n-1} \qquad \text{(VI.6)}$$

and the *second difference* (not the second power!) by

$$\nabla_n^2 = \nabla_n - \nabla_{n-1}.$$

We then have the difference table

x_{n-1}	y_{n-2}	hf_{n-2}		
			∇_{n-1}	
x_{n-1}	y_{n-1}	hf_{n-1}		∇_n^2
			∇_n	
x_n	y_n	hf_n		∇_{n+1}^2
			∇_{n+1}	
x_{n+1}	y_{n+1}	hf_{n+1}		

The new function value is calculated from the four underlined quantities by means of the formula

$$y_{n+1} = y_{n-1} + 2hf_n + \tfrac{1}{3}\,\nabla^2_{n+1}. \tag{VI.7}$$

Since, however, at the start, only the terms in the table which stand above the broken line are known, the value of ∇^2_{n+1} still has to be found. For this purpose we extrapolate this second difference, at first from the behaviour of the preceding second differences. (To start with, we put, say, $\nabla^2_{n+1} \approx \nabla^2_{n}$.) With this estimate we calculate a 'rough value' $y^{(0)}_{n+1}$. If the νth calculated approximation is denoted by the superscript (ν), then we calculate according to

$$y^{(\nu+1)}_{n+1} = y_{n-1} + 2hf_n + \tfrac{1}{3}\,\nabla^{2(\nu)}_{n+1}. \tag{VI.8}$$

Since we form the differences by working both forwards and backwards, the method already provides in itself a good check on the calculations.

The first values of the solution can be determined in various ways:

1. by using the Runge–Kutta method with a small step-width;
2. by substituting a power series;
3. by using a 'zeroth approximation' in the form of a tangent;

$$y^{(0)}_j = y_0 + jhf_0; \qquad j = -1, 1, 2;$$

a 'first approximation' is obtained from the iteration formulae

$$\left.\begin{aligned} y^{(1)}_2 &= y_0 + \frac{h}{3}\,(f^{(0)}_2 + 4f^{(0)}_1 + f_0).\\ y^{(1)}_1 &= y_0 + \frac{h}{24}\,(-f^{(0)}_2 + 13f^{(0)}_1 + 13f_0 - f^{(0)}_{-1}),\\ y^{(1)}_{-1} &= y^{(1)}_1 - \frac{h}{3}\,(f^{(0)}_1 + 4f_0 + f^{(0)}_{-1}). \end{aligned}\right\} \tag{VI.9}$$

If necessary, this iteration is repeated. These initial values have to be calculated as accurately as possible.

An example

We have chosen the same example as we did for the Runge-Kutta method. Let the step-width be $h = 0.1$. To obtain the three initial values needed, we shall here set up a power series and with it calculate the values $y(-1.1)$ and $y(-0.9)$.

By continued differentiation, we find from the differential equation, on replacing y' on the right-hand side by $-2xy^2$ each time,

$$\begin{aligned} y'' &= -2y^2 - 4xyy' = -2y^2 + 8x^2y^3,\\ y''' &= -4yy' + 16xy^2 + 24x^2y^2y' = 24(xy^3 - 2x^3y^4),\\ y^{\text{IV}} &= 24(y^3 - 12x^2y^4 + 16x^4y^5),\\ y^{\text{V}} &= 240(-3xy^4 + 16x^3y^5 - 16x^5y^6); \end{aligned}$$

and hence the values for $x = -1$ are

$$y'(-1) = \tfrac{1}{2}; \qquad y''(-1) = \tfrac{1}{2};$$
$$y'''(-1) = 0; \qquad y^{IV}(-1) = -3; \qquad y^{V}(-1) = -15$$

and by Taylor's theorem

$$y(-1 \pm 0{,}1) = \frac{1}{2} \pm \frac{0.1}{1!} \cdot \frac{1}{2} + \frac{(0.1)^2}{2!} \cdot \frac{1}{2} \pm \frac{(0.1)^2}{3!} \cdot 0$$
$$+ \frac{(0.1)^4}{4!}(-3) \pm \frac{(0.1)^5}{5!}(-15). + \cdots$$

Correct to six places, we find

$$y(-1{,}1) = 0.452\,489, \qquad y(-0.9) = 0.552\,486.$$

With these values we can already form the table of differences shown below as far as the double underlining. For convenience, the differences have been printed not on the level midway between lines, but one half line lower.

We obtain a rough value $y_3^{(0)}$ for the point $x = -0.8$ from

$$y_3^{(0)} = y_1 + 2hf_2 + \tfrac{1}{3}\nabla_3^2 \tag{VI.10}$$

where $y_1 = y(-1) = 1/2$, $hf_2 = 0.054\,943\,4$; and ∇_3^2 is at first roughly estimated at $h^4 y^{IV}(-1) = -0.0003$ (since $y'''(-1)$ vanishes); hence we find

$$y_3^{(0)} = 0.609\,787.$$

With this we can now continue the difference table by one line, and we obtain instead of the estimated value -0.0003 for ∇_3^2 the better value

The method of central differences. $h = 0.1$. $y(-1 + 0.1)$ *by a power series.*

x	y	y^2	$hf(x, y)$	∇	∇^2	new estimate of ∇^2	∇^3
−1.1	0.452 489	0.204 746	0.045 0441				
−1	0.5	0.25	0.05	0.004 9559			
−0.9	0.552 486	0.305 241	0.054 9434	0.004 9434	−0.000 0125	−0.000 3	
−0.8	0.609 787	0.371 840	0.059 4944	0.004 5510	−0.000 3924		
	0.609 756	802	4883	5449	3985		
	0.609 754	800	4880	5446	3988	−0.000 84	−0.000 386 3
−0.7	0.671 182	0.450 485	0.063 0679	0.003 5799	−0.000 9647		
	0.671 140	429	0601	5721	9725		
	0.671 138	426	0596	5716	9730	−0.001 74	−0.000 574 2
−0.6	0.735 293	0.540 656	0.064 8787	0.001 8191	−0.001 7525		
	0.735 289	650	8780	8184	7532	−0.00276	−0 000 780 2

$-0.000\,392\,4$, with which we repeat the calculation; after two further steps the calculation comes to a halt with the value $y(-0.8) \approx 0.609\,754$. The table shows a few further steps. The calculation can still be slightly shortened; we do not need to form all the differences anew at each iteration step; but we refer the reader to the literature for all the practical details of the calculation.

For a comparison of the accuracy, the approximation values obtained by the Runge–Kutta method and by the method of central differences may be compared with the true values in the following table. A judgement of the value of the two methods of approximation should not be made hastily on the basis of a single example, and certainly not here, because the expenditure of work in the two methods is greatly different, and the step-widths were also different. (In general, the Runge–Kutta method will tolerate a rather greater step-width than the method of central differences.)

Comparison of accuracy

	Exact solution	Runge–Kutta method		Method of central differences	
x	$y(x)$	y	Error	y	Error
−1	0.5	0.5	0	0.5	0
−0.9	0.552 486 2			0.552 486	–
−0.8	0.609 756 1	0.609 746	−0.000 010	0.609 754	−0.000 002
−0.7	0.671 140 9			0.671 138	−0.000 003
−0.6	0.735 294 1	0.735 269	−0.000 025	0.735 289	−0.000 005
−0.5	0.8			0.799 995	−0.000 005
−0.4	0.862 069 0	0.862 030	−0.000 039	0.862 064	−0.000 005

4. The method of difference quotients

Suppose we are presented with a boundary-value problem for an interval (a, b). We make a sub-division of the x-axis into intervals of length or 'mesh width' h by the points of sub-division $x_j = x_0 + jh$. Often one chooses $a = x_0$ and $b = x_n$ with a fixed $n > 0$. Let y_j be an approximation of the value $y(x_j)$ of the required solution at the point x_j. The differential quotients at the points x_j are now replaced by the difference quotients as follows:

$$\frac{dy}{dx} \quad \text{by} \quad \frac{1}{h}(y_{j+1} - y_j) \quad \text{or by} \quad \frac{1}{2h}(y_{j+1} - y_{j-1})$$

$$\frac{d^2y}{dx^2} \quad \text{by} \quad \frac{1}{h^2}(y_{j+1} - 2y_j + y_{j-1})$$

and higher differential quotients by the appropriate difference quotients. In this way the differential equation at each point x_j of the interval, and also the boundary conditions, are replaced by equations for the approximate values y_j.

Example

For the eigenvalue problem (IV.59) (IV.60)

$$y'' + \lambda x y = 0, \qquad y'(0) = y(1) = 0 \tag{VI.11}$$

it is recommended that the grid be chosen so that $x = 0$ lies midway between two grid points; a 'central difference quotient' can then be used for $y'(0)$. Taking $h = 2/5$ as the mesh width, see Fig. VI.2, we then have three unknowns y_0, y_1, y_2. The boundary condition $y'(0) = 0$ is then achieved by the symmetry $y_0 = y_1$. If we write $\Lambda h^2/5 = \mu$, where Λ is an approximate value for λ, then corresponding to the differential equation we have the equations

$$\begin{array}{ll} \text{for } x = 1/5 & y_2 - 2y_1 + y_1 + \mu y_1 = 0 \\ \text{for } x = 3/5 & 0 - 2y_2 + y_1 + \mu \cdot 3y_2 = 0. \end{array}$$

These two linear, homogeneous equations have a non-trivial solution if and only if the determinant of the coefficients vanishes:

$$\begin{vmatrix} -1+\mu & 1 \\ 1 & -2+3\mu \end{vmatrix} = 1 - 5\mu + 3\mu^2 = 0.$$

The roots of this equation are

$$\mu = \frac{1}{6}(5 \pm \sqrt{13}) = \begin{cases} 0.2324 \\ 1.4343 \end{cases}$$

and give

$$\Lambda = \frac{125}{4}\mu = \begin{cases} 7.263 & (-\ 7.4\%) \\ 44.8 & (-20.0\%). \end{cases}$$

Here in brackets are given the errors of the approximate values compared with the values (IV.66) calculated by the power-series method in Chapter IV, 13.

$$\lambda_1 = 7.837; \qquad \lambda_2 = 56.0.$$

With smaller mesh-widths better values can be expected. Taking $h = 2/7$ and the function values y_j as in Fig. VI.2a, we have, writing $\Lambda h^2/7 = \mu = 1/s$, the

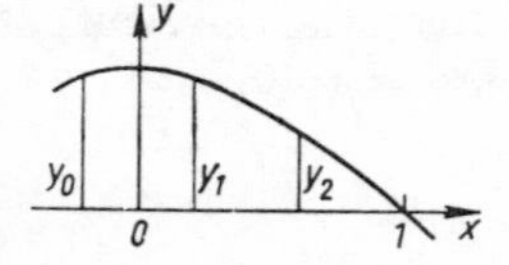

Fig. VI.2. The difference quotient method for the eigenvalue problem (V.22)

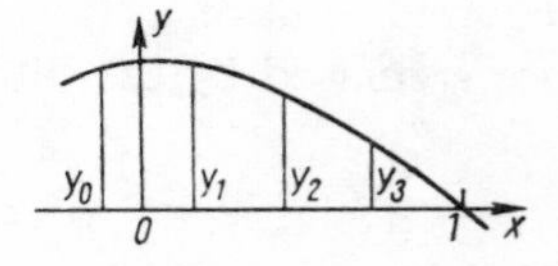

Fig. VI.2a. The same problem with a smaller mesh width

system of equations

$$y_2 - y_1 + \mu y_1 = y_3 - 2y_2 + y_1 + 3\mu y_2 = -2y_3 + y_2 + 6\mu y_2 = 0.$$

Here the determinant is

$$\begin{vmatrix} -1+\mu & 1 & 0 \\ 1 & -2+3\mu & 1 \\ 0 & 1 & -2+5\mu \end{vmatrix} = 0$$

or

$$s^3 - 14s^2 + 31s - 15 = 0.$$

We obtain

$$s = \begin{cases} 11.395\,0 \\ 1.919\,0 \\ 0.686\,0 \end{cases} \quad \text{and} \quad \Lambda = \frac{343}{4s} = \begin{cases} 7.525(-\ 4.0\%) \\ 44.68\ (-20.2\%). \\ 125.0 \end{cases}$$

5. Multi-point methods

If more accurate values are required, it is not recommended that the difference-quotient method should be repeated with a very small mesh-width, because the amount of calculation would then increase very considerably; a more exact method should be used instead. One of these is the 'multi-point method', in which, for each individual approximation equation written down for the point x_j, the differential equation at several points is brought into use. Here we shall merely illustrate the idea of the method by means of the example in the previous section. For the 'ordinary' difference-quotient method described in the previous section the second difference quotient was used instead of the second derivative; if all the values appearing therein are expanded by Taylor's theorem in terms of the value of y and its derivatives at the point x_j, then we find

$$\begin{aligned} &h^2 y''(x_j) - [y(x_{j+1}) - 2y(x_j) + y(x_{j-1})] \\ &\qquad = \text{a remainder term of the fourth order.} \end{aligned} \tag{VI.12}$$

We can now arrange, by bringing in the second derivatives at the points x_{j-1} and x_{j+1} multiplied by suitable coefficients, that the remainder term shall be of higher order:

$$\begin{aligned} &h^2[y''(x_{j+1}) + 10y''(x_j) + y''(x_{j-1})] - 12[y(x_{j+1}) - 2y(x_j) + y(x_{j-1})] \\ &\qquad = \text{a remainder term of the sixth order.} \end{aligned} \tag{VI.13}$$

We now write down this equation for each point x_j in the interval considered, omitting the remainder term, and writing the approximation y_j instead of $y(x_j)$, and replacing y'' by means of the given differential equation. (Similar multi-point formulae can also be set up for other derivatives.)

If we work the example of the previous section with the same mesh-width $h = 2/5$, it would not now be sensible to put $y_0 = y_1$ again, because by using such a crude equivalent for the boundary condition $y'(0) = 0$, the higher accuracy attained by the formula (VI.13) would be forfeited. We think of, let us say, a cubical parabola $y = p(x)$ going through the four points y_0, y_1, y_2, y_3, and calculate its slope at the point $x = 0$:

$$p'(0) = \frac{1}{24h}(-23y_0 + 21y_1 + 3y_2 - y_3). \tag{VI.14}$$

We require that $y_3 = 0$ and $p'(0) = 0$, and so

$$23y_0 - 21y_1 - 3y_2 = 0. \tag{VI.15}$$

The multi-point equations here are, for the points

$$x = \frac{1}{5}: \qquad m(3y_2 + 10y_1 - y_0) + y_2 - 2y_1 + y_0 = 0$$

$$x = \frac{3}{5}: \qquad m(30y_2 + y_1) - 2y_2 + y_1 = 0.$$

where

$$\frac{\Lambda h^2}{12 \cdot 5} = m = \frac{1}{s}.$$

If we eliminate y_0 by means of (VI.15), we obtain two equations which have a non-trivial solution if and only if

$$\begin{vmatrix} -12 + 209\,m & 26 + 66\,m \\ 1 + m & -2 + 30\,m \end{vmatrix} = 0 \quad \text{or} \quad s^2 - 52.5\,s + 258.5 = 0.$$

We obtain

$$s = \begin{cases} 47 \\ 5.5 \end{cases} \quad \text{and} \quad \Lambda = \begin{cases} 7.978\,7 & (+1.8\%) \\ 68.182 & (+22\%). \end{cases}$$

6. THE RITZ METHOD (cf. CHAPTER V, §5)

Here too only the idea of the method can be indicated without proof (a detailed presentation can be found, e.g., in the author's book *Numerische Behandlung von Differential Gleichungen*.

The considerations in Chapter III, No. 24 for ordinary differential equations of the second order will be extended here to higher orders, and the results will be given without proof.

In the calculus of variations the problem is treated of imparting to an integral of the form

$$J[\varphi] = \int_a^b F(x, \varphi(x), \varphi'(x), \ldots, \varphi^{(m)}(x))\,dx, \tag{VI.16}$$

where F is a given function of its arguments, the smallest possible value, if the function $\varphi(x)$ is allowed to traverse the set of all m times differentiable functions, which also have to satisfy certain boundary conditions. Such a problem is in no way bound to have a solution. But if it does have a solution, then the solution must satisfy the 'Euler differential equation'

$$\frac{\partial F}{\partial \varphi} - \frac{\mathrm{d}}{\mathrm{d}x}\left(\frac{\partial F}{\partial \phi'}\right) + \frac{\mathrm{d}^2}{\mathrm{d}x^2}\left(\frac{\partial F}{\partial \phi''}\right) - + \cdots + (-1)^m \frac{\mathrm{d}^m}{\mathrm{d}x^m}\left(\frac{\partial F}{\partial \phi(m)}\right) = 0. \qquad \text{(VI.17)}$$

Conversely, in the Ritz method, we try to write the given differential equation as the Euler equation for some variational problem, and to set up that variational problem (VI.16). We then make a substitution

$$\varphi(x) \approx w(x) = \sum_{n=0}^{p} a_\nu w_\nu(x) \qquad \text{(VI.18)}$$

into the expression $J[\varphi]$, i.e., we form $J[w]$ and try to determine the a_ν so that $J[w]$ shall turn out to be as small as possible; the a_ν are therefore determined from the equations

$$\frac{\partial J[w]}{\partial a_\nu} = 0, \qquad \nu = 0, \ldots, p. \qquad \text{(VI.19)}$$

The success of the Ritz method depends very strongly on how well the substitution functions $w_\nu(x)$ have been chosen.

Example

In the eigenvalues problem (VI.11) the differential equation is the Euler equation for the integral

$$J[\varphi] = \int_0^{13} (-\varphi'^2 + \lambda x \varphi^2)\, \mathrm{d}x. \qquad \text{(VI.20)}$$

The boundary conditions require, in general, special consideration; in the present case it suffices to demand of the substitution functions that they shall satisfy the boundary condition $\varphi(1) = 0$; better results are to be expected if they also satisfy the other boundary condition $\varphi'(0) = 0$.

First substitution: the simplest substitution is $w(x) = a_0(1 - x^2)$ (with $a_0 \neq 0$). Then

$$J[w] = \int_0^1 [-(2a_0x)^2 + \lambda x a_0^2(1 - x^2)^2]\, \mathrm{d}x = a_0^2\left[-\frac{4}{3} + \lambda \frac{1}{6}\right];$$

(VI.19) gives

$$\frac{\partial J}{\partial a_0} = 2a_0\left[-\frac{4}{3} + \Lambda \frac{1}{6}\right] = 0$$

or $\Lambda = 8$ (error + 2.1%), where Λ again denotes an approximation to λ.

Second substitution: $w(x) = a_0(1 - x^2) + a_1(1 - x^3)$. Then we have

$$J[w] = -\left[\frac{4}{3}a^2 + 2\cdot\frac{3}{2}ab + \frac{9}{5}b^2\right] + \lambda\left[\frac{1}{6}a^2 + 2\cdot\frac{27}{140}ab + \frac{9}{40}b^2\right],$$

where, for brevity, a, b have been written instead of a_1, a_2. The equations (VI.19) for this subsitution

$$-\frac{4}{3}a - \frac{3}{2}b + \lambda\left[\frac{1}{6}a + \frac{27}{140}\mathrm{b}\right] = 0$$

$$-\frac{3}{2}a - \frac{9}{5}b + \lambda\left[\frac{27}{140}a + \frac{9}{40}b\right] = 0$$

have a non-trivial solution if and only if the determinant vanishes:

$$\begin{vmatrix} -\frac{4}{3} + \frac{1}{6}\Lambda & -\frac{3}{2} + \frac{27}{140}\Lambda \\ -\frac{3}{2} + \frac{27}{140}\Lambda & -\frac{9}{5} + \frac{9}{40}\Lambda \end{vmatrix} = 0$$

or $\Lambda^2 - 70\Lambda + 490 = 0$. The approximation values are therefore

$$\Lambda = \begin{cases} 7.8891 & \text{(error } +0.66\%) \\ 62.111 & \text{(error } +11.0\%). \end{cases}$$

§2. FURTHER PROBLEMS, AND THEIR SOLUTIONS

7. Miscellaneous problems on chapter I

1. Find the general solution of:

(a) $y - xy' = \dfrac{x^2y'^2}{1 - 4y}$ (b) $y' = \dfrac{y^2 - x + x^3}{y}$

(c) $xy' - y = x^2 - y^2$ (d) $x^2y' + xy + ay = 0$

(e) $(x + y)\,y' + 1 = 0$ (f) $(x + y)^2y' + 1 = 0$

(g) $y - y' = ay'^3$ (h) $(x - 2x^2 - 7y)y' = 6x - 2xy - y - 3$

(i) $y + \dfrac{x}{y'} - \dfrac{a}{\sqrt{1 + y'^2}} = 0.$ (j) $y^2 - 2xyy' + (yy')^3 = 0$

(k) $xy'^2 + x - y = 0$ (l) $2\sqrt{y(1 - y)} + \sqrt{1 - x^2}\,y' = 0$

(m) $(\mathrm{e}^x - 1)\,\mathrm{e}^y y' + \mathrm{e}^x = 0$ (n) $y' = \dfrac{2x - 5}{4y + 3}$

(o) $y' \sin x = 3 \sin y$ (p) $y - xy' + \dfrac{x^2 + by^2}{ax} = 0$

(q) $y' = \dfrac{ay \sin x - \sin y}{x \cos y + a \cos x}$.

2. Set up differential equations of the first order for the following families of curves (c is the parameter of the family):

(a) $\sqrt{1+x^2} + \sqrt{1+y^2} = c$ (b) $(x - ac)^2 + y^2 = c^2$

(c) $y = cx + e^{(x/c)}$ (d) confocal conics $\dfrac{x^2}{a^2+c} + \dfrac{y^2}{c} = 1$.

3. Find the solution of the initial-value problem $xy'(x) - 2y(x) = x^3 f(x)$, with $y(a) = 0$ for $a > 0$, and $f(x)$ a given function.

4. Suppose the differential equation is given, whose general solution is, as in problem 2(c), given by

$$y = cx + e^{(x/c)}$$

Let k be the number of solutions which go through a given point. Sketch in the x,y-plane the regions in which $k = 0, 1, 2, \ldots$.

5. Discuss in the $x,\dot{x}$-phase-plane the behaviour of solutions of the differential equation for an oscillator with a non-linear restoring force and non-linear damping:

$$\ddot{x} + \dot{x}^2 \operatorname{sgn} \dot{x} + x - x^2 = 0.$$

with the initial conditions $x(0) = 0$, $\dot{x}(0) = v_0$.

6. In the following, questions are asked only about continuous solutions of the functional equation

$$y\left(\frac{x+\xi}{1+x\xi}\right) = y(x)\, y(\xi).$$

(a) What solution has the functional equation in common with the differential equation $y' = y/(1 - x^2)$?

(b) Give a differential equation which is equivalent to the functional equation.

7. Make a power-series substitution

$$y(x) = \sum_{\nu=0}^{\infty} a_\nu x^\nu$$

for the solution $y(x)$ of the non-linear initial-value problem

$$y'(x) = x^2 + xy^2, \qquad y(0) = 1,$$

and calculate some of the constants a_ν.

8. The function $y(x)$ determined in the previous problem increases beyond all bounds at a point $x = p$. Obtain bounds for p by calculating a sub-function $q(x) \leq y(x)$ and a super-function $z(x) \geq y(x)$. This can be done by including the x^2 in the differential equation between bounds so that the differential equations obtained can be solved by elementary means.

9. Consider the following very crude model for the diurnal variation of temperature in a shallow body of water. To simplify the calculations, let the unit of time be chosen so that the period of one day comprises 2π units of time. The time $t=0$ is taken to be at 12 noon, $t=\pi$ at midnight. Let the outside temperature be given by $v(t)$:

$$v(t)=\begin{cases} a & \text{for } \frac{\pi}{2}\le |t| \le \frac{3\pi}{2} \\ a+c\cos t & \text{for } |t|\le\frac{\pi}{2} \\ v(t+2\pi) & \text{for all } t \end{cases}$$

Here the expression $a+c\cos t$ corresponds to the solar insolation. Let the heating-up $\dot{u}$ of the water be given vy the difference $v-u$, i.e.,

$$\frac{du}{dt}=\dot{u}=v-u.$$

Calculate the continuous, periodic solution $u(t)$, and the time τ, and the value U, of the maximum temperature:

$$\max_{0\le t\le 2\pi} u(t)=U=u(\tau).$$

8. Solutions

1. The general solution (with c as a free parameter) is:

(a) $y=\dfrac{cx-x^2}{c^2}$ (b) $y^2=ce^{2x}-\left(x^3+\dfrac{3}{2}x^2+\dfrac{x}{2}+\dfrac{1}{4}\right)$

(c) $y=x\tanh(x+c)$ (d) $\ln(xy)-\dfrac{a}{x}=c$

(e) $(x+y-1)\,e^y=c$ (f) $y=c+\dfrac{1}{2}\ln\left|\dfrac{x+y+1}{x+y-1}\right|$

(g) With s as parameter
$$x=c+\ln s+\frac{3a}{2}s^2$$
$$y=s+as^3$$

(h) $y=\dfrac{7}{22}x-\dfrac{1}{11}+c\sqrt{7x^2-x+6}\exp\left[\dfrac{3}{\sqrt{167}}\arctan\left(\dfrac{1}{\sqrt{167}}(14x-1)\right)\right]$

(i) The solution in parametric form with $p=\tan\alpha$ as parameter:

$$x=\left(c+\frac{a}{2\sqrt{1+p^2}}\right)\frac{p}{\sqrt{1+p^2}}=c\sin\alpha+\frac{a}{2}\cos^2\alpha$$

$$y=\frac{2(a-c)p^2+a-2c}{2(1+p^2)^{3/2}}=-c\cos\alpha+\frac{a}{2}\cos\alpha[1+\sin^2\alpha]$$

(j) $y^2 = cx - \frac{1}{8}c^3$

(k) The solution is obtained by separating the variables with p as the parameter:

$$x = \frac{c}{p^2 - p + 1} \exp\left[-\frac{2}{3}\sqrt{3} \arctan\left(\frac{2p-1}{\sqrt{3}}\right)\right]$$

$$y = \frac{c(1+p^2)}{p^2 - p + 1} \exp\left[-\frac{2}{3}\sqrt{3} \arctan\left(\frac{2p-1}{\sqrt{3}}\right)\right]$$

(l) $\arcsin x + \arcsin \sqrt{y} = c$ (m) $y = \ln(c - \ln(1 - e^x))$

(n) $2y^2 + 3y = x^2 - 5x + c$ (o) $\tan \frac{y}{2} = c\left(\tan \frac{x}{2}\right)^3$

(p) $\ln x + \frac{a}{\sqrt{b}} \operatorname{arccot} \frac{y\sqrt{b}}{x} = c$ (q) $x \sin y + ay \cos x = c.$

2. (a) $x\sqrt{1+y^2} + yy'\sqrt{1+x^2} = 0$

(b) $ay\sqrt{1+y'^2} = x + yy'$

(c) $y = \frac{x^2}{\psi(xy'-y)} + e^{\psi(xy'-y)}$; where $s = \psi(t)$ is the function inverse to $t = (s-1)e^s$; $\psi(t)$ is a transcendental function for which no special symbol has yet been introduced

(d) $xyy'^2 + (x^2 - y^2 - a^2)y' = xy.$

3. $y(x) = \int_0^x x^2/(s)\, ds.$

4. Figure VI.3 (which is not drawn to scale but only qualitative) shows the regions where $k = 0, 1, 2, 3$; there is no region where $k = 4$. The boundary curves between the regions are the x-axis and the curve

$$x = 2c \ln c, \qquad y = c^2(1 + 2 \ln c)$$

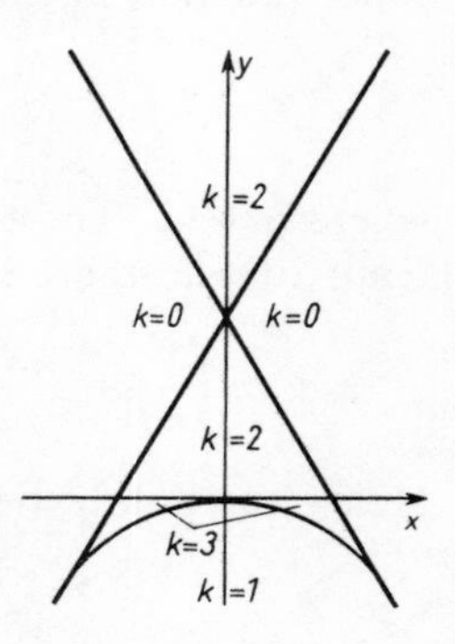

Fig. VI.3. The regions where there are $k = 0, 1, 2, 3$ solutions

(with c as a parameter), which has cusps at the two points $x = \pm 2e^{-1}$, $y = -e^{-2}$.

5. Discussion reveals the three cases:

(a) $v_0^2 > \frac{1}{4}(e^2 - 1)$, then $\dot{x}(t) \neq 0$ for all t, and there is no oscillation.

(b) $v_0^2 = \frac{1}{4}(e^2 - 1)$, say $v_0 > 0$; $x(t)$ has a fixed sign and remains bounded for $t > 0$.

(c) $v_0^2 < \frac{1}{4}(e^2 - 1)$; damped oscillations, with infinitely many zeros, take place.

6. (a) $y = \left(\frac{1+x}{1-x}\right)^{1/2}$.

(b) The general continuous solution $y = \left(\frac{1+x}{1-x}\right)^c$ of the functional equation (with c as a free constant) is also the solution of the differential equation

$$y' = \frac{2y}{1-x^2} \frac{\ln y}{\ln(1+x) - \ln(1-x)}.$$

7. Substitution of the series and comparison of coefficients gives:

$$y(x) = 1 + \frac{x^2}{2} + \frac{x^3}{3} + \frac{x^4}{4} + \frac{2}{15}x^5 + \frac{1}{8}x^6 + \frac{13}{210}x^7 + \cdots.$$

8. The function x^2 is first omitted, and a sub-function $q(x)$ is calculated from

$$q' = xq^2, \qquad q(0) = 1 \quad \text{to be} \quad q(x) = \frac{2}{2-x^2}.$$

This function has a pole at $x = x_0 = \sqrt{2} \approx 1.414$, and so we have $p < \sqrt{2}$.

Next we replace x^2 by $x\sqrt{2}$ in the interval $0 \le x \le \sqrt{2}$, and we determine a super-function in the interval $0 \le x < \sqrt{2}$ from

$$z' = x\sqrt{2} + xz^2, \qquad z(0) = 1 \quad \text{to be} \quad z(x) = \alpha \tan\left(\arctan\left(\frac{1}{\alpha}\right) + \frac{ax^2}{2}\right)$$

where $\alpha = \sqrt[4]{2}$. The tangent increases beyond all bounds when the argument tends to $\pi/2$; there is therefore a pole at $x = x_1$ with

$$\arctan\left(\frac{1}{\alpha}\right) + \alpha\frac{x_1^2}{2} = \frac{\pi}{2} \quad \text{or} \quad x_1 = \sqrt{\frac{2}{\alpha}\left(\frac{\pi}{2} - \arctan\left(\frac{1}{\alpha}\right)\right)} \approx 1.211.$$

Hence the pole p lies in the interval

$$1.211 \le p \le \sqrt{2} < 1.415.$$

9. Since the number a enters into v only additively, we can introduce $u - a$ and $v - a$ as new variables; then a drops out, and so we can from the outset put $a = 0$ for the calculation, and finally re-introduce a in the result. So we have

$$\dot{u} + u = \begin{cases} c \cos t & \text{for } |t| \leq \dfrac{\pi}{2} \\ 0 & \text{for } \dfrac{\pi}{2} \leq t \leq \dfrac{3\pi}{2} \end{cases}$$

Hence

$$u = k\mathrm{e}^{-t} \quad \text{for } \frac{\pi}{2} \leq t \leq \frac{3\pi}{2}, \text{ where } k \text{ is as yet undetermined.}$$

Also

$$u = \frac{c}{2} (\cos t + \sin t) + d\mathrm{e}^{-t} \quad \text{for } |t| \leq \frac{\pi}{2}, \quad \text{where } d \text{ also is not yet known.}$$

Putting $\varepsilon = \mathrm{e}^{-\pi/2}$, we then have the conditions

$$u\left(-\frac{\pi}{2}\right) = u\left(\frac{3\pi}{2}\right) = -\frac{c}{2} + \frac{d}{\varepsilon} = k\varepsilon^3$$

$$u\left(\frac{\pi}{2}\right) = \frac{c}{2} + d\varepsilon = k\varepsilon.$$

From these two equations we calculate

$$d = \frac{c}{2} \frac{\varepsilon}{1 - \varepsilon^2} = \frac{c}{2} \frac{1}{2 \sinh(\pi/2)}$$

$$k = \frac{c}{2} \frac{1}{\varepsilon(1 - \varepsilon^2)}.$$

The point $t = \tau$ of the maximum of u is calculated from the equation

$$\dot{u} = 0 \quad \text{or} \quad \cos t - \sin t - \frac{\varepsilon}{1 - \varepsilon^2} \mathrm{e}^{-t} = 0$$

to be $\tau = 0.7098$ (this corresponds to about 14.43 hours). Figure VI.4 shows the graphs of $u(t)$ and $v(t)$ for $c = 10^\circ\mathrm{C}$. If $v(t) < u(t)$, then $u(t)$ falls; if $v(t) > u(t)$, then $u(t)$ rises. The maximum and minimum occur when $v(t) = y(t)$, Fig. VI.4.

This model can, of course, be refined to suit the actual circumstances better.

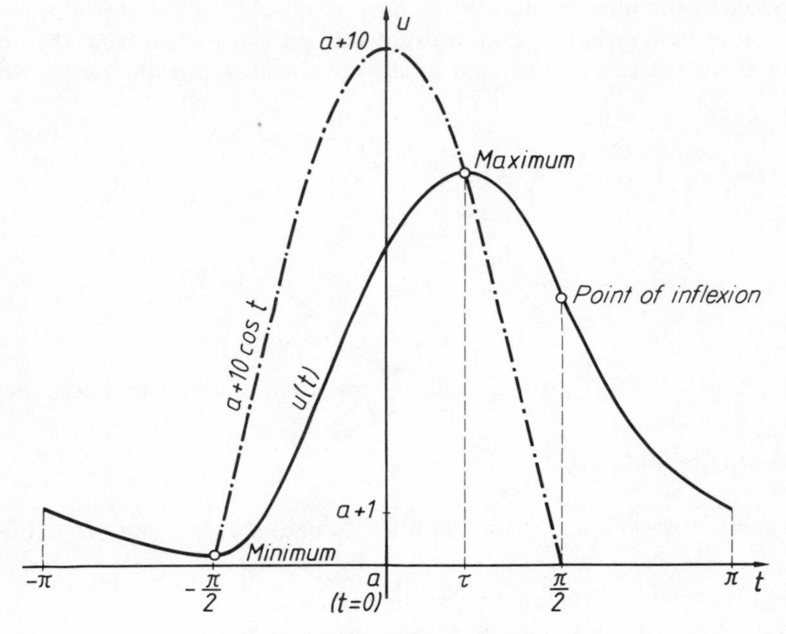

Fig. VI.4. Diurnal variation of tmeperature

9. Miscellaneous problems on Chapters II and III

1. Find the general solutions of the following non-linear differential equations.

(a) $y''^2 + 2yy'' - y'^2 = 0$ (b) $y'' = ay'^2$

(c) $y'' = f(x)(a + y')^2$ (d) $y'' = 2axy'^2$

(e) $y'' + (1 + y')^3 = 0$ (f) $yy'' = y'^2(y' - 2)$

(g) $(yy'' - y'^2)\, x^2 + y^2 = 0$ (h) $y'' = -2y'^3$

(i) $yy'' - y'(y' + xy) - y^2 e^{x^2/2} = 0.$

2. Find the general solutions of the following linear differential equations:

(a) $y'' = 3y + 1 + x$ (b) $y''' - y'' - 4y' + 4y = 4x^3$

(c) $y'' - 3y' - 10y = e^x$ (d) $y'' - 2y' = 6x^2$

(e) $y''' - 5y'' + 5y' - y = 0$ (f) $y''' - 5y'' + 7y' - 3y = e^{2x} + 2$

(g) $y'' - 2y' + y = x$ (h) $y^{IV} - 2y'' + y = \cos x + 2 \sin x$

(i) $y^{VI} - 3y^{IV} + 3y'' - y = -\dfrac{x^2}{3}$ (j) $y'' + 5y' + 7y = a\, e^{-x}$

(k) $y'' + y = 5x \cos x + 4 \sin x$ (l) $y'' + 4y' + 4y = 6\, e^{-2x}$

(m) $y^{IV} + 4y = e^x \cos x$ (n) $x^2 y'' - 2xy' + (a^2x^2 + 2)\, y = 0$

(o) $(1-x^2)\,y''-xy'+a^2y=0$ (p) $x^4y''+2x^3y'+a^2y=1$

(q) $x(1-x)\,y''+[(a+1)-2x]\;y'+a(a-1)\,y=0$

(r) $x^2y''+xy'-y=0$

(s) $y'''-y''-y'+y=(1+x)\,\mathrm{e}^x$

(t) $6y^{\mathrm{IV}}-35y'''+62y''-35y'+6y=x^3$

(u) $y''+a^2y=\mathrm{e}^{bx}\cos ax$.

3. Set up a second-order differential equation for each of the following families of curves (c_1 and c_2 are free parameters):

(a) The set of all circles which touch the unit circle centred at the origin.

(b) $y=x^x(\mathrm{e}^{c_1+c_2x})$.

4. Give a particular solution of the following differential equations, where $r(x)$ is a given function, and a, b are given constants:

(a) $y''+ay'=r(x)$ with $a\neq 0$.

(b) $y'''+ay''+by'=r(x)$ with $a^2-4b\neq 0$, $a\neq 0$.

(c) Solve the boundary-value problem $y''(x)=r(x)$ with the boundary conditions

$$y'(0)+y'(1)=0$$
$$y(0)+y(1)+y'(0)=0.$$

5. Express the function $y(x)$ uniquely determined by

$$xy''+(x-b)\,y'+ay=0,\; y(0)=1$$

as a power series, and, for $-b>a>0$ express it as a definite integral, and determine whether $y(1)$ is positive.

6. Expand the following functions as power series:

(a) $f(x)=(\arcsin x)^2$ (b) $f(x)=(\arcsin x)^3$

(c) $f(x)=(\arctan x)^2$

(d) Give a closed expression for the function

$$f(x)=\sum_{n=0}^{\infty}\frac{n!}{(2n)!}x^{2n}=1+\frac{x^2}{2}+\frac{x^4}{2\cdot 6}+\frac{x^6}{2\cdot 6\cdot 10}+\frac{x^8}{2\cdot 6\cdot 10\cdot 14}+\cdots.$$

7. Determine the periodic solutions of $y''=y\cos x$ by expanding the solution in powers of $\cos x$.

8. (a) A thin circular plate is subjected to symmetrical loading of density $p(r)$, where r is the distance from the centre of the circle, and its plate stiffness is N; the deflections $w(r)$ satisfy the differential equation

$$w^{\mathrm{IV}}+\frac{2}{r}w'''-\frac{1}{r^2}w''+\frac{1}{r^3}w'=\frac{1}{N}p(r).$$

Find the deflection $w(r)$ if the boundary of the plate is encastré (i.e., $w(a)=w'(a)=0$) and the load density is $p(r)=a^2-r^2$ (see Fig. VI.5).

(b) Find the general solution for the case where $p(r)=r^k$ $(k\geq 0)$.

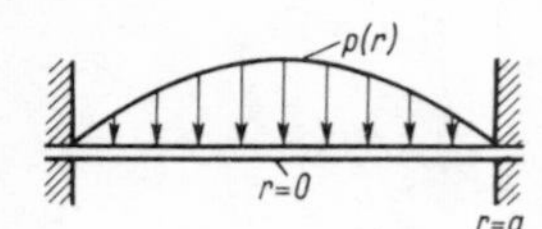

Fig. VI.5. A symmetrically loaded circular platew

9. Transform the system

$$y_j'(x) = \sum_{k=1}^{2} a_{jk}(x)\, y_k(x) + r_j(x) \qquad (j = 1, 2)$$

into a second-order differential equation for y_1 alone.

10. (a) Find the general solution $Y = \begin{Bmatrix} x(t) \\ y(t) \end{Bmatrix}$ of the following systems:

$$\begin{aligned} \dot{x} + x - 2y &= \sin t \\ \dot{y} + 5x + 3y &= \cos t. \end{aligned}$$

(b)

$$\begin{aligned} \dot{x} &= 2x + 3y - 3z \\ \dot{y} &= \qquad 2y + 2z \\ \dot{z} &+ 5x + 4y. \end{aligned}$$

(c) Integrate the Euler system

$$\begin{aligned} x^2u' + xu + 2x^2v + 3x^2w &= 0 \\ x^3v' - (4xu + 6x^2v + 10x^3w &= 0 \\ x^4w' + 2xu + 3x^2v + 6x^3w &= 0 \end{aligned} \quad \text{for} \quad \begin{cases} u(1) = 3 \\ v(1) = -2 \\ w(1) = -1. \end{cases}$$

11. A railway waggon W_1 (weight 15T) moving with velocity v_0 collides with another waggon W_2 (weight 10T), immediately behind which a third waggon W_3 (weight 15T) is standing.

The spring constant of each buffer mechanism is $c = 4\text{T/cm}$; Fig. VI.6. Discuss the subsequent motion.

12. The coefficient functions $a(t)$, $b(t)$, $c(t)$, $A(t)$, $B(t)$, $C(t)$ are each continuously differentiable (once) in a closed t-interval. What further necessary condition must they satisfy in order that the two differential equations

$$\begin{aligned} ax + b\dot{x} + c\ddot{x} &= 0 \\ Ax + B\dot{x} + C\ddot{x} &= 0 \end{aligned}$$

shall have a common solution $x(t)$? Is this condition also sufficient?

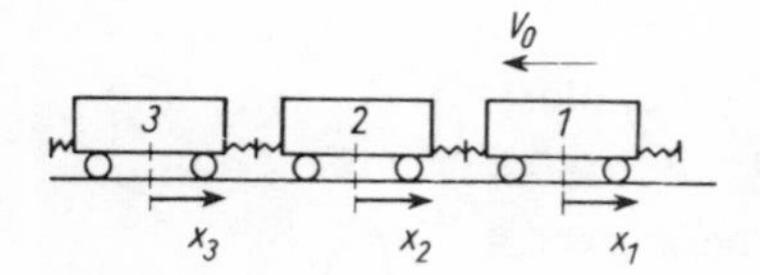

Fig. VI.6. Motion of three railway waggons after impact

13. Study the behaviour of solutions of the Schrödinger equation for a one-dimensional periodic force-field (band model in the theory of metal electrons):

$$y'' + (\mu - p(x))\, y = 0 \quad \text{with } p(x) = \begin{cases} 1 & \text{for } 0 \le x < 1 \\ 0 & \text{for } 1 \le x < 3 \\ p(x+3) & \text{for all } x. \end{cases}$$

14. What length a of a cord (length $l = 1$ m, cross-sectional area F, specific weight γ) lying on a table must overhang from the table in order that the cord can begin to slip without any initial velocity? (See Fig. VI.7.) Let the frictional force R be proportional to the normal pressure N of the end of the cord lying on the table (Coulomb friction, $R = fN$, coefficient of friction $f = 0.1$). Discuss the time $T(a)$ to the beginning of free fall as a function of the overhanging length a.

15. For each of the following differential equations, discuss when the boundary-value problem with the given differential equation and the boundary conditions $y(0) = a$, $y(h) = b$ is soluble and how many solutions it then has, i.e., give the region of the point h, b in the h,b-plane for which solutions of the boundary-value problem exist:

(a) $y'' = \alpha y'^3$ with $\alpha > 0$ (b) $y'' + 6yy'^3 = 0$ with $a = 0$

(c) $y'' = (1 + y')^3$.

16. (a) The ends of a bar of length a are maintained at the constant temperatures $y = 0$ at $x = 0$ and $y = b$ at $x = a$ $(a > 0)$. Due to chemical reactions a quantity of heat Q per unit length and unit time is developed in the bar; in dimensionless form this quantity Q is given by e^y. What is the steady temperature distribution, i.e., what is the solution of the boundary-value problem $y' = e^y$, $y(0) = 0$, $y(a) = b$? For which pairs (a, b) are there two, one, or no solutions?

(b) In the preceding problem let $b = 0$, i.e., both ends of the bar are kept at the same temperature. How can the solution now be calculated? What is the greatest length a of the bar for which a state of equilibrium is possible?

17. Set up the transcendental equation for the eigenvalues λ for

(a) $$-y'' = \lambda y = k^2 y$$

and

$$\alpha y(0) + l\beta y'(0) + \gamma y(l) + l\delta y'(l) = 0$$
$$a y(0) + lby'(0) + cy(l) + l\, \mathrm{d}y'(l) = 0.$$

(b) $$y^{\mathrm{IV}} = -\lambda y'' = -k^2 y''$$

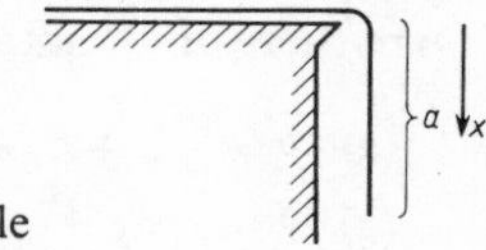

Fig. VI.7. A cord sliding off a table

with

$$y(0) = y'(0) = y''(0) + ay'''(0) = y(1) + by'(1) = 0.$$

18. A two parameter eigenvalue problem:

$$-y'' = \lambda y(x) + \mu y'(0), \qquad y(0) = y(1) = 0.$$

Give in the λ, μ-plane the 'eigencurves' on which those pairs λ, μ lie for which there is an 'eigenfunction', i.e., a solution $y(x)$ which does not vanish identically. Let k be the number of 'nodes' (zeros) of $y(x)$ in the interior of the interval (0, 1); let the various eigencurves be marked according to the values of k. Discuss, in particular, the eigenvalue problems which result from the following relations between λ and μ:

(a) $\mu = \text{const}, \ \mu < 2$ (b) $\mu = \text{const}, \ \mu > 2$
(c) $\mu = \lambda^2$ (d) $\lambda = \mu^2$
(e) $\lambda = 4\pi^2 + \mu^2$.

19. Find the eigencurves, as in question 18, for

$$y'' = \lambda y', \ y(0) = y(1) - \mu y'(1) = 0.$$

20. Consider the system

$$\begin{aligned} u' &= \alpha u + \beta v \\ v' &= -\beta u - \alpha v \end{aligned} \quad \text{with the boundary conditions} \quad \begin{aligned} u(0) + v(1) &= 0 \\ v(0) - u(1) &= 0. \end{aligned}$$

For what real eigen-pairs α, β are there non-trival solutions? (Draw a diagram in the α,β-plane.)

10. Solutions

1. In the solutions c_1 and c_2 are free parameters.

(a) $y = c_1[1 - (x + c_2)^2]$ (b) $y = -\dfrac{1}{a}\ln|x + c_1| + c_2$

(c) $y = -\displaystyle\int^x \frac{ds}{F(s) + c_1} - ax + c_2$ with $F(x) = \displaystyle\int_{x_0}^x f(s)\,ds$

(d) The solutions are

$$\frac{1}{2ac_1}\ln\left|\frac{x + c_1}{x - c_1}\right| + c_2, \qquad \frac{1}{ax} + c_2, \qquad -\frac{1}{ac_1}\left(\arctan\frac{x}{c_1}\right) + c_2.$$

Are there any relations between these forms? Are there any further solutions?

(e) $y = -x + c_1 + (2x + c_2)^{1/2}$ (f) $y + c_1 y^3 = 2x + c_2$
(g) $y = c_1 x\, e^{c_2 x}$ (h) $x = y^2 + c_1 y + c_2$

(i) $y = c_1 \exp\left[\int_0^x e^{t^2/2}\,(c_2 + t)\,dt\right]$.

2. (a) $y = c_1\,e^{x\sqrt{3}} + c_2\,e^{-x\sqrt{3}} - \dfrac{1+x}{3}$

(b) $y = c_1\,e^x + c_2\,e^{2x} + c_3\,e^{-2x} + x^3 + 3x^2 + \dfrac{15}{2}\,x + \dfrac{15}{2}$

(c) $y = c_1\,e^{5x} + c_1\,e^{-2x} - \dfrac{1}{12}\,e^x$ (d) $y = c_1 + c_2\,e^{2x} - \left(x^3 + \dfrac{3}{2}\,x^2 + \dfrac{3}{2}\,x\right)$

(e) $y = c_1\,e^x + e^2 x(c_2\,e^{x\sqrt{3}} + c_3\,e^{-x\sqrt{3}})$

(f) $y = (c_1 + c_2 x)\,e^x + c_3\,e^{3x} - e^{2x} - \dfrac{2}{3}$

(g) $y = (c_1 + c_2 x)\,e^x + x + 2$

(h) $y = (c_1 + c_2 x)\,e^x + (c_3 + c_4 x)\,e^{-x} + \dfrac{1}{4}\,(\cos x + 2\sin x)$

(i) $y = (c_1 + c_2 x + c_3 x^2)\,e^x + (c_4 + c_5 x + c_6 x^3)\,e^{-x} + \dfrac{x^2}{3} + 2$

(j) $y = \left(c_1 \cos\left(\dfrac{\sqrt{3}}{2}\,x\right) + c_2 \sin\left(\dfrac{\sqrt{3}}{2}\,x\right)\right)\,e^{-5x/2} + \dfrac{a}{3}\,e^{-x}$

(k) $y = c_1 \cos x + c_2 \sin x - \dfrac{3}{4}\,x\cos x + \dfrac{5}{4}\,x^2 \sin x$

(l) $y = (c_1 + c_2 x + 3x^2)\,e^{-2x}$

(m) $y = (c_1 \cos x + c_2 \sin x)\,e^x + (c_3 \cos x + c_4 \sin x)\,e^{-x}$

$$+ \frac{1}{16}\,x\,e^x(\sin x - \cos x)$$

(n) $y = (c_1 \cos ax + c_2 \sin ax)\,x$

(o) $y = c_1 \cos(a \arcsin x) + c_2 \sin(a \arcsin x)$

(p) $y = c_1 \cos\dfrac{a}{x} + c_2 \sin\dfrac{a}{x} + \dfrac{1}{a^2}$

(q) $y = x^{-a}\left(c_1 + c_2 \displaystyle\int_{x_0}^{x} [s(1-s)]^{a-1}\,ds\right)$

(r) $y = c_1 x + \dfrac{c_2}{x}$

(s) $y = e^x\left(c_1 + c_2 x + \frac{1}{8}x^2 + \frac{1}{12}x^3\right) + c_3\, e^{-x}$

(t) $y = c_1 e^{3x} + c_2 e^{x/3} + c_2 e^{2x} + c_4 e^{x/2} + \frac{18\,095}{216} + \frac{853}{36}x + \frac{35}{12}x^2 + \frac{1}{6}x^3$

(u) $y = c_1 \cos ax + c_2 \sin ax + \frac{1}{b^2 + 4a^2} e^{bx}\left(\cos ax + \frac{2a}{b}\sin ax\right)$.

3. (a) $y'' = \frac{2(1+y'^2)[(1+y'^2)^{1/2} + xy' - y]}{x^2 + y^2 - 1}$ (b) $yy'' - y'^2 = \frac{1}{x}y^3$.

4. (a) $y = \int_{x_0}^{x} \frac{1}{a}(1 - e^{-a(x-s)})\, r(s)\, ds$

(b) $y = \int_{x_0}^{x}\left[\frac{1}{k_1 k_2} + \frac{1}{k_2(k_2 - k_1)} e^{k_2(x-s)} + \frac{1}{k_1(k_1 - k_2)} e^{k_1(x-s)}\right] r(s)\, ds;$

here k_1 and k_2 are the roots of the equation $k^2 + ak + b = 0$. What does the solution look like if this equation has a double root, i.e., if $a^2 = 4b$?

(c) $y(x) = \frac{1}{2}\int_0^1 |x - s|\, r(s)\, ds.$

5. $y(x) = 1 + \frac{a}{1 \cdot b}x + \frac{a(a+1)}{2!b(b-1)}x^2 + \frac{a(a+1)(a+2)}{3!b(b-1)(b-2)}x^3 + \cdots,$

where b must not be a positive integer. The series converges for all x; and

$$y(x) = \frac{z(x)}{z(0)} \quad \text{with} \quad z(x) = \int_0^1 u^{a-1}(1-u)^{-b-a-1} e^{-xu}\, du.$$

By differentiating $z(x)$ and substituting into the differential equation it can be seen that $z(x)$ is indeed a solution of the differential equation. The integral exists for $-b > a > 0$, and then $z(0)$ and $z(1)$, and so also $y(1)$, are positive. But $y(x)$ is not positive for all real a, b; e.g., for $a = 1$ and $b = 0.9$ we find $y(1) < -2$.

6. (a) If we start from the series

$$\arcsin x = x + \frac{1}{2}\frac{x^3}{3} + \frac{1\cdot 3}{2\cdot 4}\frac{x^5}{5} + \frac{1\cdot 3\cdot 5}{2\cdot 4\cdot 6}\frac{x^7}{7} + \cdots$$

and square it, the law of formation of the result cannot easily be seen. But here the differential equation can be used to good effect; $f(x)$ satisfies the initial-value problem

$$(1 - x^2)\, f'' - xf' - 2 = 0, \qquad f(0) = f'(0) = 0;$$

and the substitution

$$f(x) = \sum_{n=0}^{\infty} a_n x^n \quad \text{gives} \quad (n+2)(n+1)\, a_{n+2} - n^2 a_n = 0$$

$$\text{for } n = 1, 2, \cdots; \; a_0 = a_1 = 0, \; a_2 = 1$$

and so the required series is

$$f(x) = (\arcsin x)^2 = \frac{2}{3}\frac{x^4}{2} + \frac{2\cdot 4}{3\cdot 5}\frac{x^6}{3} + \frac{2\cdot 4\cdot 6}{3\cdot 5\cdot 7}\frac{x^8}{4} + \cdots$$

$$+ \frac{2\cdot 4\cdots(2n-2)}{3\cdot 5\cdots(2n-1)}\frac{x^{2n}}{n} + \cdots$$

(b) From $(1 - x^2)f'' - xf' = 6 \arcsin x$; $f(0) = f'(0) = 0$ we obtain

$$f(x) = x^3 + \frac{3!}{5!}3^2\left(1 + \frac{1}{3^2}\right)x^5 + \frac{3!}{7!}3^2\cdot 5^2\left(1 + \frac{1}{3^2} + \frac{1}{5^2}\right)x^7 + \cdots$$

$$+ \frac{3!}{(2k+1)!}3^2\cdot 5^2\cdots(2k-1)^2\left(1 + \frac{1}{3^2} + \cdots + \frac{1}{(2k-1)^2}\right)x^{2k+1} + \cdots$$

(c) From $f'(x) = \dfrac{2 \arctan x}{1 + x^2}$, $f(0) = f'(0) = 0$ we obtain

$$f(x) = x^2 - \frac{x^4}{2}\left(1 + \frac{1}{3}\right) + \frac{x^6}{3}\left(1 + \frac{1}{3} + \frac{1}{5}\right) - \frac{x^8}{4}\left(1 + \frac{1}{3} + \frac{1}{5} + \frac{1}{7}\right) + \cdots.$$

(d) The solution of the initial-value problem

$$2xf'(x) - (2 + x^2)f(x) + 2 = 0, \qquad f(0) = 1, \qquad f'(0) = 0$$

is

$$f(x) = 1 + \frac{x}{2}e^{x^2/4}\int_0^x e^{-t^2/4}\,dt = 1 + \frac{1}{2}\sqrt{\pi}\,x\,e^{x^2/4}\,\Phi\left(\frac{x}{2}\right),$$

where $\Phi(x)$ is the error integral.

7. The substitution

$$y = \sum_{n=0}^{\infty} a_n \cos^n x$$

gives the recursion formula for the a_n

$$(n+3)(n+2)\,a_{n+3} = (n+1)^2 a_{n+1} + a_n \quad \text{for } n = 0.1, 2, \ldots \text{ and } a_2 = 0$$

and hence we have the two particular solutions

$$y_1(x) = 1 + \frac{1}{6}\cos^3 x + \frac{3}{40}\cos^5 x + \frac{1}{180}\cos^6 x + \cdots$$

$$y_2(x) = \cos x + \frac{1}{6}\cos^3 x + \frac{1}{12}\cos^4 x + \frac{3}{40}\cos^5 x + \cdots$$

From the recursion formula we can prove by induction that

$$0 \leq a_n \leq \frac{1}{n}$$

if this is true for $n = 0, 1, 2$; hence one can prove the convergence of the series for all real x except $x = k\pi \qquad (k = 0, \pm 1, \pm 2, \ldots)$.

8. (a) $w(r) = \dfrac{1}{576\,N}(a^2 - r^2)^2(7a^2 - r^2)$

(b) $w(r) = c_1 + c_2 r^2 + c_2 \ln r + c_4 r^2 \ln r + \dfrac{1}{N}\dfrac{1}{(k+2)^2}\dfrac{1}{(k+4)^2} r^{k+4}$

with $c_1, \ldots, c_4$ as free constants.

9. $b_2 y_1'' + b_1 y_1' + b_0 y_1 + r_0(x) = 0$ with
$b_2 = a_{12}$, $b_1 = -(a_{12}' + a_{12}(a_{11} + a_{22}))$.
b_0 and r_0 are easily obtainable, but are rather complicated expressions.

10.

(a) $Y = e^{-2t}\begin{pmatrix} \sin 3t - \cos 3t & \sin 3t + \cos 3t \\ \sin 3t + 2\cos 3t & -2\sin 3t + \cos 3t \end{pmatrix}\begin{pmatrix} c_1 \\ c_2 \end{pmatrix} + \begin{pmatrix} \dfrac{6}{40}\cos t + \dfrac{12}{40}\sin t \\ \dfrac{9}{40}\cos t - \dfrac{17}{40}\sin t \end{pmatrix}$

(b) $Y = \begin{pmatrix} x(t) \\ y(t) \\ z(t) \end{pmatrix} = c_1 e^{4t}\begin{pmatrix} 0 \\ 1 \\ 1 \end{pmatrix} + c_2 \begin{pmatrix} \frac{1}{5}(19\cos\sqrt{11}\,t - 2\sqrt{11}\sin\sqrt{11}\,t) \\ -2\cos\sqrt{11}\,\mathrm{t} \\ 2\cos\sqrt{11}\,t + \sqrt{11}\sin\sqrt{11}\,t \end{pmatrix}$

$$+ c_2 \begin{pmatrix} \frac{1}{5}(19\sin\sqrt{t} + 2\sqrt{11}\cos\sqrt{11}\,t) \\ -2\sin\sqrt{11}\,t \\ 2\sin\sqrt{11}\,t - \sqrt{11}\cos\sqrt{11}\,t \end{pmatrix}$$

(c) The substitutions $xu = z_1$, $x^2 v = z_2$, $x^3 w = z_3$, $x = e^t$ give

$$\dot{z} = Bz \quad \text{with} \quad B = \begin{pmatrix} 0 & -2 & -3 \\ 4 & 8 & 10 \\ -2 & -3 & -3 \end{pmatrix}.$$

The secular equation has the roots $\mu_1 = 1$, $\mu_2 = \mu_3 = 2$. The corresponding eigenvectors are

$$\begin{pmatrix} 1 \\ -2 \\ 1 \end{pmatrix}, \begin{pmatrix} 1 \\ -4 \\ 2 \end{pmatrix}.$$

A principal vector of the second stage corresponding to $\mu = 2$ is $\begin{pmatrix} 1 \\ -1 \\ -1 \end{pmatrix}$;

So the general solution of the system is

$$\begin{pmatrix} xu \\ x^2 v \\ x^2 w \end{pmatrix} = c_1 \begin{pmatrix} 1 \\ -2 \\ 1 \end{pmatrix} x + c_2 \begin{pmatrix} 1 \\ -4 \\ 2 \end{pmatrix} x^2 + c_2 \left[\begin{pmatrix} -1 \\ 1 \\ 1 \end{pmatrix} x^2 + \begin{pmatrix} -3 \\ 12 \\ -6 \end{pmatrix} x^2 \ln |x| \right]$$

and the solution which fits the initial conditions is

$$u = 3 + 4x \ln |x| \qquad v = -\frac{6}{x} + 4 - 16 \ln |x| \qquad w = \frac{3}{x^4} - \frac{4}{x} + \frac{8}{x} \ln |x|.$$

11. Let $x_j(t)$ be the distance travelled by the waggon $W_j (j = 1, 2, 3)$. Let m_j be the mass of the waggon W_j. Then for the vector x we have the system of differential equations

$$\ddot{x} = Ax, \qquad x = \begin{pmatrix} x_1 \\ x_2 \\ x^3 \end{pmatrix}; \qquad A = \begin{pmatrix} -k_1 & k_1 & 0 \\ k_2 & -2k_2 & k_2 \\ 0 & k_1 & -k_1 \end{pmatrix}; \qquad k_j = \frac{c}{m_j}.$$

The substitution $x = e^t$ leads to the characteristic equation $\det(A - \mu^2 E) = 0$ which has the roots

$$\mu_1 = i\sqrt{k_1}, \; \mu_2 = -i\sqrt{k_1}, \; \mu_3 = i\sqrt{k_1 + 2k_2}, \; \mu_4 = -i\sqrt{k_1 + 2k_2}, \; \mu_5 = \mu_4 = 0,$$

Putting $\begin{cases} \omega_1 = \sqrt{k_1}, \\ \omega_2 = \sqrt{k_1 + 2k_2} \end{cases}$ we obtain for the initial conditions

$$x_1(0) = x_2(0) = x_3(0) = 0, \qquad \dot{x}_1(0) = -v_0, \; \dot{x}_2(0) = \dot{x}_3(0) = 0$$

the solution

$$x_1 = -\frac{v_0}{2} \left[\frac{1}{\omega_1} \sin \omega_1 t + \frac{1}{8\omega_1} \sin 2\omega_1 t + \frac{3}{4} t \right]$$

$$x_2 = -\frac{v_0}{2} \left[-\frac{3}{8\omega_1} \sin 2\omega_1 t + \frac{3}{4} t \right]$$

$$x_3 = -\frac{v_0}{2} \left[-\frac{1}{\omega_1} \sin \omega_1 t + \frac{1}{8\omega_1} \sin 2\omega_1 t + \frac{3}{4} t \right].$$

This holds only while the buffer springs are compressed. At the time $t = \pi/\omega_1$ the buffer mechanisms between waggons 1 and 2, and also between waggons 2 and 3, are for the first time unstressed. The velocities at the time $t = \pi/\omega_1$, are

$$\dot{x}_1 = \frac{v_0}{16}, \qquad \dot{x}_2 = 0, \qquad \dot{x}_3 = -\frac{15}{16} v_0.$$

Thereafter, waggon 2 remains at rest; waggons 1 and 3 move away from it in opposite directions.

12. After differentiating once we obtain the following four equations:

$$\begin{aligned}
&ax + b\dot{x} && + c\ddot{x} && = 0\\
&Ax + B\dot{x} && + C\ddot{x} && = 0\\
&\dot{a}x + (a + \dot{b})\dot{x} && + (b + \dot{c})\ddot{x} + c\dddot{x} && = 0\\
&\dot{A}x + (A + \dot{B})\dot{x} && + (B + \dot{C})\ddot{x} + C\dddot{x} && = 0.
\end{aligned}$$

This system of homogeneous linear equations has non-trivial solutions $x, \dot{x}, \ddot{x}, \dddot{x}$ if the determinant of the coefficients vanishes:

$$\begin{vmatrix} a & b & c & 0 \\ A & B & C & 0 \\ \dot{a} & a+\dot{b} & b+\dot{c} & c \\ \dot{A} & A+\dot{B} & B+\dot{C} & C \end{vmatrix} = 0.$$

This condition is necessary and sufficient, as can be seen by reversing the argument.

13. The given equation is equivalent to the system

$$w' = Aw \quad \text{with } \omega = \begin{pmatrix} y \\ y' \end{pmatrix}, \qquad A = \begin{pmatrix} 0 & 1 \\ -(\mu - p) & 0 \end{pmatrix};$$

A fundamental system in the interval $0 \leq x < 1$ is

$$W_a = \begin{pmatrix} \cos ax & \frac{1}{a}\sin ax \\ -a\sin ax & \cos ax \end{pmatrix} \quad \text{with } a^2 = \mu - 1$$

and in the interval $1 \leq x < 3$ is

$$W_b = \begin{pmatrix} \cos b(x-1) & \frac{1}{b}\sin b(x-1) \\ -b\sin b(x-1) & \cos b(x-1) \end{pmatrix} \text{with } b^2 = \mu;$$

if we write

$$y_0 = y(0), \qquad y_0' = y'(0), \qquad y_3 = y(3), \qquad y_3' = y'(3) \quad \text{and} \quad \begin{pmatrix} y_3 \\ y_3' \end{pmatrix} = B\begin{pmatrix} y_0 \\ y_0' \end{pmatrix},$$

then the transition matrix $B = (b_{jk})$ can be calculated from

$$B = W_b(3)\, W_a(1)$$

$$B = \begin{pmatrix} \cos 2b & \frac{1}{b}\sin 2b \\ -b\sin 2b & \cos 2b \end{pmatrix} \begin{pmatrix} \cos a & \frac{1}{a}\sin a \\ -a\sin a & \cos a \end{pmatrix}$$

$$= \begin{pmatrix} (\cos a)(\cos 2b) - \frac{a}{b}(\sin a)(\sin 2b) & \frac{1}{a}(\sin a)(\cos 2b) + \frac{1}{b}(\cos a)(\sin 2b) \\ -b(\cos a)(\sin 2b) - a(\sin a)(\cos 2b) & -\frac{b}{a}(\sin a)(\sin 2b) + (\cos a)(\cos 2b \end{pmatrix}$$

Let $w(x+3) = kw(x)$, then k is obtained from $\det(B - kE) = 0$ or

$$\begin{vmatrix} b_{11} - k & b_{12} \\ b_{21} & b_{22} - k \end{vmatrix} = 0 = k^2 - k(b_{11} + b_{22}) + \underbrace{b_{11}b_{22} - b_{12}b_{21}}_{=1}$$

$$b_{11} + b_{22} = 2(\cos\sqrt{\mu - 1})(\cos 2\sqrt{\mu}) - (\sin\sqrt{\mu - 1})(\sin 2\sqrt{\mu})\,\frac{2\mu - 1}{\sqrt{\mu(\mu - 1)}}.$$

Distinction of cases:

$|b_{11} + b_{22}| \geq 2$, k real, in particular for $|b_{11} + b_{22}| = 2$ we have $k = \pm 1$;
$|b_{11} + b_{22}| < 2$, k complex.

In Fig. VI.8 $(b_{11} + b_{22})$ has been plotted as a function of μ. Periodic solutions exist for $\mu_1 = 0.30$, $\mu_4 = 4.58$, $\mu_5 = 4.88, \ldots$; semi-periodic solutions exist for $\mu_2 = 1.17$, $\mu_3 = 1.68, \ldots$.

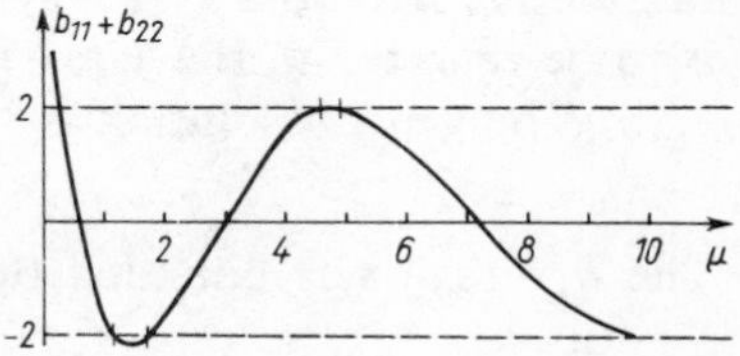

Fig. VI.8. On the construction of periodic solutions of the Schrödinger equation in Example 13

14. The equation of motion is

$$\frac{\gamma F l}{g}\ddot{x} = \gamma F x - f\gamma F(l - x)$$

or

$$\frac{l}{g}\ddot{x} = x(1 + f) - fl.$$

If the motion is to begin at $t = 0$, the relation

$$a(1 + f) - fl > 0$$

must hold for the force acting. The solution of the differential equation which satisfies the initial conditions $x(0) = a$, $\dot{x}(0) = 0$ is

$$x - \frac{fl}{1 + f} = \left(a - \frac{fl}{1 + f}\right)\cosh\sqrt{\frac{g}{l}(1 + f)}\,t.$$

Solved for the time t it gives

$$t(x, a) = \sqrt{\frac{l}{g}\frac{1}{1 + f}}\,\operatorname{arcosh}\left[\frac{x - \dfrac{fl}{1 + f}}{a - \dfrac{fl}{1 + f}}\right],$$

or, with the values substituted and with x and a measured in metres,

$$t(x, a) = 0.3044 \operatorname{arcosh} \frac{x - \frac{1}{11}}{a - \frac{1}{11}} .$$

15. (a) The general solution

$$y = \left[\frac{2}{\alpha}(c_1 - x)\right]^{1/2} + c_2$$

is a two-parameter family of parabolas, with c_1, c_2 as free parameters. The boundary-value problem has precisely one solution if the point $Q = (h, b)$ lies in the interior of a parabola P which is the locus of the vertices of the parabolas going through the point $(0, a)$ which are solutions of the differential equation, see Fig. VI.9. The boundary-value problem has no solution if Q lies outside the parabola P (always in the region $h > 0$; if $h < 0$, then the figure is to be reflected in the y-axis).

(b) The general solution is

$$x = y^3 + c_1 y + c_2$$

The h, b-region is bounded by the two curves $x = y^2$ and $x = -2y^2$, Fig. VI.10.

(c) The solutions of the differential equation are

$$y = [2(c_1 - x)]^{1/2} - x + c_2 \quad \text{and} \quad y = c - x.$$

The h,b-region is bounded by the curves

$$y + x = (\pm 2x)^{1/2}, \text{ Fig. VI.11.}$$

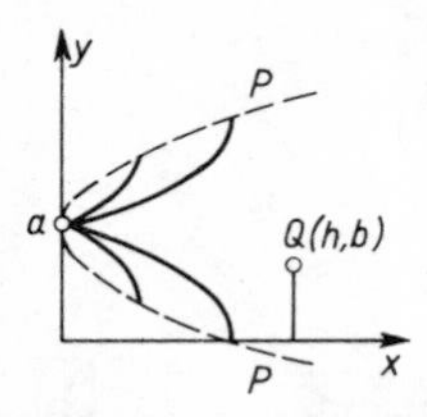

Fig. VI.9. The domains of solubility of the boundary-value problem No. 15(a)

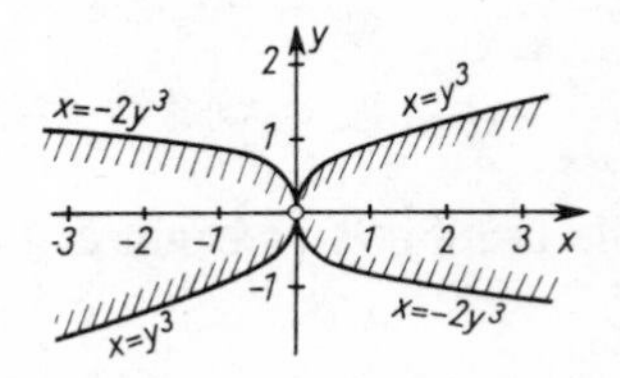

Fig. VI.10. The domain of solubility of the boundary-value problem No. 15 (b)

16. (a) The general solution of the differential equation is

$$y = \ln(2c_1^2) - 2 \ln \cosh(c_1 x + c_2),$$

with c_1, c_2 as free parameters. If we fit the parameters to the initial values $y(0) = 0$, $y'(0) = s$ we obtain

$$y = \ln\left(1 + \frac{s^2}{2}\right) - 2 \ln \cosh\left[\left(\frac{1}{2} + \frac{s^2}{4}\right)^{1/2} x + \operatorname{arsinh}(s/\sqrt{2})\right];$$

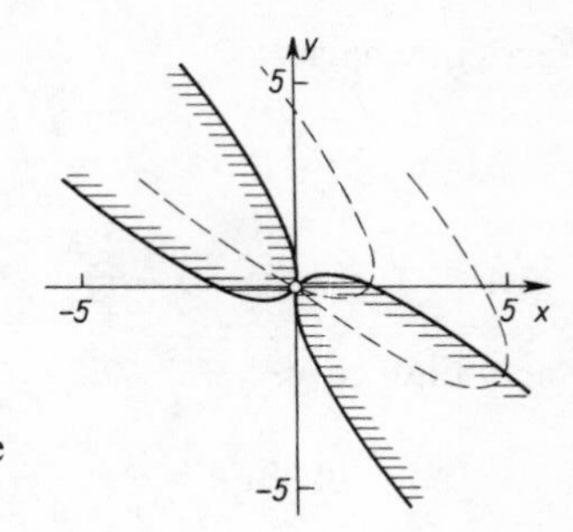

Fig. VI.11. The domain of solubility of the boundary-value problem No. 15(c)

for various s this gives the family of curves sketched in Fig. VI.12, with an envelope C. If the point $P = (a, b)$ lies between the y-axis and the curve C, the boundary-value problem has two solutions, of which the one with the smaller value of y corresponds to the stable temperature distribution, and the one with the larger value of y corresponds to the unstable temperature distribution. If P lies on C, the boundary-value problem has one solution only; and if P lies to 'the right' of C, the boundary-value problem is insoluble.

(b) In the general solution (see the previous problem) c_2 has to be found from the equation

$$a = 2\sqrt{2}\,\frac{c_2}{\cosh c_2};$$

we then have

$$c_1 = -\frac{2}{a}\,c_2.$$

The maximum value a is assumed for about $c_2 = 1.2$ and has a value of about 1.87.

17. (a) For $l\sqrt{\lambda} = lk = p$ we obtain

$$a\beta - b\alpha + c\delta - d\gamma + [a\delta - d\alpha + c\beta - b\gamma]\cos p + \left[\frac{a\gamma - c\alpha}{p} + (b\delta - d\beta)p\right]\sin p = 0.$$

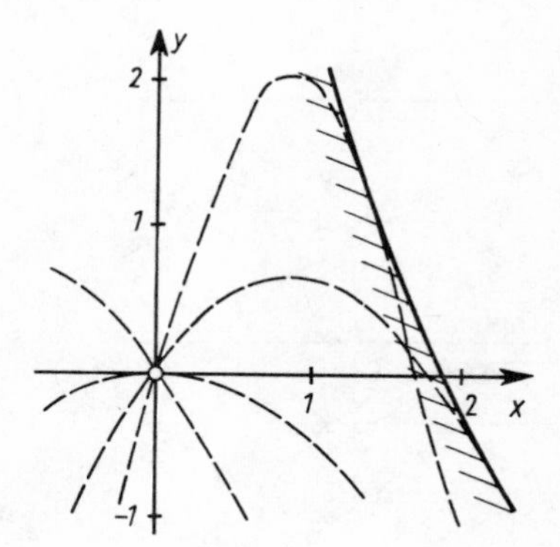

Fig. VI.12. For the temperature-distribution problem in No. 16

Particular cases:

$$\gamma = \delta = a = b = 0\colon \tan p = \frac{(c\beta - d\alpha)p}{d\beta p^2 + c\alpha}; \quad \beta = \delta = a = c = 0\colon \cos p = -\frac{b\alpha + d\gamma}{b\gamma + d\alpha}.$$

(b) $$k(a - b - 1) + (1 + abk^2)\sin k + (b - a)k \cos k = 0.$$

Particular cases:

$$a = b\colon \sin k = \frac{k}{1 + a^2k^2}; \qquad a = b + 1\colon \tan k = \frac{k}{1 + (a^2 - a)k^2}.$$

18. The eigencurves have the equations

$$\sin \frac{\rho}{2} = 0, \quad \text{where } \lambda = \rho^2,$$

i.e.,

$$\rho = 2k\pi \qquad (k = 1, 2, \ldots) \quad \text{and} \quad \mu \tan\left(\frac{\rho}{2}\right) = \rho.$$

See Fig. VI.13 in which the corresponding values of k are given.

(a) There is a countable sequence of positive eigenvalues λ.

(b) As in (a), but there is, additionally, a negative eigenvalue.

(c) As in (b); here there are two eigenfunctions, which have no nodes in a basic interval (the eigencurve with $k = 0$ is intersected twice by the dotted curve).

(d) there is a countable sequence of both positive and negative eigenvalues μ.

(e) As in (d), but this time there is no eigenfunction without nodes.

19. The general solution of the differential equation is $y = c_1 e^{\lambda x} + c_2$. The eigencurves are $e^{\lambda}(1 - \mu\lambda) = 1$; for each (complex) λ there corresponds one definite μ, but to each (complex) μ there corresponds a countable infinity of values $\lambda = \lambda_n$, which lie on the easily discussed curve $|e^{\lambda}|\,|1 - \mu\lambda| = 1$.

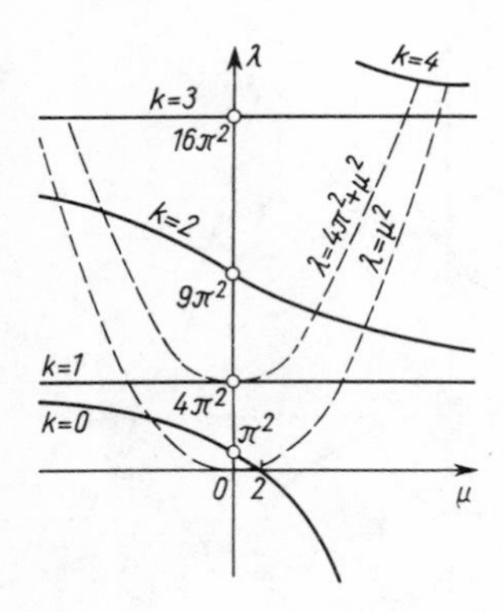

Fig. VI.13. 2-parameter eigenvalue problems

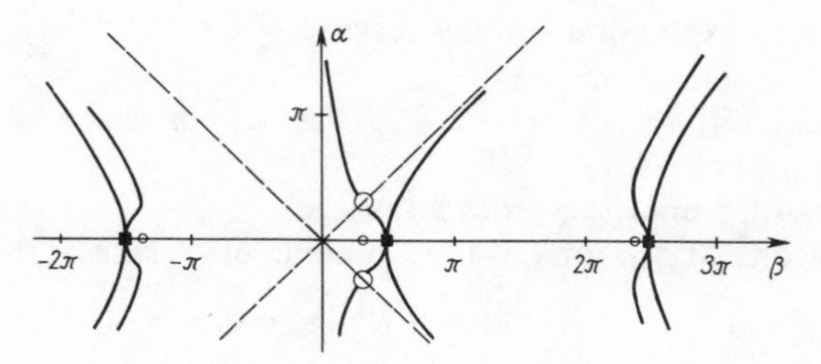

Fig. VI.14. The eigenvalue curves

20. The condition for eigenvalue pairs α, β:

$$|\alpha| > |\beta| \qquad \beta \sinh \sqrt{\alpha^2 - \beta^2} = \sqrt{\alpha^2 - \beta^2},$$
$$|\alpha| < |\beta| \qquad \beta \sinh \sqrt{\beta^2 - \alpha^2} = \sqrt{\beta^2 - \alpha^2}.$$

Limit cases: $\beta = 1$, $\alpha = \pm 1$ Figure VI.14 shows the position of the first few eigencurves.

11. Miscellaneous problems on chapters IV and V

1. Give the sum in closed form (as in (IV.79) of the following hypergeometric series:

(a) $F(1, 1, 1, z)$ (b) $F(1,1,2, -z)$

(c) $F(2, 2, 2, z)$ (d) $F(1, 3, 2, z)$

2. Show that the Legendre polynomial $P_n(x)$ can be expressed by means of a hypergeometric series as

$$\Phi(z) = F\left(-\frac{n}{2}, \frac{1-n}{2}, \frac{1-2n}{2}, z^{-2}\right)$$

3. For the spherical harmonic $P_n(z)$, as in (IV.29), calculate the integral

$$J = \int_{-1}^{1} z P_n(z) P_n'(z)\, dz.$$

4. (a) Check that the series (IV.35) for the Bessel functions converges rapidly for small values of $|z|$ by calculating the series for $J_0(2)$ to 12 decimal places.

(b) Show that for large $|z|$ the rate of convergence falls off, and calculate $J_0(10)$ to seven decimal places.

5. (a) Evaluate the sum $f(z) = \sum_{n=0}^{\infty} (2n+1) J_{2n+1}(z)$ by differentiating with respect to φ the formula for sin (z sin φ) obtained after (IV.51), and then putting $\varphi = 0$.

(b) In the same way obtain an expression for $g(z) = \sum_{n=1}^{\infty} (-1)^{n+1} (2n)^2 J_{2n}(z)$ from the formula for cos (z sin φ).

6. Give a closed expression for the series

$$f(z) = J_0(-i) + 2\sum_{n=1}^{\infty} i^n J_n(-i)\cos nz.$$

For what values of z does the series converge?

7. The vibrations of a homogeneous circular plate satisfy the equation

$$(\Delta - \lambda^2)[(\Delta + \lambda^2)w] = 0,$$

where, for rotationally symmetric vibrations, w is to be regarded as a function of the distance r from the centre of the disc, and then

$$\Delta = \frac{\partial^2}{\partial r^2} + \frac{1}{r}\frac{\partial}{\partial r}.$$

λ^2 is proportional to the circular frequency. Calculate the first few eigenvalues λ

(a) for a circular plate with its circumference clamped (at the boundary $r = a$, the conditions are

$$w = \frac{\partial w}{\partial r} = 0)$$

(b) for a circular plate secured pivotally round its circumference (at the boundary $r = a$ the conditions are

$$w = \frac{\partial^2 w}{\partial_r^2} + \frac{v}{r}\frac{\partial w}{\partial r} = 0;$$

discuss the case $\nu = 0$).

8. Expand the function

$$f(z) = \frac{1}{\sqrt{1 - z^2}}$$

in $|z| < 1$ in terms of the Legendre polynomials, $f(z) = \sum_{n=0}^{\infty} a_n P_n(z)$.

9. Set up partial differential equations for the following families of functions, where $w_1, w_2, \ldots$ are arbitrary functions of the given arguments:

(a) $z(x, y) = w_1(x)w_2(y)$

(b) $z(x, y) = w_1(x)w_2(x + y)$

(c) $z(x, y) = w_1(x) + w_2(xy)$

(d) $z(x, y) = w_1(x + y)w_2(x - y)$

(e) $z(x, y) = [w_1(x) + w_2(y)]^{1/2}$

(f) $z(x, y) = \varphi(y)w_1(x) + w_2(x\varphi(y))$ with a fixed function $\varphi(y) \neq$ constant.

10. In the system

$$u_t = au + Av_x$$
$$v_t = bu + Bw_x$$
$$w_t = cu + Cu_x,$$

let a, b, c, A, B, C be constants. Eliminate two of the three unknown functions

$$u(x, t), v(x, t), w(x, t).$$

11. Solve the Tricomi differential equation

$$u_{xx} - xu_{yy} = 0,$$

which arises in fluid dynamics (on passage through the velocity of sound), by substituting into it a power series $u(x, y) = \sum_{p,q} a_{pq} x^p y^q$.

12. Find the general solutions (including two arbitrary functions) of the partial differential equations:

$$\text{(a)}\ yu_{yy} - u_y = 2x, \qquad \text{(b)}\ xu_{xy} + yu_{yy} + u_y = 12xy^2.$$

13. For the propagation of a one-dimensional surface wave in shallow water the non-linear differential equation

$$h_t + \sqrt{gH}\left(1 + \frac{3h}{2H}\right) h_x = 0$$

arises; here g is the acceleration of gravity, $h = h(x, t)$ is the equation for the surface of the water at time t in the co-ordinate system shown in Fig. VI.15. If the initial state is

$$h(x,0) = \begin{cases} -a \sin x/L & \text{for } |x/L| \leq \pi \\ 0 & \text{for } |x/L| \geq \pi, \end{cases}$$

then after a time T the wave breaks, see Fig. VI.16. Calculate T.

14. By a product substitution solve the eigenvalue problem $\Delta u + \lambda u = 0$ for the interior of a quadrant of a cylinder with the boundary value u. In cylindrical co-ordinates the quadrant of the cylinder is given by:

$$0 \leq r \leq 1, \qquad 0 \leq z \leq 1, \qquad 0 \leq \theta \leq \pi/2$$

(see Fig. VI.17).

15. Solve the telegraphy equation $u_{xx} = u_{tt} + 2u_t + u$ for the region

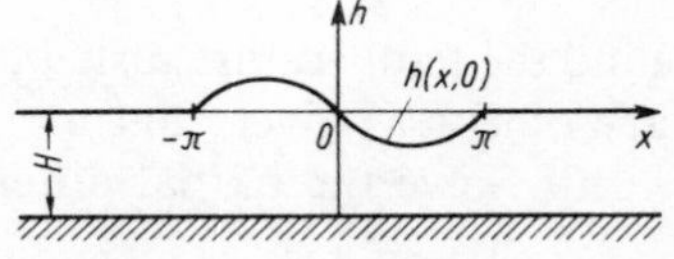

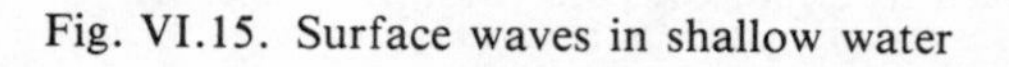
Fig. VI.15. Surface waves in shallow water

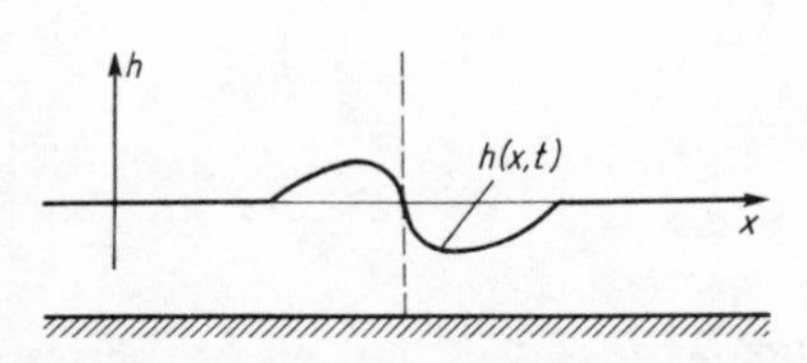

Fig. VI.16. Breaking of a wave

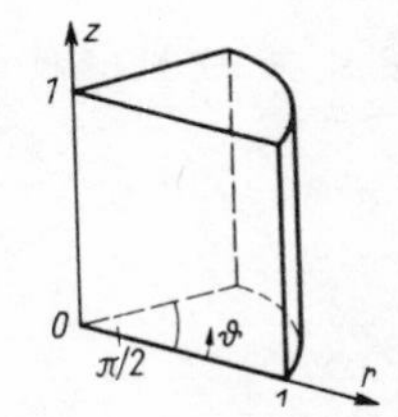

Fig. VI.17. An eigenvalue problem for a quadrant of a cylinder

$0 \leq x \leq n\pi$ ($n = 1, 2, \ldots$) with the initial and boundary conditions

$$u(0,t) = u(n\pi, t) = 0 \quad \text{for } t \geq 0$$
$$\left.\begin{aligned} u(x,0) &= \sin x \\ u_t(x,0) &= 0 \end{aligned}\right\} \quad \text{for } 0 \leq x \leq n\pi.$$

16. Find the eigenvalues and eigenfunctions for the eigenvalue problem

$$\Delta u + \lambda u = u_{xx} + u_{yy} + \lambda u = 0$$

for the right-angled isosceles triangle given by $x \geq 0$, $y \geq 0$, $x + y \leq 1$ (boundary conditions $u = 0$). Find the solution of the eigenvalue problem by superimposition of eigenfunctions for the square.

17. Find the differential equations of the family of spheres which touch the planes $u = \pm x$.

There are two different possibilities for the position of the sphere (see Fig. VI.18):

(a) $u > x$ (b) $u < x$.

18. Find the solutions $u(x, y)$ of the partial differential equations

(a) $xyu_x - u_y = 1$ (b) $u_{xyy} - u_x + xy = 0$
(c) $u_x - 2u_y + 2u = x$ (d) $u_x u_{xy} = x + y$.

19. The temperature distribution in the interior of the region bounded by the ellipse E:

$$\frac{x^2}{4} + y^2 = 1$$

satisfies Laplace's equation

$$u_{xx} + u_{yy} = 0.$$

Find the temperature distribution in the interior of the ellipse if on the boundary of E it is given that $u = x^n$ ($n = 2, ,3, 4, 5$).

20. Solve the partial differential equation (the torsion problem)

$$\Delta u = -1; \qquad u = 0 \text{ on the boundary}$$

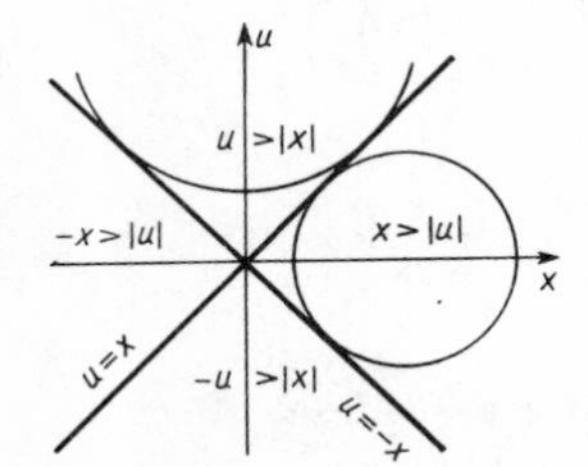

Fig. VI.18. The different positions in space of spheres touching two planes

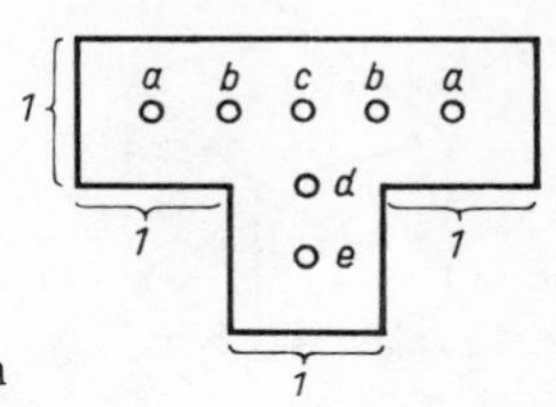

Fig. VI.19. A torsion problem

for the region shown in Fig. VI.19 by using the method of difference-quotients, with a mesh-width of

(a) $h = 1/2$ (b) $h = 1/3$.

21. Solve the boundary-value problem

$$(1 + x)y'' + y = x, y(0) = y(1) = 0$$

(a) exactly, by means of the Bessel function J_1 and the Neumann function N_1;

(b) by the Ritz method, using the substitution $y \approx u = (a_1 + a_2 x)x(x - 1)$.

22. Find an approximation solution of the problem $y' + \sinh y = \sin x, y(0) = 2$ by the Runge–Kutta method, with the step-width $h = 0.1$ up to the point $x = 0.8$ and from there onwards with the step-width $h = 0.2$.

23. Treat the vibration problem $\ddot{x} + \dot{x} + \sin x = 0, x(0) = \pi/2, \dot{x}(0) = 0$ by the Runge–Kutta method.

24. In the treatment of vehicle suspensions the differential equation

$$\ddot{x} + 2D\dot{x} + \frac{1}{K}(e^{Kx} - 1) = 0$$

arises; if we put $\dot{x}^2 = y$ this becomes

$$y' = \frac{dy}{dx} = -4Dy^{1/2} - \frac{2}{K}(e^{Kx} - 1);$$

Take $D = \frac{1}{4}, k = 1, y(0) = 1$, and apply the method of central differences with the step-width $h = 0.1$.

25. Solve the polar eigenvalue problem

$$y'' = \lambda y \sin x, y(0) = y(2\pi) = 0$$

by the method of difference-quotients in the improved form of the multi-point method, as in (VI.13); use the step-width $h = \pi/2$.

26. Find an analytic solution (i.e., the solution which can be expanded as a convergent power-series) of the functional-differential equation

$$w'(x) = w(x^2)$$

with the initial condition $w(0) = a$, and examine its convergence behaviour.

27. In the eigenvalue problem (IV.59) for the buckling of a bar under its

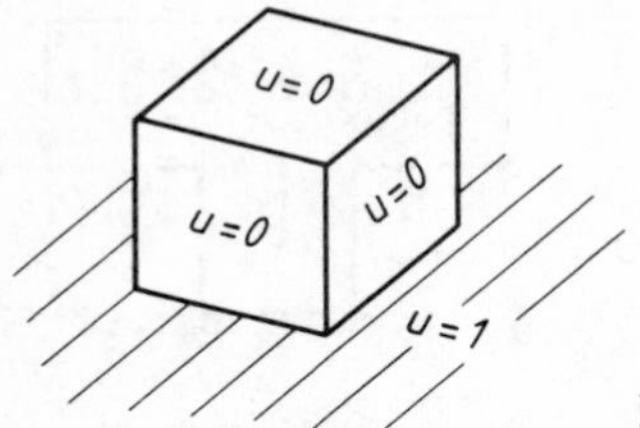

Fig. VI.20. Temperature distribution inside a cube

own weight

$$y'' + \lambda xy = 0, \qquad y'(0) = y(1) = 0$$

apply the Ritz method of §1, No. 6 with the substitution

$$w(x) = a_0(1 - x^3) + a_1(1 - x^6)$$

and the integral $I[\varphi]$ of (VI.20).

28. In an x, y, z co-ordinate system let B be the interior of the cube

$$0 < x < 1, 0 < y < 1, 0 < z < 1.$$

Suppose the base $z = 0$ is kept at the temperature $u = 1$, while the external temperature $u = 0$ prevails on the other five faces (see Fig. VI.20). The steady-state temperature in the interior satisfies Laplace's equation $\Delta u = 0$. Calculate the temperature u at the centre P of the cube ($x = y = z = 1/2$).

29. In problem 28 calculate the temperature distribution by the usual method of difference-quotients, taking the mesh-width to be: (a) $h = 1/2$ and (b) $h = 1/4$.

30. Calculate the Fourier transforms (spectral functions)

$$F(\omega) = \frac{1}{2\pi} \int_{-\infty}^{\infty} e^{-ix\omega} f(x)\, dx$$

for the functions:

(a) $f(x) = \begin{cases} e^{-x} & \text{for } x > 0 \\ 0 & \text{for } x < 0 \end{cases}$ (b) $f(x) = \dfrac{1}{\sqrt{|x|}}$

(c) $f(x) = \dfrac{1}{\sqrt{|x + c|}} - \dfrac{1}{\sqrt{|x - c|}}$ (d) $f(x) = \begin{cases} (c - |x|/2) & \text{for } |x| < 2c \\ 0 & \text{for } |x| > 2c \end{cases}$

12. Solutions

1. (a) $\dfrac{1}{1 - z}$ (b) $\dfrac{1}{z} \ln(1 + z)$ (c) $\dfrac{1}{(1 - z)^2}$ (d) $\dfrac{1}{2} \dfrac{2 - z}{(1 - z)^2}$.

2. $P_n(z) = a_n z^n \Phi(z)$ with $a_n = \dfrac{(2^n)!}{2^n (n!)^2}$.

3. $J = \dfrac{2n}{2n+1}$.

4. (a) $J_0(2) = 1 - \dfrac{1}{(1!)^2} + \dfrac{1}{(2!)^2} - \dfrac{1}{(3!)^2} + \ldots$

The calculation of the successive terms of the series, which are obtained from one another by division by $1^2, 2^2, 3^2, \ldots$ gives

$$
\begin{array}{l}
1 \\
-1 \\
+0.25 \\
-0.027\ 777\ 777\ 778 \\
+0.001\ 736\ 111\ 111 \\
-\ \ldots .\ 069\ 444\ 444 \\
+\ \ldots ..1\ 929\ 012 \\
-\ \ldots \ldots .39\ 368 \\
+\ \ldots \ldots\ 615 \\
-\ \ldots \ldots ..8 \\
\hline
J_0(2) = 0.223\ 890\ 779\ 140
\end{array}
$$

Missing decimal places are to be filled by zeros; the last decimal place is, of course, uncertain.

(b) Similarly one obtains

$$
\begin{array}{l}
1 \\
-25 \\
+156.25 \\
-434.027\ 777\ 778 \\
+678.168\ 402\ 778 \\
-678.168\ 402\ 778 \\
+470.950\ 279\ 707 \\
+240.280\ 754\ 953 \\
+.93.859\ 669\ 904 \\
-.28.969\ 033\ 922 \\
+..7.242\ 258\ 481 \\
-..1.496\ 334\ 397 \\
+....259\ 780\ 277 \\
-.....38\ 429\ 035 \\
+\ \ldots 4\ 901\ 663 \\
+\ \ldots .\ 544\ 629 \\
+\ \ldots ..53\ 186 \\
-\ \ldots ...4\ 601 \\
+\ \ldots \ldots\ 355 \\
+\ \ldots \ldots .25 \\
+\ \ldots \ldots ..2 \\
\hline
J_0(10) = -0.245\ 935\ 765
\end{array}
$$

The last two decimal places are now uncertain.

5. (a) $f(z) = \frac{1}{2} z$, (b) $g(z) = \frac{1}{2} z \sin z$.

6. $f(z) = e^{\cos z}$; the series converges for all complex z.

7. (a) The first few zeros of the equation

$$J_0(i\lambda a)\, J_0'(1a) - iJ_0(\lambda a)\, J_0'(i\lambda a) = 0$$

(I. Szabo, Höhere technische Mechanik, 4th ed. (Berlin–Heidelberg, New York, 1964, p. 192) are $\lambda a = 3.196$; 6.306; 9.439.

(b) Putting $I_0(\dot{z}) = J_0(iz), iI_1(z) = J_1(iz)$ one obtains the equation (Szabo, loc. cit, p. 238)

$$J_0(vI_1 + \lambda a I_0 - I_1) + I_0(v\, J_1 + \lambda a J_0 - J_1) = 0$$

(the argument is always λa); for $v = 0$ the first few zeros are

$$\lambda a + 2.108;\ 5.419;\ 8.592.$$

8. $$a_0 = \frac{\pi}{2}, \qquad a_1 = 0, \qquad a_{2n} = \frac{\pi}{2}(4n+1)\frac{[1 \cdot 3 \cdot 5 \cdot \cdot (2n-1)]^4}{(2n)!(2n)!},$$
$$a_{2n+1} = 0 \ (n = 1, 2, \ldots).$$

9. (a) $zz_{xy} = z_x z_y$ or $\dfrac{\partial^2 \ln z}{\partial x \partial y} = 0$ (b) $z(z_{xy} - z_{yy}) = z_y(z_x - z_y)$

(c) $xz_{xy} - yz_{yy} - z_y = 0$ (d) $z(z_{xx} - z_{yy}) = {z_x}^2 - {z_y}^2$

(e) $zz_{xy} + z_x z_y = 0$

(f) $\varphi' \varphi^2 z_{yy} - x\varphi\varphi'^2 z_{xy} + x\varphi'^3 z_x - \varphi^2\varphi'' z_y = 0.$

10. u, v, w all satisfy the same differential equation of the third order

$$\varphi_{ttt} = a\varphi_{tt} + Ab\varphi_{xt} + ABc\varphi_{xx} + ABC\varphi_{xxx}.$$

11. The terms of the series are polynomials, which themselves are solutions of the differential equation. The first few terms are:

$$\begin{aligned}
u(x, y) &= a_{0,0} && + a_{0,1} y \\
&+ a_{1,0} x && + a_{1,1} xy \\
&+ a_{3,0}\left(x^3 + 3!\,\frac{y^2}{2!}\right) && + a_{3,1}\left(x^3 y + 3!\,\frac{y^3}{3!}\right) \\
&+ a_{4,0}\left(x^4 + 4!\,\frac{xy^2}{1!2!}\,\frac{1}{2}\right) && + a_{4,1}\left(x^4 y + 4!\,\frac{xy^2}{1!3!}\,\frac{1}{2}\right) \\
&+ a_{6,0}\left[x^6 + 6!\left(\frac{x^3 y^2}{3!2!}\,\frac{1}{4} + \frac{y^4}{4!}\right)\right] && + a_{6,1}\left[x^6 y + 6!\left(\frac{x^3 y^3}{3!3!}\,\frac{1}{4} + \frac{y^5}{5!}\right)\right] + \cdots
\end{aligned}$$

12. (a) $u = w_1(x)y^2 + w_2(x) + 2xy$.

(b) By putting $z = y/x$ one obtains the differential equation $u_{xz} = 12t^3z^2$. Its general solution is $u = w_1(x) + w_2(x) + t^4z^3$. So the solution of the given equation is

$$u = w_1\left(\frac{y}{x}\right) + w_2(x) + xy^2.$$

13. First we set up the differential equation for $x/L = \varphi(h, t)$:

$$\dot{\varphi} = \sqrt{gH}\left(1 + \frac{3h}{2H}\right).$$

The general solution of this differential equation is

$$x/L = \varphi(h, t) = \sqrt{gH}\left(1 + \frac{3h}{2H}\right)t + w(h).$$

$w(h)$ is to be chosen so that the initial conditions are satisfied. For the region

$$\left|\arcsin\frac{h}{a}\right| \leq \pi$$

the solution is

$$x/L = \varphi(h, t) = \sqrt{gH}\left(1 + \frac{3h}{2H}\right)t - L\arcsin\left(\frac{h}{a}\right).$$

The time when the wave breaks is found from $t = T$ for

$$h = 0, \frac{\partial x}{\partial h} = 0: \frac{\partial x}{\partial h} = \frac{3}{2}\sqrt{\frac{g}{H}}\,t - \frac{L}{\sqrt{a_2 - h^2}};$$

for $h = 0$ this gives

$$h = 0: \frac{3}{2}\sqrt{\frac{g}{H}}\,t - \frac{L}{a} = 0.$$

Hence,

$$T = \frac{2}{3}\frac{L}{a}\sqrt{\frac{H}{g}}.$$

14. If $\alpha_{n,k}$ are the zeros of the Bessel function $J_{2k}(x)$ for $k = 0, 1, 2, \ldots$, $n = 1, 2, \ldots$, we obtain as the eigenfunctions

$$u_{nkm}(r, \theta, z) = J_{2k}(\alpha_{nk}^2 r)(\sin 2k\theta)(\sin \pi mz)$$

and the corresponding eigenvalues are

$$\lambda = \lambda_{nkm} = \pi^2 m^2 + \alpha_{nk}^2 \qquad (m, n, k = 1, 2, \ldots)$$

No. of the eigenvalue	$n\ k\ m$	α_{nk}	λ_{nkm}	$r=r_0$	$\theta=\theta_0$	$z=z_0$
1	1 1 1	5.135	36.24	—	—	—
2	1 1 2	5.135	65.84	—	—	1/2
3	1 2 1	7.586	67.42	—	$\pi/4$	—
4	2 1 1	8.417	80.71	0.61	—	—
5	1 2 2	7.586	97.02	—	$\pi/4$	1/2
6	1 3 1	9.98	109.47	—	$\pi/6$, $\pi/3$	—
7	2 1 2	8.417	110.32	0.61	—	1/2
8	1 1 3	5.135	115.19	—	—	1/3, 2/3
9	2 2 1	11.064	132.28	0.69	$\pi/4$	—
10	1 3 2	9.98	139.1	—	$\pi/6$, $\pi/3$	1/2

15. $u = e^{-t} \sin x(\sin t + \cos t)$.

16. For the square $0 \le x \le 1, 0 \le y \le 1$ the eigenfunction corresponding to the eigenvalue $\lambda_{mn} = \pi^2(m^2+n^2)$ $(m, n = 1, 2, \ldots)$ is $u_{mn} = (\sin m\pi x)(\sin n\pi y)$. The first few eigenfunctions for the right-angled isosceles triangle are then

$$\begin{aligned} u_{1,2}^* &= \sin \pi x \sin 2\pi y + \sin 2\pi x \sin \pi y && \lambda_{mn} \\ &= 2 \sin \pi x \sin \pi y \cos \frac{\pi}{2}(x+y) \cos \frac{\pi}{2}(x-y) && 5\pi^2 \\ u_{1,3}^* &= \sin \pi x \sin 3\pi y - \sin 3\pi x \sin \pi y && 10\pi^2 \\ u_{2,3}^* &= \sin 2\pi x \sin 3\pi y + \sin 3\pi x \sin 2\pi y && 13\pi^2 \end{aligned}$$

etc. Multiple eigenvalues also occur. For example, the eigenfunctions $u_{2,9}^*$ and $u_{6,7}^*$ correspond to the eigenvalue $85\pi^2$.

17. (a) $2x^2(u_x^2 + u_y^2 + 1) = (x + uu_x)^2$ (b) $2u^2(u_x^2 + u_y^2 + 1) = (x + uu_x)^2$.

18. In the following answers $w, w_1, w_2, \ldots$ denote arbitrary functions.

(a) $u = w\left(\ln|x| + \frac{y^2}{2}\right) - y$

(b) $u = \frac{1}{2}x^2 y + w_1(x)\, e^y + w_2(x)\, e^{-y} + w_3(y)$

(c) $u = \frac{1}{2}x - \frac{1}{4} + e^{-2x} w(2x+y)$

(d) $u = \int^x [2sy + y^2 + w_1(s)]^{1/2}\, ds + w_2(y)$.

19. $n = 2$: $u(x, y) = \frac{4}{5}(x^2 - y^2 + 1)$

$n = 3$: $u(x, y) = \frac{4}{7}x^3 - \frac{12}{7}xy^2 + \frac{12}{7}x$

$n = 4$: $u(x, y) = \frac{16}{41}x^4 + \frac{16}{41}y^4 - \frac{96}{41}x^2y^2 + \frac{416}{205}x^2 - \frac{416}{205}y^2 + \frac{336}{205}$

$n = 5$: $u(x, y) = \frac{16}{61}x^5 - \frac{160}{61}x^3y^2 + \frac{800}{427}x^3 - \frac{2400}{427}xy^2 + \frac{1840}{427}x$
$+ \frac{80}{61}xy^4.$

20. (a) $h = 1/2$ (notation for function values as in Fig. VI.19):

$$4a - b = \frac{1}{4} \qquad \text{Result:} \quad a = e = \frac{3}{32} = 0.093\,75$$
$$-a + 4b - c = \frac{1}{4} \qquad b = d = \frac{4}{32} = 0.125\,0$$
$$-2b + 4c - d = \frac{1}{4} \qquad c = \frac{5}{32} = 0.156\,25.$$
$$-c + 4d - e = \frac{1}{4}$$
$$-d + 4e = \frac{1}{4}$$

(b) $h = 1/3$ (notation for function values as in Fig. VI.21).

Result:

$a = \frac{5124}{72855} = 0.070\,33$ $\qquad e = \frac{5145}{72855} = 0.070\,62$

$b = 0.099\,60$ $\qquad f = 0.101\,04$

$c = 0.115\,90$ $\qquad g = 0.122\,82$

$d = 0.130\,08$ $\qquad h = 0.163\,23$

$i = 0.125\,67$ $\qquad j = 0.102\,68$ $\qquad k = 0.071\,26$

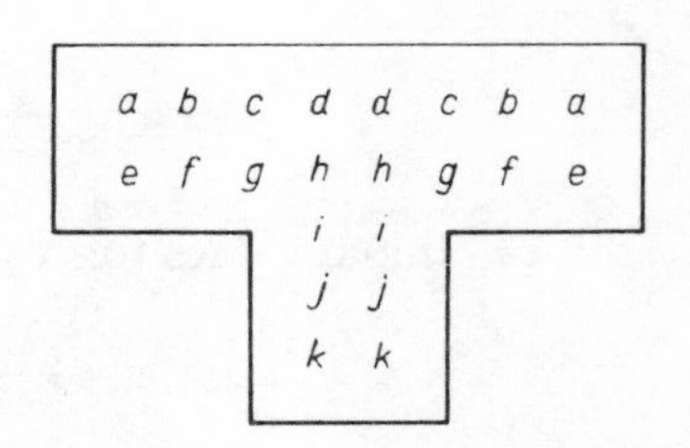

Fig. VI.21. Diagram for the method of differences with $h = 1/3$

21. (a) $y = x - \sqrt{1+x}\,[0.377\,7\,J_1(2\sqrt{1+x}) + 2.035\,8N_1(2\sqrt{1+x})]$

(b) $a_1 = 0.131\,7$; $a_2 = 0.077\,3$

Some numerical values:

x	y	u
0	0	0
0.25	−0.028 27	−0.025 0
0.50	−0.043 21	−0.042 6
0.75	−0.038 10	−0.036 5
1	0	0

22.

x	y	x	y
0	2	1	0.874 7
0.1	1.697 07	1.2	0.857 5
0.2	1.478 5	1.4	0.856 9
0.3	1.314 2	1.6	0.862 3
0.4	1.188 4	1.8	0.865 1
0.5	1.092 0	2	0.859 2
0.6	1.017 9	2.2	0.839 9
0.7	0.962 6	.	
0.8	0.922 1	.	
		3.6	0.236

23.

t	x	$\dot{x}$
0	1.570 80	0
0.3	1.529 95	−0.079 70
0.6	1.418 05	−0.136 53
0.9	1.257 8	−0.179 9
1.2	1.062 1	−0.206 9
1.5	0.849 5	−0.216 0
1.8	0.635 6	−0.208 3

24. Initial values for $x = \pm 0.1$ are obtained by means of the power series

$$y = 1 - x - \frac{3}{4}x^2 - \frac{1}{6}x^3 + \frac{1}{24}x^5 + \dots.$$

The following results are then obtained:

x	y
	1.092 67
0	1
0.1	0.892 33
0.2	0.768 68
0.3	0.628 15
0.4	0.470 07
0.5	0.294 51
0.6	0.103 55

25. Let y_j and y_j'' be the approximation values for $y(j\pi/2)$ and $y''(j\pi/2)$ $(j = 0, 1, 2, 3, 4)$ with $y_0 = y_4 = 0$.

Three equations at the points $\pi/2$, π, $3\pi/2$ are obtained, of which the first is

$$y_0'' + 10y_1'' + y_2'' - \frac{12}{h^2}(y_0 - 2y_1 + y_2) = 0.$$

If we substitute the differential equation for y'', then we have three linear homogeneous equations for y_j with the determinant

$$\begin{vmatrix} 10\Lambda + 24 & -12 & 0 \\ \Lambda - 12 & 24 & -\Lambda - 12 \\ 0 & -12 & -10\Lambda + 24 \end{vmatrix} = 0,$$

where $h^2\lambda = \Lambda$.

It follows that

$$\Lambda^2 = \frac{144}{55} \quad \text{or } \lambda = \pm 0.656.$$

26. The substitution

$$w(x) = \sum_{n=0}^{\infty} a_n x_n \quad \text{with } a_0 = a$$

gives

$$w(x) = a\left(1 + \frac{x}{1} + \frac{x^3}{1 \cdot 3} + \frac{x^7}{1 \cdot 3 \cdot 7} + \frac{x^{15}}{1 \cdot 3 \cdot 7 \cdot 15} + \dots\right)$$

$$= a\left[1 + \sum_{k=1}^{\infty} \frac{1}{1 \cdot 3 \cdot 7 \cdot \dots (2^k - 1)} x^{(2^k - 1)}\right],$$

This series converges for $|x| < 1$ and has the unit circle as its 'natural boundary'.

27. Professor J. Albrecht has supplied me with the following results:

$$I[w] = A_{00}a_0^2 + A_{01}a_0a_1 + A_{11}a_1^2$$

where

$$A_{00} = \frac{9}{5} - \frac{9}{40}\lambda, \qquad 2A_{01} = \frac{9}{2} - \frac{117}{220}\lambda, \qquad A_{11} = \frac{36}{11} - \frac{9}{28}\lambda.$$

The zeros Λ of

$$\psi(\lambda) = \begin{vmatrix} A_{00} & A_{01} \\ A_{01} & A_{11} \end{vmatrix} = 0 \qquad \text{i.e., of } 3\lambda^2 - 220\lambda - 1540 = 0,$$

are approximation values for the first two eigenvalues (error $\pm 0.005\%$)

$$\Lambda = \frac{2}{3}\{55 \mp \sqrt{1870}\} = \begin{cases} 7.837\,668 \\ 65.495\,664 \end{cases}$$

28. Let F_j $(j = 1, 2, \ldots, 6)$ be the six sides of the cube, and let $u_{(j)}$ be the temperature distribution when the side F_j has the temperature 1 and all the other five sides have the temperature 0 (see Fig. VI.22). If all these six temperature distributions $u_{(j)}$ are superimposed, then we have a cube with the temperature constant and $= 1$ on its whole surface. For the sum $v = \sum_{j=1}^{6} u_{(j)}$ we therefore have $v = 1$. Since, by symmetry, the value at P is the same for all the distribution $u_{(j)}$, it follows that $6u_{(j)}(P) = 1$, or $u(P) = 1/6$.

This deduction is possible for this special case because of the high symmetry of the region B, but is not possible for arbitrary regions.

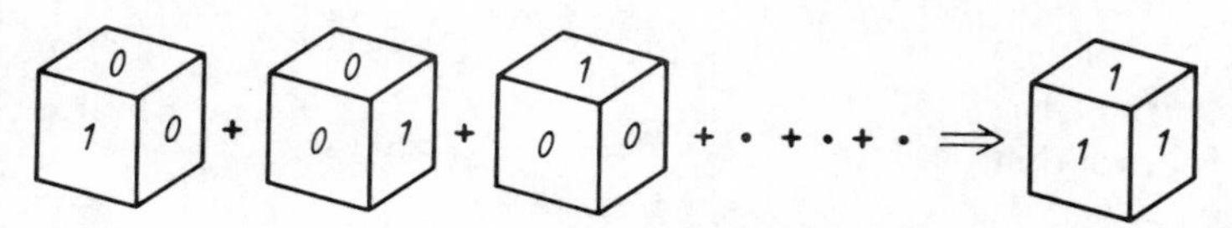

Fig. VI.22. For the symmetry of the temperature distribution

29. Analogously to the two-dimensional case in problem 20, the ordinary method of differences, with Laplace's equation, uses the approximation equation for each interior lattice point

$$u(Q) = \frac{1}{6} \cdot \text{(sum of the values at the neighbouring points)}.$$

With a mesh-width $h = 1/2$ we have only one unknown function value a in the interior, and for this we obtain the value

$$a = \frac{1}{6}(1 + 0 + 0 + 0 + 0 + 0) = \frac{1}{6}.$$

With a mesh-width $h = 1/4$, we have, taking into account the symmetry, nine unknown function values. Let $a, b, c, d, e, f, g, h, i$ be the approximation

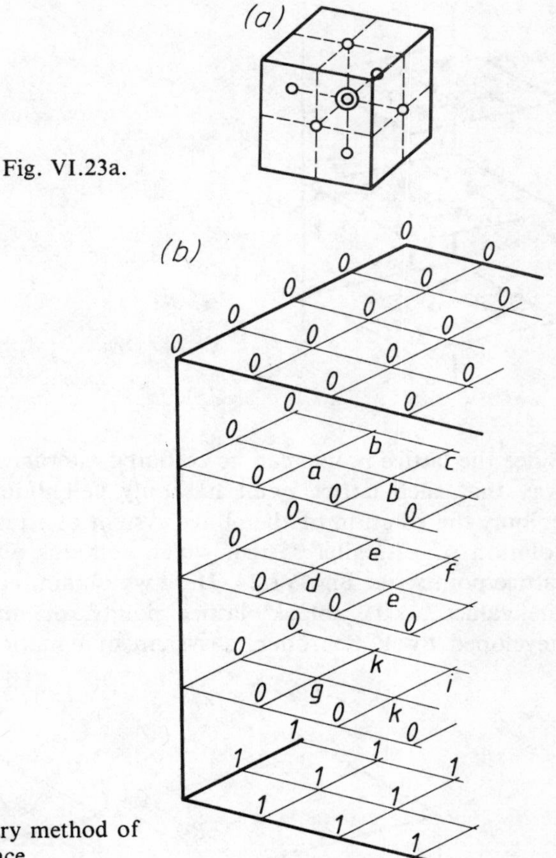

Fig. VI.23a.

Fig. VI.23b. For the ordinary method of differences in space

values for these function values, as shown in Fig. VI.23, which is probably understandable without further explanation. We then obtain the system of equations $Av = q$, where

$$A = \left[\begin{array}{ccc|ccc|ccc} 6 & -2 & 0 & -1 & 0 & 0 & 0 & 0 & 0 \\ -2 & 6 & -1 & 0 & -1 & 0 & 0 & 0 & 0 \\ 0 & -4 & 6 & 0 & 0 & -1 & 0 & 0 & 0 \\ \hline -1 & 0 & 0 & 6 & -2 & 0 & -1 & 0 & 0 \\ 0 & -1 & 0 & -2 & 6 & -1 & 0 & -1 & 0 \\ 0 & 0 & -1 & 0 & -4 & 6 & 0 & 0 & -1 \\ \hline 0 & 0 & 0 & -1 & 0 & 0 & 6 & -2 & 0 \\ 0 & 0 & 0 & 0 & -1 & 0 & -2 & 6 & -1 \\ 0 & 0 & 0 & 0 & 0 & -1 & 0 & -4 & 6 \end{array}\right], \quad v = \begin{bmatrix} a \\ b \\ c \\ d \\ e \\ f \\ g \\ k \\ i \end{bmatrix}, \quad e = \begin{bmatrix} 0 \\ 0 \\ 0 \\ 0 \\ 0 \\ 0 \\ 1 \\ 1 \\ 1 \end{bmatrix}$$

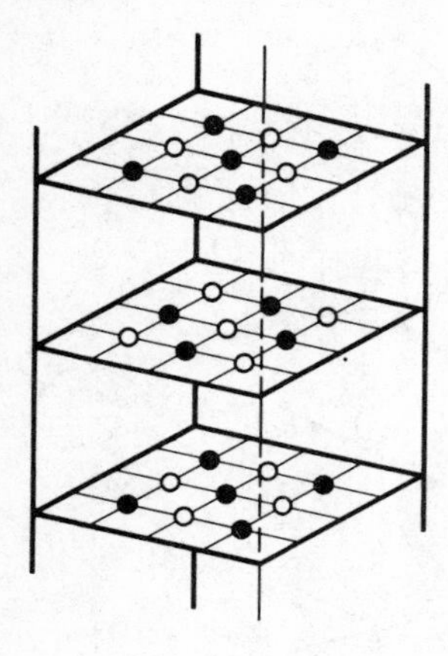

Fig. VI.24. Division of the lattice points into two classes

Since the lattice points can be coloured alternately black and white in such a way that each lattice point has only neighbouring points of the opposite colour, the solution of the above system of equations can be reduced to the solution of a smaller system which contains only the values at the 'white' lattice points; see Fig. VI.24. Here we obtain, either by direct elimination of the values at the 'black' lattice points, or, in accordance with a theory developed by J. Schröder, a system of equations in only four unknowns:

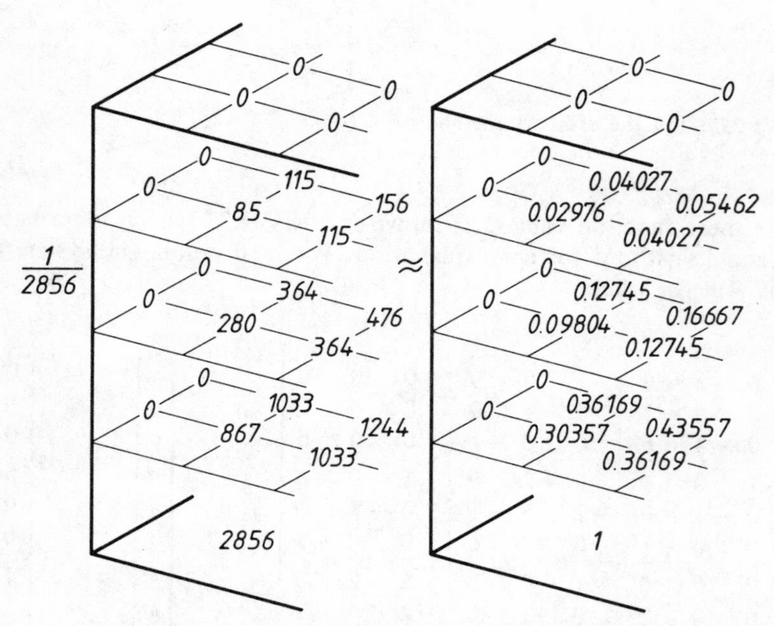

Fig. VI.25. Representation of the solution of the difference equations

$Bw = c$, where

$$B = \begin{pmatrix} -7 & 10 & 5 & -7 \\ -4 & 30 & -2 & -4 \\ -8 & -8 & 30 & -8 \\ 27 & -4 & -2 & -1 \end{pmatrix}, \qquad w = \begin{pmatrix} b \\ d \\ f \\ k \end{pmatrix} \qquad c = \begin{pmatrix} -1 \\ 1 \\ 1 \\ 0 \end{pmatrix}.$$

Here again we obtain at the centre point P of the cube the exact value

$$u = f = 1/6.$$

The solution of the difference equations with $a = 1/2856$ is shown pictorially in Fig. VI.25, with the same arrangement as in Fig. VI.23 (on the left with rational numbers, and on the right, rounded off to five decimal places).

30. (a) $F(\omega) = \frac{1}{2\pi}\int_0^\infty e^{-x} e^{-i\omega x}\,dx = \frac{1}{2\pi}\frac{1 - i\omega}{1 - \omega^2}$

(b) $F(\omega) = \frac{1}{\sqrt{2\pi\,|\,\omega\,|}}$ (c) $F(\omega) = i\sqrt{\frac{2}{\pi\,|\,\omega\,|}}\sin(c\omega)$

(d) $F(\omega) = \left[\frac{\sin(c\omega)}{\omega}\right]^2.$

§3. SOME BIOGRAPHICAL DATA

13. THE DATES OF SOME MATHEMATICIANS

The following list singles out (my own selection) a few of the mathematicians who have worked on differential equations or who are mentioned in this book.

Name	Lived:	Worked in:
Johannes Kepler	1571–1630	Prague, Linz
Isaac Newton	1642–1727	Cambridge, London
Gottfried Wilhelm Leibniz	1646–1716	Hannover
Jacob I. Bernoulli	1654–1795	Basel
Jacopo F. Riccati	1676–1754	Padua
Daniel Bernoulli	1701–1784	St. Petersburg, Basel
Leonhard Euler	1707–1783	St. Petersburg, Berlin
Alexis Claude Clairaut	1713–1765	France
Jean B. le Rond d'Alembert	1717–1783	France
Joseph L. Lagrange	1736–1813	Berlin, Paris
Gaspard Monge	1746–1818	Paris
Pierre S. Laplace	1749–1827	Paris
Adrien-Marie Legendre	1752–1833	Paris
Jean B. J. Fourier	1768–1830	Paris, Grenoble
Carl Friedrich Gauss	1777–1855	Göttingen

Simeon D. Poisson	1781–1840	Paris
Friedrich W. Bessel	1784–1846	Königsberg
Louis Navier	1785–1836	Paris
Augustin L. Cauchy	1789–1857	Paris
George Green	1793–1841	Cambridge
Gabriel Lamé	1794–1870	Russia, Paris
Joseph Plateau	1801–1883	Belgium
Jacques C. F. Sturm	1803–1855	Paris
Carl G. Jacobi	1804–1851	Köningsberg, Berlin
Lejeune Dirichlet	1805–1859	Berlin, Göttingen
William R. Hamilton	1805–1865	Dublin
Joseph Liouville	1809–1882	Paris
Karl Weierstraß	1815–1897	Berlin
George Gabriel Stokes	1819–1903	Cambridge
Pafnuti L. Chebyshev	1821–1894	St. Petersburg
Bernhard Riemann	1826–1866	Göttingen
Rudolf Lipschitz	1832–1903	Bonn
Carl Neumann	1832–1905	Leipzig
Immanuel L. Fuchs	1833–1902	Heidelberg, Berlin
Edmond Laguerre	1834–1886	Paris
Gaston Darboux	1842–1917	Paris
J. W. S. Rayleigh	1842–1919	Cambridge, London
Hermann Amandus Schwarz	1843–1921	Berlin
Oliver Heaviside	1850–1925	England
Carl Runge	1856–1927	Hannover, Göttingen
David Hilbert	1862–1943	Göttingen
Ivar Fredholm	1866–1927	Uppsala
Boris G. Galerkin	1871–1945	Leningrad
Erhard Schmidt	1876–1959	Göttingen, Berlin
Walter Ritz	1878–1909	Tübingen, Göttingen
Richard Edler v. Mises	1883–1953	Berlin, Harvard
Richard Courant	1888–1972	Göttingen, New York
Erich Kamke	1890–1961	Tübingen
Stefan Banach	1892–1945	Lvov
Pawel J. Schauder	1896–1943	Lvov
Edward C. Titchmarsh	1899–1963	Cambridge, Oxford
Johann v. Neumann	1903–1957	Princeton
Franz Rellich	1906–1955	Göttingen

SOME TEXTBOOKS ON DIFFERENTIAL EQUATIONS AND BOOKS FOR FURTHER READING

AMELING, ₁84]: *Laplace Transformation*, 3rd edition, Vieweg & Sohn, 1984, 291 pp.

BIEBERBACH, L. [65]: *Theorie der gewöhnlichen Differentialgleichungen. Auf funktionentheoretischer Grundlage dargestellt*. 2. Aufl. Berlin-Heidelberg-New York: Springer 1965, 389 S.

CLEGG, J. C. [70]: *Variationsrechnung*. Stuttgart: Teubner 1970, 138 S.
COLLATZ, L. [63]: *Eigenwertaufgaben mit technischen Anwendungen*. 2. Aufl. Leipzig: Akad, Verl. Ges. 1963, 466 S.
COLLATZ, L. [55/66]: *Numerische Behandlung von Differentialgleichungen*. 2. Aufl. 1955, 526 S.; 3. Aufl. 1966 (Engl.), 568 S. Berlin-Heidelberg-New York: Springer.
COURANT, R.; HILBERT, D. [672]: *Methods of Mathematical Physics*, Vol. II. Interscience Publishers 1962, 830 S.
DRIVER, R. D. [77]: *Ordinary and Delay Differential Equations*. Berlin-Heidelberg-New York: Springer 1977. *Appl. Math. Sci.* Vol. 20, 501 S.
FRANK, Ph.; v. MISES, R. [61]: *Die Differential- und Integralgleichungen der Mechanik und Physik*. Bd. I. Nachdr. d. 2. Aufl. 1961, 916 S.; Bd. II. Nachdr. d. 2. Aufl. 1961, 1106 S. Braunschweig: Vieweg
GILBARG, D.; TRUDINGER, N.S. [77]: *Elliptic Partial Differential Equations of Second Order*. Berlin-Heidelberg-New York: Springer 1977, 401 S.
GRIGORIEFF, R. D. [72/77]: *Numerik gewöhnlicher Differentialgleichungen*. Bd. I. 1972, 202 S.; Bd. II. 1977, 411 S. Stuttgart: Teubner
HUTSON, V.; PYM, J. S. [80]: *Applications of Functional Analysis and Operator Theory*. New York-London: Acad. Press 1980, 490 S.
JAHNKE, E.; EMDE, F.; LÖSCH, F. [66]: *Tafeln höherer Funktionen*. 7. Aufl. Stuttgart: Teubner 1966, 334 S.
JORDAN, D. W.; SMITH, P. [77]: *Nonlinear Ordinary Differential Equations*. Oxford: Clarendon Press 1977, 360 S.
KAMKE, E. [62/64]: *Differentialgleichungen*. Tl. 1, 5. Aufl. 1964, 316 S.; Tl. 2, 4. Aufl. 1962, 255 S. Leipzig: Akad. Verlagsges.
KAMKE, E. [61/65]: *Differentialgleichungen, Lösungsmethoden und Lösungen*. Bd. I, 7. Aufl. 1961, 666 S.; 9. Aufl. 1977, 668 S — Bd. II, 5 Aufl. 1965, 243 S.; 6. Aufl. 1979, 243 S. Leipzig: Akad. Verlagsges; Stuttgart: Teubher.
KAUDERER, H. [58]: *Nichtlineare Mechanik*. Berlin-Göttingen-Heidelberg: Springer 1958, 684 S.
KINDERLEHRER, D.; STAMPACCHIA, G. [80]: *An Introduction to Variational Inequalities and their Applications*. New York-London: Acad. Press 1980, 336 S.
KNOBLOCH, H. W.; KAPPEL, F. [74]: *Gewöhnliche Differentialgleichungen*, Stuttgart: Teubner 1974, 332 S.
MAGNUS, W.; OBERHETTINGER, F. [48]: *Formeln und Sätze für die spezielle Funktionen der mathematischen Physik*, Springer 1948, 230pp.
MAGNUS, K. [76] *Schwingungen*, 3rd edition, Stuttgart, Teubner 1976, 251 pp.
MARCHUK, G. I. [75]: *Methods of Numerical Mathematics*. Berlin-Heidelberg-New York: Springer 1975, 316 S.
MEIS, Th.: MARCOWITZ, U. [78]: *Numerische Behandlung partieller Differentialgleichungen*. Berlin-Heidelberg-New York: Springer 1978, 452 S.
MITCHELL, A. R.: WAIT, R. [77]: *The Finite Element Method in Partial Differential Equations*. New York: John Wiley & Sons 1977, 198 S.
REISSIG, R.; SANSONE, G.; CONTI, R. [69]: *Nichtlineare Differentialgleichungen höherer Ordnung*. Edizioni Cremonese 1969, 738 S.
SCHRÖDER, J. [80]: *Operator Inequalities*. New York-London: Academic Press 1980, 384 S.
VELTE, W. [76]: *Direkte Methoden der Variationsrechnung*. Stuttgart: Teubner 1976, 198 S.
WALTER, W [70]: *Differential- and Integral Inequalities*. Berlin-Heidelberg-New York: Springer 1970. *Ergebn. Math. und Grengebiete Bd.* 55, 352 S.
WALTER, W. [72]: *Gewöhnliche Differentialgleichungen*. Berlin-Heidelberg-New York: Springer 1972. *Heidelberger Taschenbücher*, Bd. 110, 229 S.
WEINBERGER, H. F. [65]: *A First Course in Partial Differential Equations*. New York-Toronto-London: Blaisdell Publ. Comp. 1965, 446 S.

WEISE, K. H. [66]: *Differentialgleichungen*. Göttingen: Vandenhoeck und Reprecht 1966, 358 S.

WENZEL, H. [80]: *Gewöhnliche Differentialgleichungen*. Leipzig: Akad. Verlagsges. 3. Aufl. 1980, 119 S.

WERNER, H.: SCHABACK, R. [79]: *Praktische Mathematik* II. 2. Aufl. Berlin-Heidelberg-New York: Springer 1979, 388 S. (hier speziell S. 245–372 (gewöhnl. Diff. Gln.)).

WIEGHARDT, K. [65]: *Theoretische Strömungslehre*. Stuttgart: Teubner 1965, 226 S.

WYLIE, RAY; BARRETT L. C. [82]: *Advanced Engineering Mathematics*, McGraw-Hill 5th edition 1982, 1103 pp.

Index